SEMA SIMAI Springer Series

Volume 32

As of 2013, the SIMAI Springer Series opens to SEMA in order to publish a joint series aiming to publish advanced textbooks, research-level monographs and collected works that focus on applications of mathematics to social and industrial problems, including biology, medicine, engineering, environment and finance. Mathematical and numerical modeling is playing a crucial role in the solution of the complex and interrelated problems faced nowadays not only by researchers operating in the field of basic sciences, but also in more directly applied and industrial sectors. This series is meant to host selected contributions focusing on the relevance of mathematics in real life applications and to provide useful reference material to students, academic and industrial researchers at an international level. Interdisciplinary contributions, showing a fruitful collaboration of mathematicians with researchers of other fields to address complex applications, are welcomed in this series.

THE SERIES IS INDEXED IN SCOPUS

Giacomo Albi · Walter Boscheri · Mattia Zanella
Editors

Advances in Numerical Methods for Hyperbolic Balance Laws and Related Problems

Springer

Editors
Giacomo Albi
Department of Informatics
University of Verona
Verona, Italy

Walter Boscheri
Department of Mathematics
and Informatics
University of Ferrara
Ferrara, Italy

Mattia Zanella
Department of Mathematics
University of Pavia
Pavia, Italy

ISSN 2199-3041 ISSN 2199-305X (electronic)
SEMA SIMAI Springer Series
ISBN 978-3-031-29874-5 ISBN 978-3-031-29875-2 (eBook)
https://doi.org/10.1007/978-3-031-29875-2

This Springer imprint is published by the registered company Springer Nature Switzerland AG
The registered company address is: Gewerbestrasse 11, 6330 Cham, Switzerland

Preface

A wide range of phenomena in science and technology can be described by nonlinear partial differential equations characterized by systems of conservation laws with source terms. Well-known examples are hyperbolic systems with source terms, kinetic equations, and convection-reaction-diffusion equations. This class of equations fits several fundamental physical laws and plays a crucial role in real-world applications, like plasma physics, continuum mechanics, geophysics, semiconductor design, and granular gases. Yet, in the recent years, there has been an explosion of new applications of these techniques to other areas of research like biology and social sciences. Indeed, the theoretical background of hyperbolic balance laws has proven its effectiveness in the description of the collective motion of a large number of particles such as pedestrian and traffic flows, swarming dynamics, fake-news spread, tumor growth dynamics, and cardiovascular flow modeling. The investigation of observable phenomena in these new fields of applications presents many mathematical challenges at the level of modeling, mathematical analysis, and numerical methods. From the numerical point of view, these phenomena may be characterized by multiple scales and complex geometries, which render classical methods ineffective. Specifically, to deal with real-world objects, immersed boundary methods based on a level-set strategy must be designed such that the complex shape of the object is accurately represented by the numerical scheme, as well as its associated boundary conditions. Besides, moving mesh techniques must be faced when the physical phenomena require the precise tracking of material interfaces with no mass flux across element boundaries, thus solving exactly contact discontinuities in multi-material and multi-phase flows. The multiscale nature of the models, which simultaneously must deal with different time scales, like transport and diffusive terms, leads to the design and implementation of sophisticated and modern implicit-explicit time integrators that contribute to devise asymptotic preserving and accurate numerical methods.

All these aspects have been presented in the Young Researchers Workshop on Numerical Aspects of Hyperbolic Balance Laws and Related Problems organized by the editors of the present book at the University of Verona, Italy, in December 2021. During this event, 22 invited young researchers presented advances and cutting-edge

results to a broad scientific community working in the field of applied mathematics.
The editors of this book wish to express their gratitude to the INdAM National Groups of Scientific Computing (GNCS) and of Mathematical Physics (GNFM) and to the PRIN2017 Project "Innovative Numerical Methods for Evolutionary Partial Differential Equations and Applications" for their financial support in the organization of the event. We would like to thank all the speakers of this exciting event for their inspiring presentations that stimulated fruitful discussions among the participants.

Verona, Italy — Giacomo Albi
Ferrara, Italy — Walter Boscheri
Pavia, Italy — Mattia Zanella

Contents

Contributors

Alessandro Alla Department of Molecular Sciences and Nanosystems, Universitá Ca' Foscari Venezia, Venice, Italy

Giulia Bertaglia Department of Environmental and Prevention Sciences, University of Ferrara, Ferrara, Italy

Felisia Angela Chiarello Dipartimento di Ingegneria e Scienze dell'Informazione e Matematica (DISIM), University of L'Aquila, L'Aquila, Italy

Simone Chiocchetti Department of Civil, Environmental and Mechanical Engineering, University of Trento, Trento, Italy

Armando Coco Dipartimento di Matematica e Informatica, Università degli Studi di Catania, Catania, Italy

Fiammetta Conforto Department of ChiBioFarAm, University of Messina, Messina, Italy

Elena Gaburro Inria, University of Bordeaux, CNRS, Bordeaux INP, IMB, UMR 5251, Talence cedex, France

Paola Goatin Université Côte d'Azur, Inria, CNRS, LJAD, Sophia Antipolis Cedex, France

Elisa Iacomini Institut für Geometrie und Praktische Mathematik RWTH Aachen, Aachen, Germany

Liu Liu The Chinese University of Hong Kong, Hong Kong, Hong Kong

Vincent Liu Department of Electrical and Electronic Engineering, The University of Melbourne, Melbourne, Australia

Peter M. Dower Department of Electrical and Electronic Engineering, The University of Melbourne, Melbourne, Australia

Giorgio Martalò Department of SMFI, University of Parma, Parma, Italy

Gabriella Puppo Department of Mathematics, Sapienza University of Rome, Rome, Italy

Matteo Semplice Dipartimento di Scienza e Alta Tecnologia, Università degli Studi dell'Insubria, Como, Italy

Santina Chiara Stissi Istituto Nazionale di Geofisica e Vulcanologia, Osservatorio Etneo, Catania, Italy

Silvia Tozza Department of Mathematics, University of Bologna, Bologna, Italy

Giuseppe Visconti Department of Mathematics, Sapienza University of Rome, Rome, Italy

Yuhua Zhu Department of Mathematics and Halicioğlu Data Science Institute, University of California-San Diego, San Diego, CA, USA

A Tree Structure Approach to Reachability Analysis

Alessandro Alla, Peter M. Dower, and Vincent Liu

Abstract Reachability analysis is a powerful tool when it comes to capturing the behaviour, thus verifying the safety, of autonomous systems. However, general-purpose methods, such as Hamilton-Jacobi approaches, suffer from the curse of dimensionality. In this paper, we mitigate this problem for systems of moderate dimension and we propose a new algorithm based on a tree structure approach with geometric pruning. The numerical examples will include a comparison with a standard finite-difference method for linear and nonlinear problems.

Keywords Reachability analysis · Hamilton-Jacobi equations · Optimal control · Tree structure · Convex geometry

1 Introduction

The development and production of self-driving cars and the deployment of drones in industrial applications are exemplars of society's ever-increasing fascination of autonomous vehicles. Commensurate to the growing integration of these vehicles in day-to-day life is the concern regarding how safe these unmanned vehicles are. These concerns are particularly prevalent in safety-critical applications such as human-robot interactions, disaster responses, and the use of high-value machinery. In such applications, being able to characterize all possible behaviours of these autonomous

A. Alla (✉)
Department of Molecular Sciences and Nanosystems, Universitá Ca' Foscari Venezia, Venice, Italy
e-mail: alessandro.alla@unive.it

P. M. Dower · V. Liu
Department of Electrical and Electronic Engineering, The University of Melbourne, Melbourne, Australia
e-mail: pdower@unimelb.edu.au

V. Liu
e-mail: liuv2@student.unimelb.edu.au

G. Albi et al. (eds.), *Advances in Numerical Methods for Hyperbolic Balance Laws and Related Problems*, SEMA SIMAI Springer Series 32,
https://doi.org/10.1007/978-3-031-29875-2_1

vehicles would be a rigorous way to verify their safety. Reachable sets lend themselves well towards this goal. When computed forwards in time, they characterise all possible states that can be reached using a constraint admissible control from some initial set of states. Similarly, when computed backwards in time, they characterise all possible states that are able to reach a terminal set of states using a constraint admissible control.

The computation of reachable sets may be done via the Hamilton-Jacobi-Bellman (HJB) equations, i.e. one of the most powerful formal verification tools for guaranteeing performance and safety properties of systems. This approach is rather general and works for controlled nonlinear systems that involve disturbances or adversarial behaviors, and despite this, characterizes the exact reachable set rather than approximations. However, this method suffers from the curse of dimensionality and it is hard to build numerical methods for high dimensional problems. In the last two decades several contributions on the mitigitation of the curse of dimensionality have been investigated mainly for optimal control problems such as, e.g. model order reduction [1, 2], tree structure algorithms [3, 4], spectral methods [5], max-plus algebra [6, 7], Hopf-Lax approaches [8, 9], neural networks [10, 11], tensor decomposition [12, 13] and sparse grids method [14].

For the approximation of the reachable sets in [15], the authors provides a formulation that requires numerically solving a Hamilton-Jacobi partial differential equation. This is a grid-based approach which is typically limited to systems of no more than 4 states on standard computers. Therefore, the study of higher dimensional problems remains an open research area. Under certain assumption on the system, one could decompose it appropriately, and obtain efficient algorithms for computing reachable sets, see e.g. [16]. Other approaches have been studied in [17] where reachable sets for nonlinear systems are computed via results on reachability for uncertain linear system. Linearisation error is explicitly accounted for in [17] using an iterative algorithm to bound this error in an over-approximative manner. Similarly, in [18], uncertain linear systems are considered with a focus on producing zonotopic under-approximations of reachable sets. In their work, the representational complexity of the reachable sets grows as the algorithm iterates in time, thus motivating the use of a 'pruning' step, which reduces the order of the zonotopic sets.

In this paper, we present an algorithm based on a tree structure to approximate the HJB equation for backwards reachable sets. The idea of the algorithm is based on the paper [3] and approximates the value function using the Dynamic Programming Principle (DPP) on an unstructured mesh for optimal control problems. The idea of our proposed algorithm is as follows.

We start from a discretization of the terminal set and compute the value function on those points. Then, we neglect the interior points, say those nodes for which the value function is strictly negative. This is a pruning strategy that aims to mitigate the exponential increase in the cardinality of the tree. We then evolve these nodes backwards in time, making use of a result provided in Sect. 4, where the forward controlled dynamical system is equivalent, up to minor changes in the problem, to the backward system. This is the main difference with respect to the method in [3]. We are able, in this work, to compute the tree backwards in time and to prune the tree using the information from the value function. Our pruning is based on geometric considerations where the interior of the reachable set is pruned. This comes from the observation, which is demonstrated in this work, that the boundary of the reachable set at a particular time cannot evolve from the interior of the reachable set at a prior time. Thus, it is wasteful to propagate the tree structure of [3] for nodes that lie interior to the reachable set. When the value function is convex, interior nodes can be easily identified using off-the-shelf algorithms for computing convex hulls.

The outline of the paper is the following. In Sect. 2 we present the control problem setup and in Sect. 3 we provide the relevant background for the characterization of the backwards reachable set. In Sect. 4 we show the equivalence between backwards and forwards reachable sets. Our algorithm is introduced and discussed in Sect. 5. Numerical examples are then shown in Sect. 6. Finally, conclusions and future works are discussed in Sect. 7.

Notation

- Let $\mathbb{I}_n$ denote an n-by-n identity matrix.
- Let $\langle \cdot, \cdot \rangle$ denote the Euclidean inner product.
- Let $\|w\|$ denote any norm of a vector w.
- Let $int\ (\mathcal{A})$, $cl\ (\mathcal{A})$, and $\partial\mathcal{A}$ denote the interior, closure, and boundary of a set $\mathcal{A}$, respectively.
- Let $C^k(\Omega; \mathbb{R})$ denote the space of k-times continuously differentiable functions from Ω to $\mathbb{R}$ with $C \doteq C^0$.
- Let $\mathbb{B}_R(x_0) \doteq \{x \in \mathbb{R}^n \mid \|x - x_0\| < R\}$ denote a ball of radius $R \geq 0$ with centre $x_0 \in \mathbb{R}^n$.
- An ellipsoidal set with centre q and shape Q is defined as

$$\mathcal{E}(q, Q) \doteq \left\{x \in \mathbb{R}^n \mid (x - q)^T Q^{-1}(x - q) \leq 1\right\}, \tag{1}$$

  where $q \in \mathbb{R}^n$ and $Q \in \mathbb{R}^{n \times n}$ is a symmetric, positive definite matrix. The axes of $\mathcal{E}(q, Q)$ are aligned with the eigenvectors of Q with lengths along these axes being equal to the square root of the corresponding eigenvalues.
- Let conv $(\mathcal{A})$ denote the convex hull of a finite set of points $\mathcal{A} = \{a_i\}_{i \in \{1, \cdots, n_p\}}$, defined by

$$\text{conv}\ (\mathcal{A}) = \left\{ \sum_{i=1}^{n_p} \lambda_i a_i \,\middle|\, \lambda_i \geq 0 \text{ for all } i \in \{1, \cdots, n_p\} \text{ and } \sum_{i=1}^{n_p} \lambda_i = 1 \right\}. \tag{2}$$

2 Problem Setup

In this section we begin with a system description and review relevant background relating to optimal control. Consider the continuous-time nonlinear system described by

$$\dot{x}(t) = f(x(t), u(t)), \quad \forall t \in (0, T), \tag{3}$$

where $T \geq 0$, $x(t) \in \mathbb{R}^n$ is the state and $u(t) \in \mathbb{U}$ is the input at time t, with $\mathbb{U} \subset \mathbb{R}^m$ being compact. The input is selected such that $u \in \mathcal{U}$, where

$$\mathcal{U} \doteq \{u : [0, T] \to \mathbb{U} \mid u \text{ measurable}\}. \tag{4}$$

The following flow field conditions are assumed throughout.

Assumption 1 The function $f : \mathbb{R}^n \times \mathbb{U} \to \mathbb{R}^n$ satisfies

(i) $f \in C(\mathbb{R}^n \times \mathbb{U}; \mathbb{R}^n)$; and
(ii) f is locally Lipschitz continuous in x, uniformly in u, i.e. for any $R > 0$, there exists a Lipschitz constant $C_R^f > 0$ such that $\|f(x, u) - f(y, u)\| \leq C_R^f \|x - y\|$, $\forall (x, y, u) \in \mathbb{B}_R(0) \times \mathbb{B}_R(0) \times \mathbb{U}$.

Under Assumption 1, the system described by (3) admits a unique and continuous solution for any initial condition $x(0) = x_0$ and fixed control $u(\cdot) \in \mathcal{U}$. We denote these solutions at time $0 \leq t \leq T$ by $\varphi(t; 0, x_0, u(\cdot))$.

Attach to (3) the value function $V : [0, T] \times \mathbb{R}^n$ corresponding to the optimal control problem

$$V(t, x) \doteq \inf_{u(\cdot)\in\mathcal{U}} J(t, x, u(\cdot)) \doteq \inf_{u(\cdot)\in\mathcal{U}} \left\{ \int_t^T h(\varphi(s; t, x, u(\cdot)), u(s))\, ds + g\left(\varphi(T; t, x, u(\cdot))\right) \right\}, \tag{5}$$

where the running cost $h : \mathbb{R}^n \times \mathbb{U} \to \mathbb{R}$ and the terminal state cost $g : \mathbb{R}^n \to \mathbb{R}$ satisfy the assumptions below.

Assumption 2 The functions $h : \mathbb{R}^n \times \mathbb{U} \to \mathbb{R}$ and $g : \mathbb{R}^n \to \mathbb{R}$ satisfy

(i) $h \in C(\mathbb{R}^n \times \mathbb{U}; \mathbb{R})$ and $g \in C(\mathbb{R}^n; \mathbb{R})$; and
(ii) h and g are locally Lipschitz continuous in x (uniformly in u for h), i.e. for any $R > 0$, there exists a Lipschitz constant $C_R^h > 0$ such that $|h(x, u) - h(y, u)| \leq C_R^h \|x - y\|$, $\forall (x, y, u) \in \mathbb{B}_R(0) \times \mathbb{B}_R(0) \times \mathbb{U}$ and there exists a $C_R^g > 0$ such that $|g(x) - g(y)| \leq C_R^g \|x - y\|$, $\forall (x, y) \in \mathbb{B}_R(0) \times \mathbb{B}_R(0)$.

The value function defined by (5) satisfies the Dynamic Programming Principle (see e.g. [19]), which is presented below.

Theorem 1 *For any $t \in [0, T]$, $s \in [0, t]$, and $x \in \mathbb{R}^n$, the value function V : $[0, T] \to \mathbb{R}^n$ in (5) satisfies the Dynamic Programming Principle*

$$V(s,x) = \inf_{u(\cdot)\in\mathcal{U}} \left\{ \int_s^t h(\varphi(\tau;s,x,u(\cdot)),u(\tau))\,d\tau + V\left(t,\varphi(t;s,x,u(\cdot))\right) \right\}. \quad (6)$$

The value function can be characterized via the viscosity solution of a HJB equation. A detailed discussion on viscosity solutions can be found in [20].

Theorem 2 *Let Assumptions 1 and 2 hold. Then, $V = V(t,x)$ in (5) is the unique, and locally Lipschitz continuous viscosity solution of the HJB equation given by*

$$\begin{aligned} -V_t + H(x,\nabla V) = 0, \quad \forall (t,x)\in(0,T)\times\mathbb{R}^n, \\ V(T,x) = g(x), \quad \forall x\in\mathbb{R}^n, \end{aligned} \quad (7)$$

where the Hamiltonian $H : \mathbb{R}^n \times \mathbb{R}^n \to \mathbb{R}$ is given by

$$H(x,p) \doteq \max_{u\in\mathbb{U}} \left\{ -\langle p, f(x,u)\rangle - h(x,u) \right\}. \quad (8)$$

3 Backwards Reachability

Let us now consider a special case of the optimal control problem in (5), which can be used to characterize a *backwards* reachable set. This set characterizes all states $x\in\mathbb{R}^n$ for which there exists an admissible control $u(\cdot)\in\mathcal{U}$ leading to some terminal set of states

$$\mathcal{X}_T \doteq \left\{ x\in\mathbb{R}^n \,|\, g(x) \le 0 \right\}, \quad (9)$$

in time $T \ge 0$, where $g : \mathbb{R}^n \to \mathbb{R}$ is a bounded and locally Lipschitz continuous function. The backwards reachable set is more precisely defined below.

Definition 1 The backwards reachable set at time $T \ge 0$ is defined as the set

$$\mathcal{G}(T) \doteq \left\{ x\in\mathbb{R}^n \,|\, \exists\, u(\cdot)\in\mathcal{U} \text{ such that } \varphi(T;0,x,u(\cdot))\in\mathcal{X}_T \right\}, \quad (10)$$

where $\varphi(t;0,x_0,u(\cdot))$ denotes solutions of (3) at time t from an initial state $x(0)=x_0$ with an admissible control $u(\cdot)\in\mathcal{U}$, and $\mathcal{X}_T$ is the terminal set described by (9).

The backwards reachable set can be characterized via the viscosity solution of the HJB equation given in (7)–(8), which is described in the following theorem (see Sect. 2.2.2 of [21]).

Theorem 3 *Let Assumption 1 hold for (3) and let the function $g : \mathbb{R}^n \to \mathbb{R}$, which defines the terminal set $\mathcal{X}_T$ in (9), satisfy Assumption 2. Let $v \in C([0,T]\times\mathbb{R}^n;\,\mathbb{R})$ be the unique and locally Lipschitz continuous viscosity solution of the HJB equation given by*

$$\begin{aligned} -v_t + H(x, \nabla v) &= 0, \quad \forall (t, x) \in (0, T) \times \mathbb{R}^n, \\ v(T, x) &= g(x), \quad \forall x \in \mathbb{R}^n, \end{aligned} \tag{11}$$

where the Hamiltonian $H : \mathbb{R}^n \times \mathbb{R}^n \to \mathbb{R}$ *is given by*

$$H(x, p) \doteq \max_{u \in \mathbb{U}} \{-\langle p, f(x, u)\rangle\}. \tag{12}$$

Then, the backwards reachable set $\mathcal{G}(T)$ *for* (3) *is*

$$\mathcal{G}(T) = \left\{x \in \mathbb{R}^n \mid v(0, x) \leq 0\right\}. \tag{13}$$

Next, we will demonstrate that the boundary of the backwards reachable set $\mathcal{G}(T)$ can only reach the boundary of the terminal set $\mathcal{X}_T$. This result will motivate the use of a geometric pruning criterion that will be introduced in our proposed algorithm for computing $\mathcal{G}(T)$.

Lemma 1 *Let* $\mathcal{G}(T)$ *denote the backwards reachable set of* (3) *as defined in* (10)*. Then,*

$$\partial \mathcal{G}(T) \subseteq \left\{x \in \mathbb{R}^n \mid \exists\, u(\cdot) \in \mathcal{U} \text{ such that } \varphi(T; 0, x, u(\cdot)) \in \partial \mathcal{X}_T\right\}. \tag{14}$$

Moreover, there does not exist a control $u(\cdot) \in \mathcal{U}$ *such that* $\varphi(T; 0, x, u(\cdot)) \in int\,(\mathcal{X}_T)$ *for any* $x \in \partial \mathcal{G}(T)$*.*

Proof Let us denote the right hand side of (14) by $\bar{\mathcal{B}}$. We will first show that the boundary of $\mathcal{G}(T)$ can not reach the interior of $\mathcal{X}_T$. Suppose this is not true. That is, there exists $\hat{x} \in \partial \mathcal{G}(T)$ and $\hat{u}(\cdot) \in \mathcal{U}$ such that $\varphi(T; 0, \hat{x}, \hat{u}(\cdot)) \in int\,(\mathcal{X}_T)$. Under Assumption 1, φ is continuous in x, thus there exists a sufficiently small neighbourhood $\mathcal{N}$ of $\hat{x}$ such that $\varphi(T; 0, \bar{x}, \hat{u}(\cdot)) \in int\,(\mathcal{X}_T)$ for all $\bar{x} \in \mathcal{N}$. This would then imply $\mathcal{N} \subset \mathcal{G}(T)$, but this leads to a contradiction as $\hat{x}$ lies on the boundary of $\mathcal{G}(T)$, i.e. there exists a $\bar{x} \in \mathcal{N} \setminus \mathcal{G}(T)$ such that $\varphi(T; 0, \bar{x}, \hat{u}(\cdot)) \in int\,(\mathcal{X}_T)$. Furthermore, since $\hat{x} \in \partial \mathcal{G}(T) \subset \mathcal{G}(T)$, there exists a $u(\cdot) \in \mathcal{U}$ such that $\varphi(T; 0, \hat{x}, u(\cdot)) \in \mathcal{X}_T$. Thus, for all $\hat{x} \in \partial \mathcal{G}(T)$, there exists a $u(\cdot) \in \mathcal{U}$ such that $\varphi(T; 0, \hat{x}, u(\cdot)) \in \mathcal{X}_T \setminus int\,(\mathcal{X}_T) = \partial \mathcal{X}_T$, and hence $\partial \mathcal{G}(T) \subseteq \bar{\mathcal{B}}$.

4 Forwards Reachability

An analogous set to the backwards reachable set is the forwards reachable set, which characterizes all states that can be reached from some initial set of states under the influence of an admissible control. We will verify that the forwards reachable set of a time-reversed system is exactly the backwards reachable set of (3). This will prove to be convenient for us as the algorithm in Sect. 5 is more intuitively described when considering a forwards reachability problem. The aforementioned time-reversed system is described by

$$\dot{x}(t) = -f(x(t), u(t)), \quad \forall t \in (0, T), \tag{15}$$

where $T \geq 0$, $f : \mathbb{R}^n \times \mathbb{U} \to \mathbb{R}^n$ satisfies Assumption 1, and $u(\cdot) \in \mathcal{U}$. When referring to the forwards reachable set, we will use $\mathcal{X}_0$ defined identically to $\mathcal{X}_T$ in (9) to describe the initial set of states. The forwards reachable set for (15) is now precisely defined below.

Definition 2 The forwards reachable set of the time-reversed system in (15) at time $T \geq 0$ is defined as

$$\mathcal{F}_{-}(T) \doteq \left\{ x \in \mathbb{R}^n \mid \exists\, u(\cdot) \in \mathcal{U} \text{ and } \exists\, x_0 \in \mathcal{X}_0 \text{ such that } \varphi_{-}(T; 0, x_0, u(\cdot)) = x \right\}, \tag{16}$$

where $\varphi_{-}(t; 0, x_0, u(\cdot))$ denotes the unique solutions of (15) at time t from an initial state $x(0) = x_0$ with an admissible control $u(\cdot) \in \mathcal{U}$, and $\mathcal{X}_0$ is the initial set.

To demonstrate that $\mathcal{F}_{-}(T)$ is exactly the backwards reachable set $\mathcal{G}(T)$, we formally verify a standard result relating solutions of (3) to (15). In particular, we wish to show that an initial state x that evolves under a control $u(\cdot)$ for (3) can be recovered by setting the terminal state of (3) as the initial state for (15) and applying a time-reversed control $u_{-}(\cdot)$. It follows naturally that the other direction would hold as well. This is more precisely described below.

Lemma 2 *Let $x \in \mathbb{R}^n$ and $u(\cdot) \in \mathcal{U}$. Select $u_{-}(\cdot) \in \mathcal{U}$ such that $u_{-}(t) = u(T - t)$ for all $t \in [0, T]$, then, for any $T \geq 0$ we have*

$$\varphi_{-}(T; 0, \varphi(T; 0, x, u(\cdot)), u_{-}(\cdot)) = x, \tag{17}$$

$$\text{and} \quad \varphi(T; 0, \varphi_{-}(T; 0, x, u_{-}(\cdot)), u(\cdot)) = x, \tag{18}$$

where $\varphi(t_2; t_1, x_1, u(\cdot))$ and $\varphi_{-}(t_2; t_1, x_1, u(\cdot))$ denote solutions of (3) and (15), respectively, at time t_2 with initial state $x(t_1) = x_1$ and control input $u(\cdot)$.

Proof Since $\varphi(\cdot; 0, x, u(\cdot))$ is a solution to (3) with an initial condition $x(0) = x$ under the control $u(\cdot) \in \mathcal{U}$, we have that $\frac{\partial}{\partial s}\varphi(s; 0, x, u(\cdot)) = f(\varphi(s; 0, x, u(\cdot)), u(s))$ for all $s \in (0, T)$. Substituting $s = T - t$, we obtain $\frac{\partial}{\partial t}\varphi(T - t; 0, x, u(\cdot)) = -f(\varphi(T - t; 0, x, u(\cdot)), u_{-}(t))$ for all $t \in (0, T)$. Then, taking the initial condition of the reversed system (15) to be $x(0) = \varphi(T; 0, x, u(\cdot))$ and the control input to be $u_{-}(\cdot) \in \mathcal{U}$ where $u_{-}(t) = u(T - t)$, $\forall t \in [0, T]$, we must have that

$$\varphi_{-}(t; 0, \varphi(T; 0, x, u(\cdot)), u_{-}(\cdot)) = \varphi(T - t; 0, x, u(\cdot)), \tag{19}$$

for all $t \in [0, T]$, since (15) admits *unique* solutions only. Letting $t = T$ in (19) makes the right-hand side become $\varphi(0; 0, x, u(\cdot)) = x$, which yields (17). Finally, by alternating the signs between (3) and the time-reversed system (15), we obtain (18) from (17).

Theorem 4 *Let $\mathcal{F}_-(T)$ denote the forwards reachable set defined in (16) for the time-reversed system in (15). Let $\mathcal{G}(T)$ be the backwards reachable set as defined in (10). Then,*

$$\mathcal{G}(T) = \mathcal{F}_-(T). \tag{20}$$

Proof Let $x \in \mathcal{F}_-(T)$, which means for some $u_-(\cdot) \in \mathcal{U}$ and $x_0 \in \mathcal{X}_0$, we have $\varphi_-(T; 0, x_0, u_-(\cdot)) = x$. From Lemma 2, we have that $\varphi(T; 0, \varphi_-(T; 0, x_0, u_-(\cdot)), u(\cdot)) = x_0 \in \mathcal{X}_0 = \mathcal{X}_T$, where $u(t) \doteq u_-(T - t)$ for all $t \in [0, T]$. Since $u(\cdot) \in \mathcal{U}$ takes the state $\varphi_-(T; 0, x_0, u_-(\cdot)) = x$ to $x_0 \in \mathcal{X}_T$ for the system in (3), we have $x \in \mathcal{G}(T) \implies \mathcal{F}_-(T) \subseteq \mathcal{G}(T)$.

To demonstrate that the other direction holds, let $x \in \mathcal{G}(T)$, which means for some $u(\cdot) \in \mathcal{U}$ we have $x_T \doteq \varphi(T; 0, x, u(\cdot)) \in \mathcal{X}_T = \mathcal{X}_0$. From Lemma 2, we have $\varphi_-(T; 0, \varphi(T; 0, x, u(\cdot)), u_-(\cdot)) = x$ where $u_-(t) = u(T - t)$ for all $t \in [0, T]$. Since $u_-(\cdot) \in \mathcal{U}$, this means x can be reached under the influence of an admissible control from $\varphi(T; 0, x, u(\cdot)) = x_T \in \mathcal{X}_0$, thus $x \in \mathcal{F}_-(T) \implies \mathcal{G}(T) \subseteq \mathcal{F}_-(T)$.

Likewise with the characterization of the backwards reachable set in Theorem 3, the forwards reachable set $\mathcal{F}_-(T)$ for (15) can be also be characterized via the solution of a corresponding HJB PDE.

Corollary 1 *Let Assumption 1 hold for (15) and let the function $g : \mathbb{R}^n \to \mathbb{R}$, which defines the initial set $\mathcal{X}_0 = \mathcal{X}_T$ in (9), satisfy Assumption 2. Let $w \in C([0, T] \times \mathbb{R}^n; \mathbb{R})$ be the unique and locally Lipschitz continuous viscosity solution of the HJB equation given by*

$$\begin{aligned} w_t - H(x, \nabla w) = 0, \quad & \forall (t, x) \in (0, T) \times \mathbb{R}^n, \\ w(0, x) = g(x), \quad & \forall x \in \mathbb{R}^n, \end{aligned} \tag{21}$$

where the Hamiltonian $H : \mathbb{R}^n \times \mathbb{R}$ is given by

$$H(x, p) \doteq - \max_{u \in \mathbb{U}} \langle p, -f(x, u) \rangle = \min_{u \in \mathbb{U}} \langle p, f(x, u) \rangle. \tag{22}$$

Then, the forwards reachable set for (15) is

$$\mathcal{F}_-(T) = \left\{ x \in \mathbb{R}^n \mid w(T, x) \leq 0 \right\}. \tag{23}$$

Proof For the proof, we make use of the test function form of the definition for viscosity solutions in (5.17) of [22]. Let $w \in C([0, T] \times \mathbb{R}^n; \mathbb{R})$ be the viscosity solution of (21) and let $v \in C([0, T] \times \mathbb{R}^n; \mathbb{R})$ be defined such that $v(t, x) \doteq w(T - t, x)$, $\forall (t, x) \in [0, T] \times \mathbb{R}^n$. Suppose for some $\xi \in C^1((0, T) \times \mathbb{R}^n; \mathbb{R})$ we have that $v - \xi$ attains a local maximum at $(t_0, x_0) \in (0, T) \times \mathbb{R}^n$. Define now $\bar{\xi} \in C^1((0, T) \times \mathbb{R}^n; \mathbb{R})$ such that $\bar{\xi}(t, x) \doteq \xi(T - t, x)$, $\forall (t, x) \in [0, T] \times \mathbb{R}^n$. Then, at

$(T - t_0, x_0)$ we must also have that $w - \bar{\xi}$ attains a local maximum. Since w being a viscosity solution of (21) implies it is also a viscosity subsolution of (21), we have that

$$\bar{\xi}_t(T - t_0, x_0) + \max_{u \in \mathbb{U}} \langle \nabla \bar{\xi}(T - t_0, x_0), -f(x_0, u) \rangle \leq 0. \tag{24}$$

Substituting ξ into (24) yields

$$-\xi_t(t_0, x_0) + \max_{u \in \mathbb{U}} \langle \nabla \xi(t_0, x_0), -f(x_0, u) \rangle \leq 0, \tag{25}$$

which implies v is a viscosity subsolution of (11). The same procedure can be used to show that v is also a viscosity supersolution of (11) by considering a candidate function $\xi \in C^1((0, T) \times \mathbb{R}^n; \mathbb{R})$ such that $v - \xi$ attains a local *minimum* at $(t_0, x_0) \in (0, T) \times \mathbb{R}^n$. Thus, v must be a viscosity solution of (11) if w is a viscosity solution of (21). From Theorem 3, this implies that

$$\left\{x \in \mathbb{R}^n \mid w(T, x) \leq 0\right\} = \left\{x \in \mathbb{R}^n \mid v(0, x) \leq 0\right\} = \mathcal{G}(T). \tag{26}$$

Theorem 4 can then be used to conclude (23). The Lipschitz and uniqueness properties follow from the properties of v in Theorem 3.

5 A Tree-Based Algorithm for Computing Reachable Sets

In this section we provide the description of our new algorithm to approximate solutions of (7) with zero running cost, i.e. $h = 0$. Our algorithm is based on a tree structure as proposed in [3] that is adapted to our problem for computing the backwards reachable set as defined in (10). This tree structure is depicted in Fig. 1. In [3], the value function is computed by first constructing a tree, which represents a discretization in time and space of the forwards reachable set from a *known* initial state denoted by x_1^0 in the diagram. Using the Dynamic Programming Principle (DPP) stated in Theorem 1, the value function is then computed backwards in time starting from the nodes in the tree corresponding to some terminal time $t = T$. However, when it comes to computing the backwards reachable set, we want to compute the set of all initial states such that a *known* terminal set $\mathcal{X}_T$ can be reached. Thus, a key difference between our approach and the algorithm proposed in [3] is that we build the tree backwards in time starting from time T at the terminal set. We now move to describing how the algorithm of [3] could be adjusted to compute the backwards reachable set and state its connection to the value function in (5). A geometric condition for pruning is then introduced to reduce the computational expense.

To begin, the control set $\mathbb{U}$ is discretized into a finite set $\bar{\mathbb{U}} \doteq \{u_1, \ldots, u_{n_u}\} \subset \mathbb{U}$. Let Δ_t denote the time discretization interval with $t_k \doteq T - k\Delta_t$ being the k-th time point for a total of $N \doteq \lceil \frac{T}{\Delta_t} \rceil$ time points. We define the k-th tree level $\mathcal{T}^k$ as the set

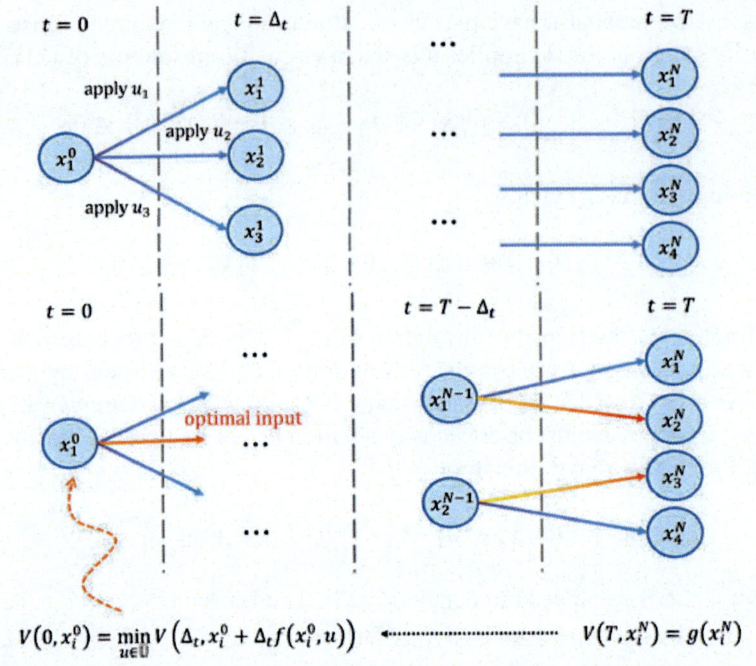

Fig. 1 Diagram depicting tree structure and value function computation as described in [3]. Upper figure depicts generation of nodes in each level of the tree from a discretized input set $\mathbb{U}$. Lower figure depicts computation of the value function starting from nodes at the terminal time T

of nodes generated at time t_k, which has a cardinality denoted by n_k. Additionally, we use x_i^k to denote the i-th node of the k-th tree level.

Then, we start to compute the nodes of our tree. The initial (zeroth) level contains nodes given by a discretization of the terminal set $\mathcal{X}_T$. Let $\{x_i^0\}_{i \in \{1,\dots,n_0\}}$ be a finite set of points in $\mathcal{X}_T$. For these points, we can directly compute the value function as follows:

$$V(T, x_i^0) = g(x_i^0), \qquad \forall i \in \{1, \dots n_0\}, \tag{27}$$

which is given by the terminal condition of (11) at time T. Equivalently, we can consider this step as initialising the value function w in (21) for the forwards reachable set of the time-reversed system in (15).

We continue generating the nodes of the tree for level 1 corresponding to the time $t_1 = T - \Delta_t$. The set of nodes $\mathcal{T}^1 \doteq \{x_i^1\}_{i \in \{1,\dots,n_1\}}$ will be computed from the time-reversed ODE (15). Although other discretizations are possible, an Euler discretization has been chosen here leading to:

$$x_{i_j}^1 = x_i^0 - \Delta_t f(x_i^0, u_j), \qquad \forall i \in \{1, \dots, n_0\},\ \forall j \in \{1, \dots, n_u\}. \tag{28}$$

Once the nodes in $\mathcal{T}^1$ are generated we approximate the value function at time $T - \Delta_t$ using a one-step discretisation of the DPP (rewritten from (6)) using the discretized input set $\bar{U}$:

$$V(T - \Delta_t, x_i^1) = \min_{u \in \{u_1, \ldots, u_{n_u}\}} V(T, x_i^1 + \Delta_t f(x_i^1, u)), \qquad \forall i \in \{1, \ldots, n_1\}. \quad (29)$$

The DPP in (29) approximates the value function for a point x_i^1 at time $T - \Delta_t$ via a minimization of the value function at time T across all points that can be reached by x_i^1 *forwards* in time. However, the value function at time T is only computed in (27) on the finite set of points $\mathcal{T}^0 = \{x_i^0\}_{i \in \{1, \ldots, n_0\}}$, which may not contain the point $x_i^1 + \Delta_t f(x_i^1, u_j)$. In general, (29) may require an interpolation of the value function.

In order to compute an approximation of the value function using pre-computed values of nodes on previous tree levels, we can instead perform the minimization in (29) over all nodes in $\mathcal{T}^0$ that can be reached by x_i^1. To be more precise, we consider an Euler discretisation of the forward system (3):

$$x[n+1] = x[n] + \Delta_t f(x[n], u[n]), \qquad \forall n \in \{0, 1, \ldots\}, \quad (30)$$

where $x[n] \approx x(T - t_n)$, $u[n] \approx u(T - t_n) \in \mathbb{U}$, $\forall n \in \{0, 1, \ldots\}$. The set of points that can be reached in a single time-step Δ_t is defined precisely below.

Definition 3 The one-step reachable set for the discrete-time system (30) from a point $\bar{x}$ is defined as

$$\mathcal{R}_1(\bar{x}) \doteq \left\{x \in \mathbb{R}^n \mid \exists u \in \mathbb{U} \text{ such that } \bar{x} + \Delta_t f(\bar{x}, u) = x\right\}. \quad (31)$$

From this, we construct a general iteration step for generating nodes on all tree levels $k \in \{1, \ldots, n_k\}$, as well as approximating the value function at these nodes:

$$x_{i_j}^k = x_i^{k-1} - \Delta_t f(x_i^{k-1}, u_j), \qquad \forall i \in \{1, \ldots, n_{k-1}\}, \ \forall j \in \{1, \ldots, n_u\}, \quad (32)$$

$$V(t_k, x_{i_j}^k) = \min_{x \in \mathcal{T}^{k-1} \cap \mathcal{R}_1(x_{i_j}^k)} V(t_{k-1}, x), \qquad \forall i_j \in \{1, \ldots, n_k\}. \quad (33)$$

In (32), we generate nodes contained in the forwards reachable set of the time-reversed system in (15) starting from an initial set $\mathcal{X}_0 = \mathcal{X}_T$. From Theorem 4, this gives us the backwards reachable set of (3). The value function is then approximated via a minimization over a subset of nodes in the previous tree level. Unlike in Fig. 1, the entire tree does not need to be generated a priori. Algorithm 1 summarises the procedure outlined above.

Remark 1 To check whether a point $x \in \mathcal{T}^{k-1}$ sits within the one-step reachable set $\mathcal{R}_1(x_{i_j}^k)$ as required by (33), we can consider a root finding problem of the form

$$x_{i_j}^k - x + \Delta_t f(x_{i_j}^k, u) = 0. \tag{34}$$

If (34) can be solved for some $u \in \mathbb{U}$, then x must be contained in $\mathcal{R}_1(x_{i_j}^k)$. Alternatively, theory on computing N-step reachable sets for linear systems is well-established (see e.g. Chapter 10 of [23]) and can be used here to compute $\mathcal{R}_1(x_{i_j}^k)$ if the system of interest is linear. Note that there must always exist an $x \in \mathcal{T}^{k-1}$ that sits within $\mathcal{R}_1(x_{i_j}^k)$ since the input u_j, which was used to generate the node $x_{i_j}^k$ can be used to reach the node $x_i^{k-1} \in \mathcal{R}_1(x_{i_j}^k)$. In particular, $\mathcal{T}^{k-1} \cap \mathcal{R}_1(x_{i_j}^k)$ is non-empty.

Algorithm 1 Tree Structure Algorithm for Computing Backwards Reachable Sets

Require: $\Delta_t, \bar{\mathbb{U}} \doteq \{u_1, \ldots, u_{n_u}\} \subset \mathbb{U}$.

1: **initialise**

  Discretize $\mathcal{X}_T$ to $\{x_i^0\}_{i \in \{1,\ldots,n_0\}}$

  $V(T, x_i^0) = g(x_i^0)$.

2: **for** $k = 1, \ldots, N$ **do**

3: $\quad x_{i_j}^k = x_i^{k-1} - \Delta_t f(x_i^{k-1}, u_j), \quad \forall i \in \{1, \ldots, n_{k-1}\}, \ \forall j \in \{1, \ldots, n_u\}$

4: $\quad V(t_k, x_{i_j}^k) = \min_{x \in \mathcal{T}^{k-1} \cap \mathcal{R}_1(x_{i_j}^k)} V(t_{k-1}, x), \quad \forall i_j \in \{1, \ldots, n_k\}$

5: **end for**

As the number of nodes grows by a factor of n_u in each tree level k of Algorithm 1, a pruning strategy becomes necessary to mitigate the exponential increase in the cardinality of the tree. In [3], a pruning strategy based on the distance of the nodes on the same level of the tree was implemented. This turned out to be efficient due to the fact that the value function is a Lipschitz continuous function. Here, in our problem, we offer an alternative pruning criterion based on storing only the nodes that lie along the boundary of the reachable set. In Lemma 1, it was shown that the boundary of the backwards reachable set cannot come from the interior of the terminal set $\mathcal{X}_T$. This applies in a recursive manner, meaning, the boundary of the backwards reachable set at time t cannot come from the interior of the set at a time $\tau > t$. Thus, propagating the tree from nodes that lie in the interior of the backwards reachable set seems to be a wasted expense.

Interior nodes can be identified by points for which $V(t, x) < 0$, and are candidates for pruning. Algorithm 1 could potentially be modified in Steps 2 to 5 with a pruning step that removes nodes in $\mathcal{T}^k$ if $V(t_k, x_{i_j}^k) < -\epsilon$ for some chosen tolerance $\epsilon > 0$. However, removing *all* such nodes may be problematic as (33) minimizes over all nodes in the previous level $k - 1$. If none of the nodes in the previous level have a value less than $-\epsilon$, then it is not possible for any node on level k to have a value less than $-\epsilon$ and hence interior nodes can no longer be identified. This issue is highlighted in Fig. 2. If nodes x_2^2 and x_3^2 can reach node x_2^1 using an input $u \in \mathbb{U}$ and $V(T - \Delta_t, x_2^1) < 0$, then it must also be the case that the value at these nodes is also strictly negative, and are hence interior nodes. However, if x_2^1 is pruned from the previous level, then x_2^2 and x_3^2 can no longer be identified as interior nodes. In

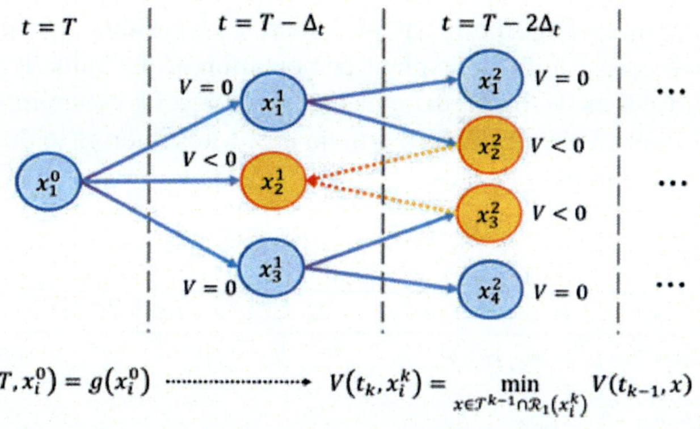

Fig. 2 Diagram depicting tree structure and value function computation as described by Algorithm 1. In orange are nodes that have a strictly negative value, which are candidates for pruning

general, some interior nodes may need to be stored so that nodes on further tree levels can be identified as being interior. These nodes can then be stored more sparsely by removing some of its members without impacting knowledge of the boundary of the reachable set.

There is, however, a case where interior nodes can be identified without needing to store interior nodes in previous tree levels. In particular, a geometric condition for identifying interior nodes can be used. To this extent, first consider the short result below.

Lemma 3 *Assume that the value function* $v \in C([0, T] \times \mathbb{R}^n ; \mathbb{R})$ *in* (11) *is convex in* x *for all* $t \in [0, T]$ *and let* $\{x_i\}_{i\in\{1,\dots,k\}}$ *be a finite set of points contained in* $\mathcal{G}(\tau)$ *for some time* $\tau \in [0, T]$. *Then,*

$$conv\left(\{x_i\}_{i\in\{1,\dots,k\}}\right) \subseteq \mathcal{G}(\tau). \tag{35}$$

Proof Let $\hat{x} \in \text{conv}\left(\{x_i\}_{i\in\{1,\dots,k\}}\right)$, then $\hat{x} = \sum_{i=1}^{k} \lambda_i x_i$ for $\lambda_i \geq 0$ and $\sum_{i=1}^{k} \lambda_i = 1$. By convexity of v, we have that for any $\tau \in [0, T]$,

$$v\left(T - \tau, \hat{x}\right) = v\left(T - \tau, \sum_{i=1}^{k} \lambda_i x_i\right) \leq \sum_{i=1}^{k} \lambda_i v\left(T - \tau, x_i\right). \tag{36}$$

Since $x_i \in \mathcal{G}(\tau) \implies v(T - \tau, x_i) \leq 0$, the right-hand side of (36) is non-positive. This then implies $v\left(T - \tau, \hat{x}\right) \leq 0$, thus $\hat{x} \in \mathcal{G}(\tau)$ and (35) holds.

It follows from Lemma 3 that for convex value functions, points that lie interior to the convex hull of the nodes on any given tree level k must also be interior to the backwards reachable set $\mathcal{G}(t_k)$ (ignoring integration errors of (32)). This implicitly identifies nodes for which $V < 0$.

A modification to Algorithm 1 is presented in Algorithm 2 where assumptions on convexity are used to avoid explicit computation of the value function for identifying interior points. Here $\tilde{\mathcal{T}}^k$ denotes the tree at level k containing a set of nodes $\{\tilde{x}_i^k\}_{i\in\{1,\ldots,\tilde{n}_k\}}$ prior to any pruning. If a node in $\tilde{\mathcal{T}}^k$ lies interior to the convex hull of $\tilde{\mathcal{T}}^k$, it is not added to the tree level $\mathcal{T}^k$, and is thus not propagated in the overall tree structure.

Algorithm 2 Tree Structure Algorithm with Convex Hull Pruning

Require: $\Delta_t, \bar{\mathbb{U}} \doteq \{u_1, \ldots, u_{n_u}\} \subset \mathbb{U}$.
1: **initialise**
  Discretize $\partial\mathcal{X}_T$ to $\{x_i^0\}_{i\in\{1,\ldots,n_0\}}$
2: **for** $k = 1, \ldots, N$ **do**
3: $\tilde{x}_{i_j}^k = x_i^{k-1} - \Delta_t f(x_i^{k-1}, u_j), \quad \forall i \in \{1, \ldots, n_{k-1}\}, \ \forall j \in \{1, \ldots, n_u\}$
4: $\tilde{\mathcal{T}}^k \leftarrow \tilde{x}_{i_j}^k, \quad \forall i \in \{1, \ldots, n_{k-1}\}, \ \forall j \in \{1, \ldots, n_u\}$
5: $n_k \leftarrow 0$
6: **for all** $\tilde{x}_{i_j}^k \in \tilde{\mathcal{T}}^k$ **do**
7: **if** $\tilde{x}_{i_j}^k \in \partial\text{conv}\left(\tilde{\mathcal{T}}^k\right)$ **then**
8: $\mathcal{T}^k \leftarrow \tilde{x}_{i_j}^k$
9: $n_k \leftarrow n_k + 1$
10: **end if**
11: **end for**
12: **end for**
13: **return** $\mathcal{T}^N$

Remark 2 The value function $v \in C([0, T] \times \mathbb{R}^n ; \mathbb{R})$ in (11) is convex in x for all $t \in [0, T]$, if $\mathbb{U}$ is a convex set, $g : \mathbb{R}^n \to \mathbb{R}$ is a convex function, and (3) is described by the linear dynamics

$$\dot{x}(t) = Ax(t) + Bu(t). \tag{37}$$

To see this, we note that the viscosity solution of (11) corresponds to the value function

$$v(t, x) \doteq \inf_{u(\cdot)\in\mathcal{U}} g(\varphi(T; t, x, u(\cdot))). \tag{38}$$

Solutions φ of (37) take the form

$$\varphi(T; t, x, u(\cdot)) = e^{A(T-t)}x + \int_t^T e^{A(T-s)}Bu(s)ds, \tag{39}$$

which is affine in both x and $u(\cdot)$. Thus, g must be convex in both arguments x and $u(\cdot)$. Furthermore, convexity of $\mathbb{U}$ implies convexity of $\mathcal{U}$, then standard results from convex analysis (see e.g. [24]) can be used to show that v must be convex in x. In particular, if a function $f : \mathcal{X} \times \mathcal{Y} \to \mathbb{R}$ is convex in both of its arguments with $\mathcal{Y}$

being a convex set, then the function $\bar{f}(x) \doteq \inf_{y \in \mathcal{Y}} f(x, y)$ is convex in x. This is a special case of Theorem 7.4.13 in [22] where the running cost h in (5) is omitted.

Remark 3 If the system (3) is control affine, i.e. the flow field $f : \mathbb{R}^n \times \mathbb{U} \to \mathbb{R}$ can be decomposed as

$$f(x(t), u(t)) = f_1(x(t)) + f_2(x(t))u,$$

and $\mathbb{U}$ is an ellipsoidal set, then, nodes that lie along the boundary $\partial \mathcal{G}(T)$ must originate from $\partial \mathcal{X}_T$ under a control satisfying $u(t) \in \partial \mathbb{U}$ for all $t \in [0, T]$. This can be demonstrated by noting that points along $x \in \partial \mathcal{G}(T)$ can only reach $\partial \mathcal{X}_T$ (see Lemma 1) and they do so under a control law satisfying $u(t) \in \operatorname{argmin}_{u \in \mathbb{U}} \langle \nabla v, f(x, u) \rangle$. If (3) is control affine, then the control is minimal with respect to a linear function in u. Thus, if $\mathbb{U}$ is ellipsoidal, the minimising control lies along $\partial \mathbb{U}$ (see standard results on support functions over ellipsoidal sets e.g. [24]).

6 Numerical Examples

We use Algorithm 2 to compute the backwards reachable set $\mathcal{G}(T)$ for two example systems. The first example looks at a linear system with two states and two inputs, whilst the second example looks at a nonlinear system model for a DC motor consisting of three states. For comparison purposes, the backwards reachable sets are also computed using an off-the-shelf toolbox provided by [25]. This toolbox contains a grid-based, finite-difference scheme for numerically evaluating the value function of Theorem 2. By omitting the running cost $h : \mathbb{R}^n \times \mathbb{U} \to \mathbb{R}$ and extracting the zero level set $\{x \in \mathbb{R}^n \mid V(0, x) = 0\}$, we obtain the boundary of the backwards reachable set. Throughout the following, we will use $\mathcal{G}_{\mathcal{T}}(T)$ to denote the backwards reachable set computed by taking the convex hull of the final tree nodes $\mathcal{T}^N$ in Algorithm 2 and $\mathcal{G}_{FD}(T)$ to denote the backwards reachable set computed via the toolbox of [25]. In the following examples, we make use of the 'qhull' algorithm (see [26]) to compute the convex hull in Step 7 of Algorithm 2, which is available via standard routines in MATLAB.

6.1 Numerical Example 1: Linear System with Two States

Consider the linear time-invariant system described by

$$\dot{x}(t) = \begin{bmatrix} 0 & 1 \\ 1 & 0 \end{bmatrix} x(t) + \begin{bmatrix} 1 & 0 \\ 0 & 1 \end{bmatrix} u(t), \quad \forall t \in (0, T), \tag{40}$$

where $x(t) = [x_1(t), x_2(t)]^T \in \mathbb{R}^2$ is the state and $u(t) \in \mathbb{U} \subset \mathbb{R}^2$ is the input at time t, with a terminal condition $x(T) \in \mathcal{X}_T \subset \mathbb{R}^2$. We take the input constraint set $\mathbb{U}$ and the terminal set $\mathcal{X}_T$ to be ellipsoidal sets given by

$$\mathbb{U} \doteq \mathcal{E}\left(\begin{bmatrix}0\\1\end{bmatrix}, \begin{bmatrix}4 & 0\\0 & 1\end{bmatrix}\right), \quad \mathcal{X}_T \doteq \mathcal{E}\left(\begin{bmatrix}0\\0\end{bmatrix}, \begin{bmatrix}0.01 & 0\\0 & 0.01\end{bmatrix}\right). \tag{41}$$

In the notation of (9), we have the terminal state cost $g(x) = 100\|x\|_2^2 - 1$.

To implement Algorithm 2 for (40), the terminal set $\mathcal{X}_T$ was discretised into a set of $n_0 = 20$ nodes, uniformly distributed along its boundary. Since (40) is control affine and $\mathbb{U}$ is ellipsoidal, the set of optimal inputs that generate nodes along the boundary of the backwards reachable set must come from the boundary of $\mathbb{U}$ (see Remark 3). Accordingly, $\mathbb{U}$ in (41) was discretised into a set of $n_u = 15$ points distributed about its boundary. In particular, the following discretised input set was used:

$$\bar{\mathbb{U}} \doteq \left\{\begin{bmatrix}2 & 0\\0 & 1\end{bmatrix} w + \begin{bmatrix}0\\1\end{bmatrix} \,\middle|\, w = \left[\sin\left(\frac{2\pi k}{n_u}\right), \cos\left(\frac{2\pi k}{n_u}\right)\right]^T, k \in \{1, \cdots, n_u\}\right\}. \tag{42}$$

To compute $\mathcal{G}_{FD}(T)$ via a finite-difference scheme, a grid of 200×200 nodes was used, noting that coarser grids were noticeably under-approximating compared to expected results. The backwards reachable sets as computed by Algorithm 2 and the finite-difference scheme are depicted in Fig. 3 for $T = 1$s. Also displayed in Fig. 4

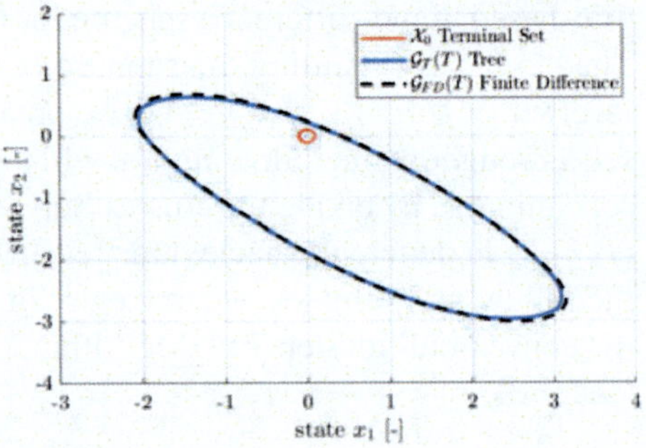

Fig. 3 Comparison of backwards reachable sets for the system dynamics in (40) computed via a finite difference approach and Algorithm 2

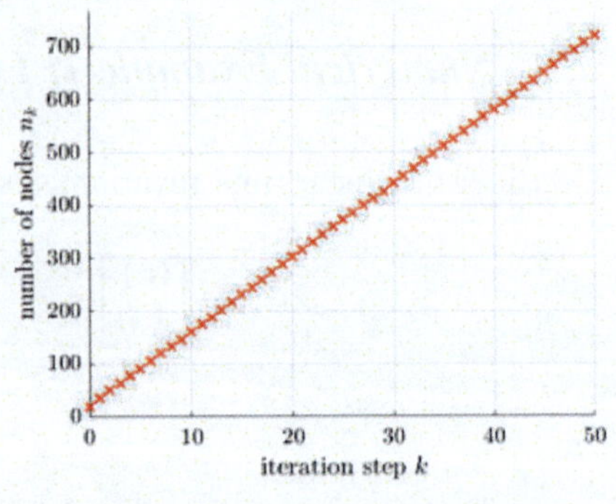

Fig. 4 Number of nodes used to describe $\mathcal{G}_\mathcal{T}(T)$ for the system dynamics (40) in each iteration of Algorithm 2

is the number of nodes n_k used to describe the boundary of the reachable set in each iteration of Algorithm 2, noting that a time-step of $\Delta_t = 0.02$s was selected.

Using a four-core Intel® Core™ i7-1065G7 CPU, computation times in MATLAB for computing $\mathcal{G}_{\mathcal{T}}(T)$ and $\mathcal{G}_{FD}(T)$ were 6.11s, and 34.1s, respectively. The internal area of the sets $\mathcal{G}_{\mathcal{T}}(T)$ and $\mathcal{G}_{FD}(T)$ were computed via a trapezoidal integration scheme and were found to be 8.50 units^2 and 8.68 units^2, respectively. We observe that Algorithm 2 offers markedly lower computation times at a fairly small expense to the captured area, which can be improved by increasing the size of the discretised set $\bar{\mathbb{U}}$ if needed. It is interesting to note that despite Step 3 of Algorithm 2 producing 15 'candidate' nodes for each node in the previous tree level, the resulting number of nodes does not grow exponentially after pruning. The final tree level returned by Algorithm 2 has only 720 nodes as compared with the 40,000 grid points used in the finite difference scheme. Moreover, if a longer horizon T was selected, then a new grid, covering a larger domain, may be required for the finite-difference scheme. This is not necessary for Algorithm 2.

In this example, the value function $v : [0, T] \times \mathbb{R}^n \to \mathbb{R}$ as described in Theorem 3 can be shown to be convex (see Remark 2) thus the backwards reachable set $\mathcal{G}(T)$ of (40) is also convex, since sub-level sets of convex functions are convex. Consequently, taking a convex hull of the nodes contained in each tree level $\{x_i^k\}_{i\in\{1,\cdots,n_k\}}$ results in the backwards reachable set computed via Algorithm 2 being an inner approximation of $\mathcal{G}(T)$ (ignoring integration errors in Step 3 of Algorithm 2).

6.2 *Numerical Example 2: DC Motor*

In our next example, we consider a typical nonlinear system model for a DC motor, which consists of three states: the rotor angle x_1, the rotor angular velocity x_2, and the armature current x_3. The input to the system is the supplied voltage u. An example DC motor model with arbitrarily selected system parameters is given by

$$\dot{x}(t) = \begin{bmatrix} x_2(t) \\ -10\sin(x_1(t)) - \text{sign}(x_2(t))x_2^2(t) + 5x_3(t) \\ -10x_2(t) + 50x_3(t) \end{bmatrix} + \begin{bmatrix} 0 \\ 0 \\ 50 \end{bmatrix} u(t), \quad \forall t \in (0, T), \tag{43}$$

where $x(t) = [x_1(t), x_2(t), x_3(t)]^T \in \mathbb{R}^3$ is the state and $u(t) \in \mathbb{U} \subset \mathbb{R}$ is the input at time t, with a terminal condition $x(T) \in \mathcal{X}_T \subset \mathbb{R}^2$. We take the input constraint set $\mathbb{U}$ and the terminal set $\mathcal{X}_T$ to be ellipsoidal sets given by

$$\mathbb{U} \doteq \mathcal{E}(0, 4), \qquad \mathcal{X}_T \doteq \mathcal{E}\left(\begin{bmatrix} \frac{\pi}{2} \\ 0 \\ 0 \end{bmatrix}, 0.04\mathbb{I}_3\right). \tag{44}$$

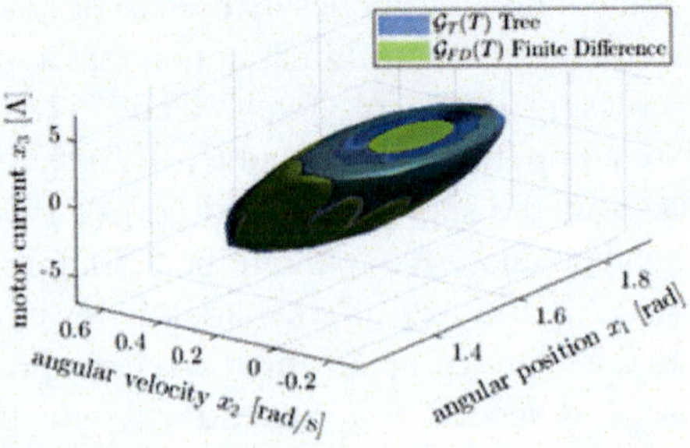

Fig. 5 Comparison of backwards reachable sets for the system dynamics in (43) computed via a finite difference approach and Algorithm 2

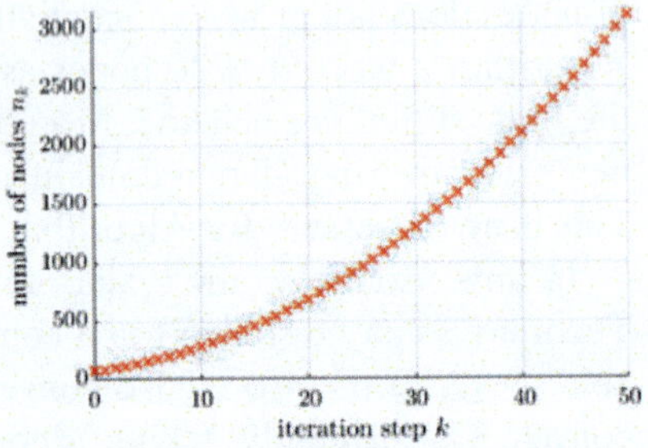

Fig. 6 Number of nodes used to describe $\mathcal{G}_{\mathcal{T}}(T)$ for the system dynamics (43) in each iteration of Algorithm 2

Here, the input constraint set is equivalent to the interval $\mathbb{U} = [-2, 2]$ and the terminal set $\mathcal{X}_T$ is a small ball around the unstable equilibrium (with zero input) $\bar{x} = \left[\frac{\pi}{2}, 0, 0\right]^T$ of (43). Since (43) is control affine and $\mathbb{U}$ is an interval constraint, optimal controls that generate nodes along the boundary of the reachable set $\mathcal{G}(T)$ must lie on the extremal points of $\mathbb{U}$ (see Remark 3 and Remark 3.1 of [3]). A natural discretization of $\mathbb{U}$ is then given by $\bar{\mathbb{U}} \doteq \{-2, 2\}$, which contains the optimal control for evolving the boundary of the backwards reachable set. Additionally, to compute $\mathcal{G}(T)$ of (43) using Algorithm 2, the terminal set $\mathcal{X}_T$ was discretised into a set of $n_0 = 84$ nodes, uniformly distributed along its boundary.

Likewise with the first numerical example, the finite-difference scheme implemented by the toolbox of [25] was used as comparison for computing the backwards reachable set. In the finite-difference scheme, a grid of $101 \times 101 \times 101$ nodes was used. The backwards reachable set as computed by Algorithm 2 and the finite-difference scheme are depicted in Fig. 5 for $T = 0.02$s with a time discretisation of $\Delta_t = 0.4$ms used in Algorithm 2. Figure 6 displays the number of nodes n_k used to describe the boundary of the reachable set in each iteration of Algorithm 2.

Computation times in MATLAB for computing $\mathcal{G}_{\mathcal{T}}(T)$ and $\mathcal{G}_{FD}(T)$ were 4.984s and 456.8s, respectively. Numerical integration of the sets $\mathcal{G}_{\mathcal{T}}(T)$ and $\mathcal{G}_{FD}(T)$ produced internal volumes of 1.059 units^3 and 1.055 units^3, respectively. Again, the computation times are noticeably smaller for Algorithm 2 whilst still producing a backwards reachable set of comparable volume and shape to the finite-difference scheme. The backwards reachable set appears to be convex, and from Lemma 3, this would suggest that $\mathcal{G}_{\mathcal{T}}(T)$ should be an inner approximation of $\mathcal{G}_{FD}(T)$, and thus should be contained inside the set in green in Fig. 5. However, integration errors in both Algorithm 2 and the finite-difference scheme may cause the boundary of $\mathcal{G}_{\mathcal{T}}(T)$ to lie outside of $\mathcal{G}_{FD}(T)$.

The number of nodes in the final tree level $\mathcal{T}^N$ of Algorithm 2 was 3111, and like the first numerical example, the nodes seem to exhibit sub-exponential growth in the number of iteration steps k due to the pruning of interior points. We note that if a long horizon is used, then the number of nodes may grow to be quite large. In such a case, additional pruning may be needed, for instance, removing nodes that lie sufficiently close to other nodes in the same tree level as is suggested in [3].

7 Conclusions and Future Works

In this work, we have proposed a tree structure algorithm to compute reachable sets using the Hamilton-Jacobi approach. Our method computes the tree backwards in time, starting from the terminal set and using a finite set of controls. To mitigate issues associated with the exponential increase in the cardinality of the tree, we have introduced a pruning strategy based on geometric considerations. In fact, at each time level of the tree, we neglect all the nodes that lie in the interior of the convex hull of that level set.

In our numerical examples, we have shown how the algorithm compares to a finite-difference approach for a 2D linear and a 3D nonlinear system. Our method provides very accurate results with a measurable speed up in terms of computational time.

This is, to the best of the authors' knowledge, the first approach which uses a tree structure algorithm for reachable set computation. To further validate our approach, we will consider higher dimensional systems driven by applications. For instance, guidance of aircraft, collision avoidance of multi-agent systems, and control of systems described by partial differential equations.

In the future, we would like to generalize our method by relaxing our convexity assumption, which was crucial for validating our approach, and instead consider semi-concave value functions. This would then allow us to retain the same degree of accuracy for a larger class of systems. Finally, it will also be of interest to be able to construct a safety-based control law so as to guarantee that we stay within the backwards reachable set for all time.

Acknowledgements AA is a member of the InDAMGNCS activity group. AA wants to acknowledge the *Overseases Mobility program* financed by Università Ca' Foscari Venezia. Funding for this research was also supported through an Australian Research Council Linkage Project grant (Grant number: LP190100104), an Asian Office of Aerospace Research and Development grant (Grant number: AOARD22IOA074), and the Australian Commonwealth Government through the Ingenium Scholarship. Acknowledgement is also given to BAE systems as a collaborator in the aforementioned research grants.

References

1. Kunisch, K., Volkwein, S., Xie, L.: HJB-POD-based feedback design for the optimal control of evolution problems. SIAM J. Appl. Dyn. Syst. **3**(4), 701–722 (2004)
2. Alla, A., Falcone, M., Volkwein, S.: Error analysis for POD approximations of infinite horizon problems via the dynamic programming approach. SIAM J. Control Optim. **55**(5), 3091–3115 (2017)
3. Alla, A., Falcone, M., Saluzzi, L.: An efficient DP algorithm on a tree-structure for finite horizon optimal control problems. SIAM J. Sci. Comput. **41**(4), A2384–A2406 (2019)
4. Alla, A., Saluzzi, L.: A HJB-POD approach for the control of nonlinear PDEs on a tree structure. Appl. Numer. Math. 155, 192–207 (2020). Structural Dynamical Systems: Computational Aspects held in Monopoli (Italy) on 12–15 June 2018. https://doi.org/10.1016/j.apnum.2019.11.023
5. Kalise, D., Kunisch, K.: Polynomial approximation of high-dimensional Hamilton-Jacobi-Bellman equations and applications to feedback control of semilinear parabolic PDEs. SIAM J. Sci. Comput. **40**(2), A629–A652 (2018)
6. McEneaney, W.M.: A curse-of-dimensionality-free numerical method for solution of certain HJB PDEs. SIAM J. Control Optim. **46**(4), 1239–1276 (2007)
7. McEneaney, W.M.: Convergence rate for a curse-of-dimensionality-free method for Hamilton-Jacobi-Bellman PDEs represented as maxima of quadratic forms. SIAM J. Control Optim. **48**(4), 2651–2685 (2009)
8. Chow, Y.T., Darbon, J., Osher, S., Yin, W.: Algorithm for overcoming the curse of dimensionality for state-dependent Hamilton-Jacobi equations. J. Comput. Phys. **387**, 376–409 (2019)
9. Yegorov, I., Dower, P.M.: Perspectives on characteristics based curse-of-dimensionality-free numerical approaches for solving Hamilton-Jacobi equations. Appl. Math. Optim. **83**(1), 1–49 (2021)
10. Darbon, J., Langlois, G.P., Meng, T.: Overcoming the curse of dimensionality for some Hamilton-Jacobi partial differential equations via neural network architectures. Res. Math. Sci. **7**(3), 1–50 (2020)
11. Darbon, J., Meng, T.: On some neural network architectures that can represent viscosity solutions of certain high dimensional Hamilton-Jacobi partial differential equations. J. Comput. Phys. **425**, 109907 (2021)
12. Dolgov, S., Kalise, D., Kunisch, K.K.: Tensor decomposition methods for high-dimensional Hamilton-Jacobi-Bellman equations. SIAM J. Sci. Comput. **43**(3), A1625–A1650 (2021)
13. Oster, M., Sallandt, L., Schneider, R.: Approximating optimal feedback controllers of finite horizon control problems using hierarchical tensor formats. SIAM J. Sci. Comput. **44**(3):B746–B770 (2022). https://doi.org/10.1137/21M1412190
14. Bokanowski, O., Garcke, J., Griebel, M., Klompmaker, I.: An adaptive sparse grid semi-Lagrangian scheme for first order Hamilton-Jacobi Bellman equations. J. Sci. Comput. **55**(3), 575–605 (2013)
15. Mitchell, B.A., Tomlin, I.M.C.: A time-dependent Hamilton-Jacobi formulation of reachable sets for continuous dynamic games. IEEE Trans. Autom. Control **50**(7), 947–957 (2005)
16. Chen, M., Herbert, S., Vashishtha, M., Bansal, S., Tomlin, C.: Decomposition of reachable sets and tubes for a class of nonlinear systems. IEEE Trans. Autom. Control **63**(11), 3675–3688 (2018)
17. Althoff, M., Krogh, B.: Reachability analysis of nonlinear differential-algebraic systems. IEEE Trans. Autom. Control **59**(2), 371–383 (2013)
18. Yang, L., Ozay, N.: Scalable zonotopic under-approximation of backward reachable sets for uncertain linear systems. IEEE Control Syst. Lett. **6**, 1555–1560 (2021)
19. Bardi, M., Capuzzo-Dolcetta, I.: Optimal Control and Viscosity Solutions of Hamilton-Jacobi-Bellman Equations. Birkhäuser, Basel (1997)
20. Crandall, M.G., Lions, P.-L.: Viscosity solutions of Hamilton-Jacobi equations. Trans. Am. Math. Soc. **277**(1), 1–42 (1983)

21. Chen, M., Tomlin, C.J.: Hamilton-Jacobi reachability: some recent theoretical advances and applications in unmanned airspace management. Ann. Rev. Control Robot. Auton. Syst. **1**(1), 333–358 (2018)
22. Cannarsa, P., Sinestrari, C.: Semiconcave Functions, Hamilton-Jacobi Equations, and Optimal Control, vol. 58. Springer Science & Business Media (2004)
23. Borrelli, F., Bemporad, A., Morari, M.: Predictive Control for Linear and Hybrid Systems. Cambridge University Press (2017)
24. Rockafellar, R.T.: Convex Analysis, vol. 18. Princeton University Press (1970)
25. Mitchell, I.M.: A toolbox of level set methods. https://www.cs.ubc.ca/~mitchell/ToolboxLS/
26. Barber, C.B., Dobkin, D.P., Huhdanpaa, H.: The quickhull algorithm for convex hulls. ACM Trans. Math. Soft. (TOMS) **22**(4), 469–483 (1996)

Asymptotic-Preserving Neural Networks for Hyperbolic Systems with Diffusive Scaling

Giulia Bertaglia

Abstract With the rapid advance of Machine Learning techniques and the deep increase of availability of scientific data, data-driven approaches have started to become progressively popular across science, causing a fundamental shift in the scientific method after proving to be powerful tools with a direct impact in many areas of society. Nevertheless, when attempting to analyze dynamics of complex multiscale systems, the usage of standard Deep Neural Networks (DNNs) and even standard Physics-Informed Neural Networks (PINNs) may lead to incorrect inferences and predictions, due to the presence of small scales leading to reduced or simplified models in the system that have to be applied consistently during the learning process. In this Chapter, we will address these issues in light of recent results obtained in the development of *Asymptotic-Preserving Neural Networks* (APNNs) for hyperbolic models with diffusive scaling. Several numerical tests show how APNNs provide considerably better results with respect to the different scales of the problem when compared with standard DNNs and PINNs, especially when analyzing scenarios in which only little and scattered information is available.

Keywords Asymptotic-preserving methods · Physics-informed neural networks · Multiscale hyperbolic systems · Discrete-velocity kinetic models · Diffusion limit

1 Introduction

The study of dynamics of complex systems described by multiscale partial differential equations (PDEs) has applications ranging from classical physics and engineering to biology, socio-economics and life sciences in general [1–5]. Despite the continuous progress achieved in the understanding of such systems, the modeling and prediction of the evolution of nonlinear multiscale phenomena using classical analytical or

G. Bertaglia (✉)
Department of Environmental and Prevention Sciences, University of Ferrara,
Corso Ercole I D'Este 32, 44121 Ferrara, Italy
e-mail: giulia.bertaglia@unife.it

G. Albi et al. (eds.), *Advances in Numerical Methods for Hyperbolic Balance Laws and Related Problems*, SEMA SIMAI Springer Series 32,
https://doi.org/10.1007/978-3-031-29875-2_2

computational tools goes through several demanding challenges. Firstly, the numerical solution of a multiscale problem requires complex and elaborate computational codes and can introduce prohibitive costs, due to the well-known *curse of dimensionality*. Secondly, especially when working in the social sciences, it is inevitable to come up against the difficulties associated with data scarcity and multiple sources of uncertainty [6–10]. Not to mention that solving real physical problems with missing or incomplete initial or boundary conditions is currently impractical with traditional approaches.

With rapid advances in the field of Machine Learning (ML) and hugely increasing amounts of scientific data, data-driven approaches have started to become increasingly popular across science, causing a fundamental shift in the scientific method after proving to be tools with innumerable potentials and with direct impact in many areas of society [11–15]. One can then attempt to gain support from ML techniques even for the study of complex systems described by multiscale dynamics. When doing so, approaching the problem with the usage of standard ML methods, it is important to recall that purely data-driven models may fit observations very well (especially when a huge amount of data is available), but predictions may be physically inconsistent or unrealistic, with extrapolations leading to erroneous generalizations. Since, in general, we possess a-priori, albeit incomplete, knowledge of the physical laws governing the phenomenon under study, the key idea in the blossoming field of physics-informed ML is to include this type of prior scientific knowledge into the machine learning work-flow, "teaching" ML models not only to match observed data, but also to respect, in the extrapolations produced, the physical laws that we know govern the dynamics of interest. In fact, this approach aims to complement any (very likely) scarcity of available data with the knowledge of mathematical physical models, even in partially-understood, uncertain and high-dimensionality contexts, guaranteeing robustness, more accurate and, above all, physically consistent predictions [16].

A recent example reflecting this new learning philosophy is represented by Physics-Informed Neural Networks (PINNs), first introduced in [17]. PINNs constitute a new class of deep neural networks (DNNs) that are trained to solve supervised learning tasks respecting physical laws described through linear or nonlinear ordinary differential equations (ODEs) or PDEs. The physical knowledge of the phenomenon under study, in addition to being incorporated into the learning process of the DNN directly through data embodying the underlying physics of the phenomenon of interest (observational bias), is introduced through an appropriate choice of the loss function that the DNN has to minimize, which forces the training phase of the neural network to converge towards solutions that adhere to the physics (learning bias) [18–21].

However, the use of a standard formulation of PINNs in the context of multiscale problems can still lead to incorrect inferences and predictions [22, 23]. This is mainly due to the presence of small scales leading to reduced or simplified models in the system that have to be applied consistently during the learning process. In these cases, a standard PINN formulation only allows an accurate description of the process at the leading order, thus losing accuracy in asymptotic limit regimes. A remedy to this problem, as recently proposed in [22, 23], is to modify the loss function to include asymptotic-preserving (AP) properties during the training process. The realization of such an AP-loss function will therefore depend on the particular problem under consideration and be based on an appropriate asymptotic analysis of the model.

In this Chapter, we will address these issues in the light of recent results obtained in the development of *asymptotic-preserving neural networks* (APNNs) for hyperbolic models with diffusive scaling. More precisely, the presentation is organized as follows: Sect. 2 is devoted to the overview of hyperbolic systems with relaxation terms leading to an asymptotic parabolic limit; in Sect. 3 we review the basic concepts of DNNs and PINNs; in Sect. 4 the construction of APNNs with AP loss functions for multiscale hyperbolic systems of interest is discussed, giving an analytical demonstration of the AP property and its importance in the context of neural networks; in Sect. 5 the different behavior of a standard DNN, a standard PINN and an APNN are analyzed and compared considering the Goldstein-Taylor model as prototype system of equations, evaluated for different regimes; finally, an application of APNNs in the context of multiscale hyperbolic transport models for the study of the propagation of infectious diseases is presented in Sect. 6 and conclusive remarks are given in Sect. 7.

2 Hyperbolic Systems with Diffusive Scaling

The prototype hyperbolic system with relaxation that we will consider is a discrete-velocity kinetic model describing the space-time evolution of two particle densities, $f^+(x,t)$ and $f^-(x,t)$, at time $t > 0$, traveling in a 1D domain, $x \in \mathcal{D} \subseteq \mathbb{R}$, with opposite velocities ± 1, respectively, which is known as Goldstein-Taylor model [2, 24, 25]:

$$\frac{\partial f^+}{\partial t} + \frac{1}{\epsilon}\frac{\partial f^+}{\partial x} = -\frac{\sigma}{2\epsilon^2}\left(f^+ - f^-\right), \tag{1}$$

$$\frac{\partial f^-}{\partial t} - \frac{1}{\epsilon}\frac{\partial f^-}{\partial x} = \frac{\sigma}{2\epsilon^2}\left(f^+ - f^-\right). \tag{2}$$

Here σ identifies a possible scattering coefficient and ϵ represents the scaling factor. Indeed, in this model, particles can collide and, at the same time, assume the opposite velocity, randomly, with a probability that depends on the scaling factor ϵ, which is strictly related to the *mean free path* of particles: the more ϵ tends to zero, the smaller the mean free path of the particles and the more collisions will occur. In this context, the role of the operator which describes the interactions with the background is to push the solution towards a universal steady state (at exponential-in-time rate proportional to the collision frequency). The typical definition used in kinetic theory for this type of operator is indeed *relaxation operator* [2].

If we define the total particles density $\rho = f^+ + f^-$ and the re-scaled flux $j = \epsilon^{-1}(f^+ - f^-)$, we can rewrite system (1)–(2) obtaining an equivalent model expressed in terms of macroscopic variables:

$$\frac{\partial \rho}{\partial t} + \frac{\partial j}{\partial x} = 0\,, \tag{3}$$

$$\frac{\partial j}{\partial t} + \frac{1}{\epsilon^2}\frac{\partial \rho}{\partial x} = -\frac{\sigma}{\epsilon^2} j\,, \tag{4}$$

With this formulation, we can analyze the behavior of the solution as $\epsilon \to 0$, i.e., in the *zero-relaxation limit*. In this asymptotic limit, the second equation relaxes to the local equilibrium

$$j = -\frac{1}{\sigma}\frac{\partial \rho}{\partial x}\,, \tag{5}$$

and substituting into the first equation gives the following parabolic, diffusive limit that recalls the standard heat equation:

$$\frac{\partial \rho}{\partial t} = \frac{\partial}{\partial x}\left[\frac{1}{\sigma}\frac{\partial \rho}{\partial x}\right]. \tag{6}$$

It is therefore clear that, depending on the scaling factor ϵ, system (1)–(2) describes different distinct physical propagation phenomena that range from that of advective transport to that of parabolic diffusion [26, 27], associated with spatio-temporal scales which differ by several orders of magnitude.

In classical kinetic theory, the space-time scaling just discussed is related to one specific hydrodynamical limit of the Boltzmann equation. In particular, when the dissipative effects become non negligible, from the Boltzmann equation we recover the incompressible Navier–Stokes scaling. We refer the reader to [28] for further details and the mathematical theory behind the hydrodynamical limits of the Boltzmann equation and to [29] for theoretical results on the diffusion limit of a system like (3)–(4).

3 Review of Deep Neural Networks and Physics-Informed Neural Networks

In this Section, we provide a brief summary of the general framework of DNNs and their extension to PINNs [17, 18]. After this overview, the relevant concepts of APNNs for the problems of interest will be discussed in Sect. 4.

3.1 Deep Neural Networks (DNNs)

Let us assume we want to evaluate the dynamic of a variable $U \in \Omega \subset \mathbb{R}^d \times \mathbb{R}$ through a data-driven approach, recurring to a classical DNN, for example an $L+1$ layered feed-forward neural network (FNN), which consists of an input layer, an

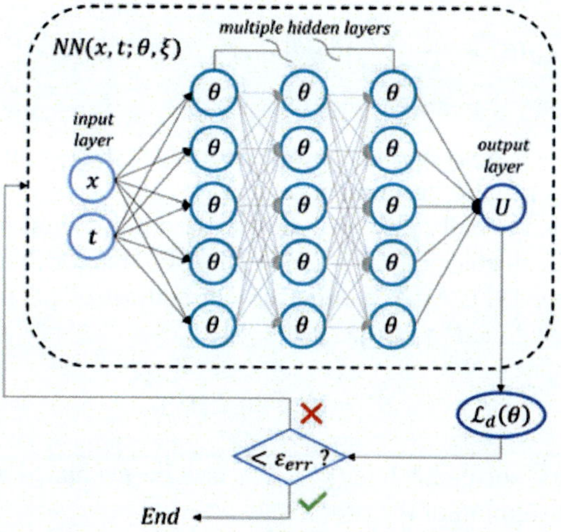

Fig. 1 NN schematic work-flow. Given the location (x, t) of the available training data as input layer, to find the optimal values for θ in the hidden layers, the neural network is trained by minimizing the loss function $\mathcal{L}_d(\theta)$ that generally consists in the mean squared error between the NN's predictions (output layer) and the training points. Iterations end when the error evaluated at the training points is less than a set threshold ε_{err}

output layer, and $L - 1$ hidden layers. Given the location of an available dataset of N_d observations $z \in \mathbb{R}^d$, $d = 2$, as input, the FNN can be used to output a prediction of the value $U \approx U_{NN}(z; \theta)$, parameterized by network parameters θ, as shown in Fig. 1 for a spatio-temporal dynamic in 1D space. To this aim, we can define the FNN as follows:

$$\begin{aligned} z^1 &= W^1 z + b^1 , \\ z^l &= \sigma \circ (W^l z^{l-1} + b^l), \quad l = 2, \dots, L-1 , \\ U_{NN}(z; \theta) = z^L &= W^L z^{L-1} + b^L , \end{aligned}$$

where $W^l \in \mathbb{R}^{m_l \times m_{l-1}}$ are the weights, $b^l \in \mathbb{R}^{m_l}$ the bias, m_l is the width of the l-th hidden layer with $m_1 = d_{in} = d$ the input dimension and $m_L = d_{out}$ the output dimension, σ is a scalar activation function (such as ReLU [30]), and "$\circ$" denotes entry-wise operation. Thus, we denote the set of network parameters $\theta = (W^1, b^1, \dots, W^L, b^L)$.

To learn the model, the networkâŁ™s free parameters θ are tuned to find the optimal values at each iteration of the neural network through a supervised learning process, so that the DNNâŁ™s predictions closely match the available experimental data. This is usually done by minimizing a loss function $\mathcal{L}_d(\theta)$ (also called cost or risk function) that consists in the mean squared error (MSE) between the DNN's predictions and the training points, which for a spatio-temporal dynamic in 1D space reads:

$$\mathcal{L}_d(\theta) = \frac{1}{N_d} \sum_{i=1}^{N_d} \left| U_{NN}(x_d^i, t_d^i; \theta) - U(x_d^i, t_d^i) \right|^2 , \tag{7}$$

where U_d^i, $i = 1, \ldots, N_d$ is the available measurement dataset and $\{(x_d^i, t_d^i)\}_{i=1}^{N_d} \subset \Omega$ are the coordinates of these training points.

This problem is typically cast as a stochastic optimization problem and generally solved using a stochastic gradient descent (SGD) algorithm such as the Adam advanced optimizer [31]. After finding the optimal set of parameter values θ^* by minimizing the loss function (7), i.e.,

$$\theta^* = \operatorname{argmin} \mathcal{L}_d(\theta),$$

the neural network surrogate $U_{NN}(x, t; \theta^*)$ can be evaluated at any point in the domain to get the solution of the problem.

Therefore, the design of a DNN can be summarized in the following three main steps [32]:

- the choice of the neural network structure,
- the choice of the loss function,
- the choice of the method to minimize the loss over the parameter space.

Notice that, in practice, the performance of the neural network is estimated on a finite set of points which is unrelated to any data used to train the model evaluating a *test error* (or *validation error*); whereas the error in the loss function, which is used for training purposes, is called the *training error*.

3.2 *Physics-Informed Neural Networks (PINNs)*

As already discussed in Sect. 1, using a purely data-driven approach like the one just presented can have significant weaknesses. Indeed, whilst a standard DNN might accurately model the physical process within the vicinity of the experimental data, it will fail to generalize away from the training points. This is because the neural network cannot actually *learn* the physical dynamic of the phenomenon [21], and recurring to standard DNNs we are essentially forgetting about our existing scientific knowledge.

To improve the efficiency and efficacy of the learning process, PINNs take advantage of the prior physical knowledge in our possession including it into the neural network work-flow, providing a powerful tool that can address both the inverse and the forward problem of data-driven learning of the dynamic and solving the problem of interest, respectively [17, 18, 33].

More specifically, let us first consider the context of inverse problems assuming that, in addition to having access to experimental data, we know that the dynamic investigated is governed by a differential operator $\mathcal{F}$ in the spatio-temporal domain $\Omega \subset \mathbb{R}^d \times \mathbb{R}$ which depends on parameters ξ related to the physics whose values are unknown. Furthermore, the dynamic is subject to a general operator $\mathcal{B}$ that prescribes arbitrary initial and boundary conditions of the system in $\partial\Omega$, which could be known at specific locations. This generic governing system will read as follows:

$$\mathcal{F}(U, x, t; \xi) = 0, \quad (x, t) \in \Omega, \tag{8}$$

$$\mathcal{B}(U, x, t; \xi) = 0, \quad (x, t) \in \partial\Omega. \tag{9}$$

To transform a standard NN into a PINN, it is sufficient to require the neural network to satisfy not only the agreement with experimental data, but also the physics we know of the system, acting directly on the loss function by adding two more terms:

$$\mathcal{L}(\theta, \xi) = w_d^T \mathcal{L}_d(\theta, \xi) + w_b^T \mathcal{L}_b(\theta, \xi) + w_r^T \mathcal{L}_r(\theta, \xi), \tag{10}$$

with

$$\mathcal{L}_r(\theta, \xi) = \frac{1}{N_r} \sum_{n=1}^{N_r} \left| \mathcal{F}(U_{NN,r}^n, x_r^n, t_r^n; \theta, \xi) \right|^2, \tag{11}$$

$$\mathcal{L}_b(\theta, \xi) = \frac{1}{N_b} \sum_{k=1}^{N_b} \left| \mathcal{B}(U_{NN,b}^k, x_b^k, t_b^k; \theta, \xi) \right|^2, \tag{12}$$

where $\{(x_r^n, t_r^n)\}_{n=1}^{N_r} \subset \Omega$ and $\{(x_b^k, t_b^k)\}_{k=1}^{N_b} \subset \partial\Omega$ are scattered points within the domain and on the boundary, respectively, called *residual points*, at which PINN is asked to satisfy the known physics; while $U_{NN,r}^n = U_{NN}(x_r^n, t_r^n)$ and $U_{NN,b}^k = U_{NN}(x_b^k, t_b^k)$. Here, the additional residual terms, $\mathcal{L}_r$ and $\mathcal{L}_b$, quantify the discrepancy of the neural network surrogate U_{NN} with respect to the underlying differential operator and its initial or boundary conditions in (9), respectively. Finally, w_r, w_b, w_d are the weights associated to each contribution. Notice that if one aims at penalizing a specific loss term with respect to the others (e.g., $\mathcal{L}_r$, in the context of a scalar $\mathcal{F}$ and $\mathcal{B}$) it is either possible to set a weight smaller than one in front of it (e.g., $\omega_r = 0.1$, $\omega_d = \omega_b = 1$) or to use weights higher than one in front of the rest of the loss terms in the loss function $\mathcal{L}$ (e.g., $\omega_d = \omega_b = 10$, $\omega_r = 1$).

In this additional process, gradients of the networkâĿ™s output with respect to its input are computed at each residual point typically recurring to automatic differentiation [34], and the residual of the underlying differential equation is evaluated using these gradients.

In the context of inverse problems, the unknown physical parameters ξ are treated as learnable parameters, and we search for those that best describe the observed data. As a result, the training process involves optimizing θ and ξ jointly:

$$(\theta^*, \xi^*) = \text{argmin}\, \mathcal{L}(\theta, \xi).$$

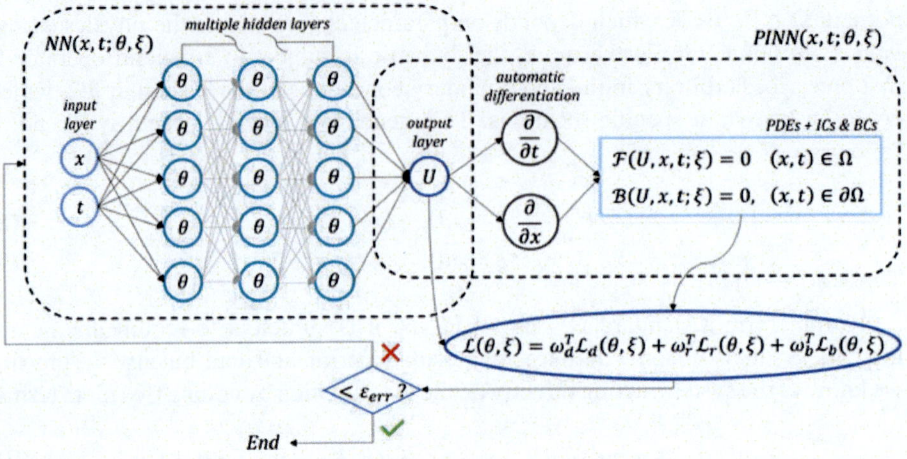

Fig. 2 PINN schematic work-flow. The NN architecture is integrated with the physical knowledge of the dynamic of interest through the inclusion of the differential operator (in this case a PDE) and the enforcement of initial and boundary conditions (and eventually conservation properties), when known, becoming a PINN. To this aim, the loss function encloses three different residual terms, related to the mismatch with respect to available data $\mathcal{L}_d(\theta)$, known physical law $\mathcal{L}_r(\theta)$ and boundary/initial conditions $\mathcal{L}_b(\theta)$, respectively

For a schematic representation of the discussed PINN work-flow the reader can refer to Fig. 2.

In the context of forward problems, the structure of the network is almost the same as in the inverse problem, except that no matching with data in the learning phase is normally considered and we assume to know ξ. In essence, we ask the PINN what can be said about the unknown hidden state U of the system, given the fixed parameters ξ of the model. Thus, the optimization task will generally read:

$$\theta^* = \text{argmin}\left[w_r^T \mathcal{L}_r(\theta) + w_b^T \mathcal{L}_b(\theta)\right].$$

The physics-informed neural networks thus constructed are able to predict the solution also away from the experimental data points by behaving much better than a standard neural network, as we will explicitly see in Sect. 5.1.

4 Asymptotic-Preserving Neural Networks

Since we aim at analyzing multiscale hyperbolic dynamics regardless of the propagation scaling, in order to obtain physically-based predictions, it is important that the PINN is able to preserve also the correct equilibrium solution in the diffusive regime of the equations, such as that of system (3)–(4) with limit (6). The neural networks satisfying this requirement are called *Asymptotic-Preserving Neural Net-*

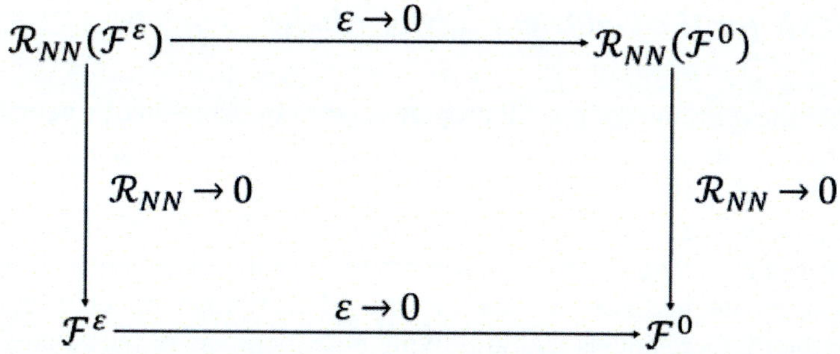

Fig. 3 AP diagram for physics-informed neural networks. $\mathcal{F}^{\epsilon}$ is the multiscale hyperbolic model that depends on the scaling parameter ϵ, while $\mathcal{F}^0$ is the corresponding formulation in the diffusive limit, which does not depend anymore on ϵ. The solution of the system $\mathcal{F}^{\epsilon}$ is approximated by the neural network through the imposition of the residual term $\mathcal{R}_{NN}(\mathcal{F}^{\epsilon}) = \mathcal{R}^{\epsilon}_{NN}$. The asymptotic limit of $\mathcal{R}_{NN}(\mathcal{F}^{\epsilon})$ as $\epsilon \to 0$ is denoted with $\mathcal{R}_{NN}\left(\mathcal{F}^0\right) = \mathcal{R}^0_{NN}$. The neural network is called AP if $\mathcal{R}_{NN}(\mathcal{F}^{\epsilon})$ represents a good approximation of $\mathcal{F}^0$

works (APNNs), and have been recently introduced in [23, 35] precisely to efficiently solve multiscale kinetic problems with scaling parameters that can have several orders of magnitude of difference. In essence, an APNN is said to be such if the neural network (already PINN) benefits from the *asymptotic preservation* (AP) property, by analogy with that of classical numerical schemes. In this analogy, the loss function is viewed as a numerical approximation of the original equation and has to benefit from the AP property. Let us recall that the so-called AP schemes constitute a class of numerical methods that aim to preserve the correct asymptotic behavior of the system without any loss of efficiency due to time step restrictions related to the small scales [6, 27, 36]. Therefore, the definition of an APNN (reported in [23] for the case of multiscale kinetic models with continuous velocity fields) can be generalized as follows (see Fig. 3).

Definition 2.1 Assume the solution of the problem is parametrized by a PINN trained by using an optimization method to minimize a loss function which includes a residual term $\mathcal{R}_{NN}$ enforcing the physics of the phenomenon $\mathcal{F}^{\epsilon}$, which depends on a scaling parameter ϵ. Then we say it is an *Asymptotic-Preserving Neural Network* (APNN) if, as the physical scaling parameter of the multiscale model tends to zero (i.e., $\epsilon \to 0$), the loss function of the full model-constraint converges to the loss function of the corresponding reduced order model $\mathcal{F}^0$.

4.1 APNN for the Goldstein-Taylor Model

To illustrate the relevance of the AP property in the construction of a neural network for hyperbolic systems with diffusive scaling, let us carry on a detailed example by directly considering the Goldstein-Taylor model (1)–(2).

We assume $f^{\pm}_{NN}(x,t;\theta,\sigma)$ to be the outputs of a PINN given inputs x and t and trainable parameters θ and σ (so we assume the scattering coefficient to be unknown), which approximate the solution of our system: $f^{\pm}(x,t;\sigma) \approx f^{\pm}_{NN}(x,t;\theta,\sigma)$. Then, we define the PDEs residuals by multiplying both members of the equations by the square of the scaling parameter ϵ^2 to allow the usage of the model also when $\epsilon \to 0$,

$$\mathcal{R}^{\epsilon,\pm}_{NN} = \epsilon^2 \frac{\partial f^{\pm}_{NN}}{\partial t} \pm \epsilon \frac{\partial f^{\pm}_{NN}}{\partial x} - \frac{\sigma}{2}\left(f^{\mp}_{NN} - f^{\pm}_{NN}\right), \tag{13}$$

and incorporate them into the loss function term $\mathcal{L}_r(\theta,\sigma)$ of the neural network in (10) by taking the mean squared error to obtain a standard PINN:

$$\omega_r^T \mathcal{L}_r(\theta,\sigma) = \frac{\omega_r^+}{N_r}\sum_{n=1}^{N_r}\left|\mathcal{R}^{\epsilon,+}_{NN}(x_r^n, t_r^n;\theta,\sigma)\right|^2 + \frac{\omega_r^-}{N_r}\sum_{n=1}^{N_r}\left|\mathcal{R}^{\epsilon,-}_{NN}(x_r^n, t_r^n;\theta,\sigma)\right|^2 . \tag{14}$$

It is easy to observe that, with this construction, the standard PINN residual (13) is not consistent with the reduced order model of the system, namely the diffusion limit (6). In fact, $\mathcal{R}^{\epsilon,\pm}_{NN}$ in the limit $\epsilon \to 0$ reduce to

$$\mathcal{R}^{0,\pm}_{NN} = -\frac{\sigma}{2}\left(f^{\mp}_{NN} - f^{\pm}_{NN}\right),$$

which basically corresponds to forcing $f^+(x,t) = f^-(x,t)$ and does not suffice to achieve the correct diffusive behavior. Therefore, the thus constructed PINN does not satisfy the sought AP property.

In contrast, using the macroscopic formulation of the model (3)–(4) results sufficient to construct an APNN, with outputs $\rho_{NN}(x,t;\theta,\sigma)$ and $j_{NN}(x,t;\theta,\sigma)$. Incorporating now in the loss function the mean squared error of the PDEs residuals of the macroscopic formulation,

$$\mathcal{R}^{\epsilon,\rho}_{NN} = \frac{\partial \rho_{NN}}{\partial t} + \frac{\partial j_{NN}}{\partial x}, \qquad \mathcal{R}^{\epsilon,j}_{NN} = \epsilon^2 \frac{\partial j_{NN}}{\partial t} + \frac{\partial \rho_{NN}}{\partial x} + \sigma j_{NN}, \tag{15}$$

the loss term of the physics residual will read

$$\omega_r^T \mathcal{L}_r(\theta,\sigma) = \frac{\omega_r^\rho}{N_r}\sum_{n=1}^{N_r}\left|\mathcal{R}^{\epsilon,\rho}_{NN}(x_r^n, t_r^n;\theta,\sigma)\right|^2 + \frac{\omega_r^j}{N_r}\sum_{n=1}^{N_r}\left|\mathcal{R}^{\epsilon,j}_{NN}(x_r^n, t_r^n;\theta,\sigma)\right|^2 . \tag{16}$$

In the limit $\epsilon \to 0$, we obtain

$$\mathcal{R}_{NN}^{0,\rho} = \frac{\partial \rho_{NN}}{\partial t} + \frac{\partial j_{NN}}{\partial x}, \qquad \mathcal{R}_{NN}^{0,j} = \frac{\partial \rho_{NN}}{\partial x} + \sigma j_{NN},$$

which analytically confirms that we are now dealing with a loss function that remains consistent also with the residual of the limiting diffusive model (6).

5 Application to the Goldstein-Taylor Model

In this Section, we will analyze and compare the behavior of a standard DNN with respect to a standard PINN and of a standard PINN and an APNN when trying to solve the Goldstein-Taylor model in a non-multiscale and a multiscale configuration, respectively.

5.1 *Standard DNN Versus Standard PINN in Hyperbolic Regime*

To first analyze the different behavior of a standard DNN with respect to a PINN, let us consider a numerical test for the Goldstein-Taylor model presented in Sect. 2 without the presence of the scaling parameter, hence fixing $\epsilon = 1.0$. Therefore, we build up a DNN that is trained only with observed data points taken from a synthetic solution obtained solving system (1)–(2) with a second-order AP-IMEX Runge-Kutta finite volume method [10, 37] that is considered as ground truth (see Fig. 4 bottom, right). On the other hand, we consider also a PINN having in the loss function the additional physics loss term as in (14). Moreover, in the PINN's loss function we enforce the positivity of the densities by adding the term

$$\begin{aligned}\mathcal{L}_P(\theta) = &\frac{1}{N_r}\sum_{n=1}^{N_r}\left|\text{abs}\left(f_{NN}^{+}(x_r^n, t_r^n; \theta)\right) - f_{NN}^{+}(x_r^n, t_r^n; \theta)\right|^2 \\ &+ \frac{1}{N_r}\sum_{n=1}^{N_r}\left|\text{abs}\left(f_{NN}^{-}(x_r^n, t_r^n; \theta)\right) - f_{NN}^{-}(x_r^n, t_r^n; \theta)\right|^2 .\end{aligned}$$

In the domain $\mathcal{D} = [-1, 1]$, we consider two initial Gaussian distributions of densities $f^{\pm}$ as

$$f^{\pm}(x, 0) = \frac{1}{s\sqrt{2\pi}} e^{-\frac{(x\pm 0.5)^2}{2s^2}}$$

with $s = 0.15$ and scattering coefficient $\sigma = 1$, which is assumed to be known. We consider periodic boundary conditions and run the simulation until $t_{end} = 0.9$.

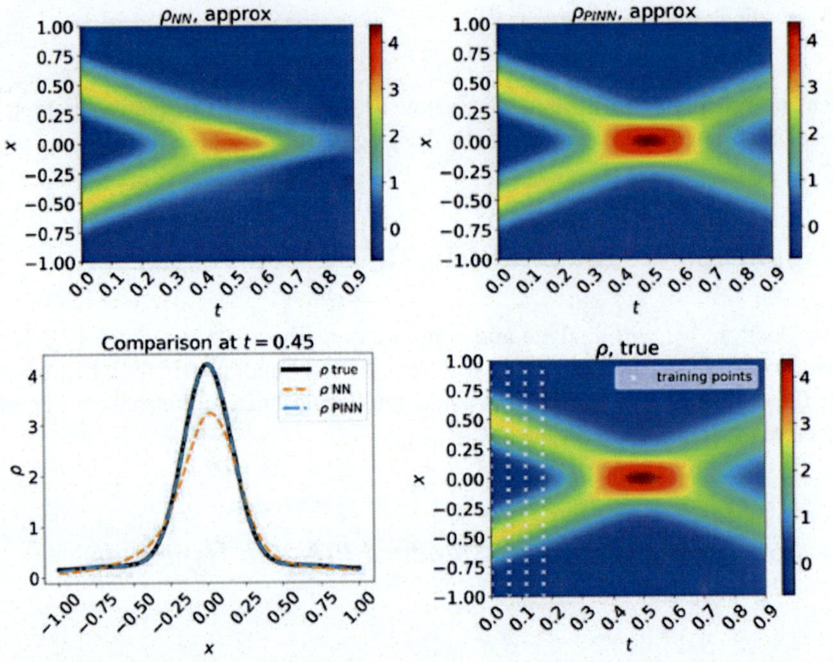

Fig. 4 Forward problem for the Goldstein-Taylor model with $\epsilon = 1.0$. Solution of the forward problem obtained with a standard DNN (top, left) or with a standard PINN (top, right) with respect to the ground truth (bottom, right) in terms of total density ρ; comparison of the DNN and PINN solution at $t = 0.45$ (bottom, left). White crosses in the bottom right plot identify the 80 training data points used for both the neural networks

In both the NNs, we consider a feed-forward network with depth 3 (i.e., number of layers, including output layer but excluding input layer) and width 32 (i.e., number of nodes in a layer). We use tanh as activation function and fix a learning rate LR=10^{-3}. Finally, the Adam method [31] is employed for the optimization process and the derivatives in the physics loss function are computed applying automatic differentiation [34]. The weights in the PINN loss function (10) are chosen in order to penalize the physics residuals. Hence, we set all of them equal to 1, except for $\omega_d = 10$. The spatio-temporal approximation of the total density ρ obtained with the standard DNN trained with $N_d = 80$ observation points of densities f^+ and f^- is compared with the one obtained using a standard PINN, which is trained not only on the same dataset but also enforcing the PDE system structure and the density positivity on $N_r = 3600$ residual points uniformly distributed in the entire domain (without enforcing boundary conditions) in Fig. 4. From this Figure, it is clearly visible that the standard DNN is not able to reproduce a correct solution of the problem when trained with limited data and does not fulfil to predict credible trends far from the dataset availability. In contrast, the PINN is able to fit the solution not only in the vicinity of the training points, but also outside of them, giving reliable predictions.

5.2 Standard PINN Versus APNN in Diffusive Regime

To emphasize also numerically the importance of choosing the correct formulation to fulfill the AP property and correctly approximate dynamics of multiscale systems even in diffusive regimes, we consider a test similar to the one discussed in the previous Section for the Goldstein-Taylor model, but choosing $\epsilon = 10^{-5}$ and evaluating the solution until $t_{end} = 0.02$. In the domain $\mathcal{D} = [-1, 1]$, centered in $x_0 = 0$, we consider a problem with an initial Gaussian distribution of the total density ρ that is the same obtained in the previous test

$$\rho(x, 0) = \frac{1}{s\sqrt{2\pi}} \left(e^{-\frac{(x-0.5)^2}{2s^2}} + e^{-\frac{(x+0.5)^2}{2s^2}} \right),$$

but evaluating the corresponding equilibrium flux j with (5), which results

$$j(x, 0) = -\frac{1}{\sigma s^3\sqrt{2\pi}} \left((x-0.5)e^{-\frac{(x-0.5)^2}{2s^2}} + (x+0.5)e^{-\frac{(x+0.5)^2}{2s^2}} \right),$$

again with $s = 0.15$, $\sigma = 1$ (assumed to be known), and periodic boundary conditions.

We solve the problem first applying the standard PINN, with the same setting previously discussed, and then, for comparison, using the APNN formulation, with residuals (15) in the AP loss function and enforcing the positivity of the total density:

$$\mathcal{L}_p(\theta) = \frac{1}{N_r} \sum_{n=1}^{N_r} \left| \text{abs}\left(\rho_{NN}(x_r^n, t_r^n; \theta)\right) - \rho_{NN}(x_r^n, t_r^n; \theta) \right|^2 .$$

Notice that the standard PINN shares the same setting with the APNN, except for the different residual terms enforced for the physics, as discussed in Sect. 4.1. In the APNN, indeed, we consider the correctly scaled system (16) instead of (14).

Results obtained with the two neural networks, trained with $N_d = 160$ data points and enforcing the two different physics residuals in $N_r = 2500$ uniformly distributed points, are presented in Fig. 5. We observe that the standard PINN is not able to recover the correct solution in this regime, especially with respect to forecast scenarios far from the dataset, because the physics imposed (and learned by the PINN) is not consistent with the relaxation limit of the system. In contrast, the APNN approximates the solution very well even at temporal points where there is no nearby data availability.

Inverse Problem

To analyze also the behavior of standard PINN and APNN when dealing with the resolution of inverse problems, we design an additional test for the Goldstein-Taylor model, for which initial conditions are

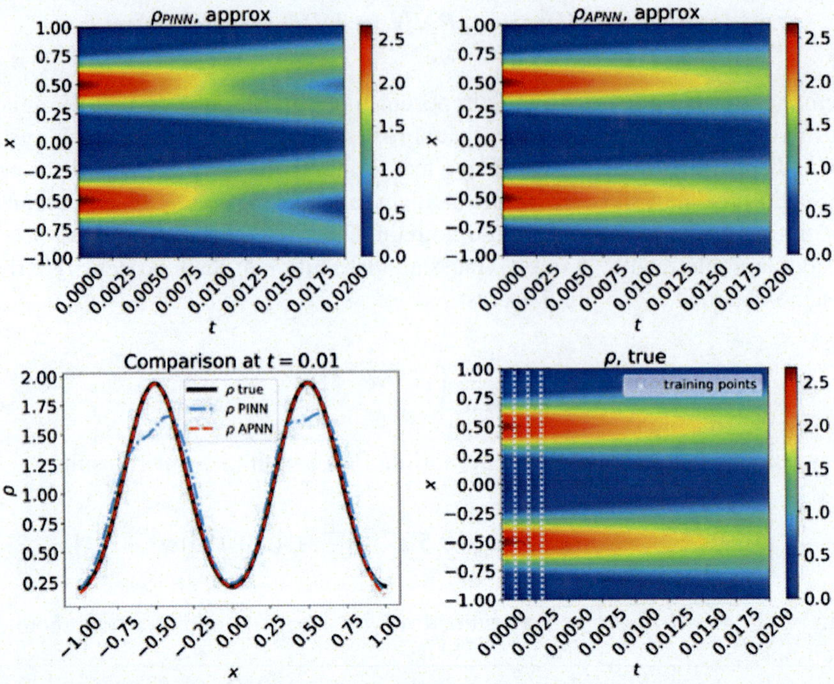

Fig. 5 Forward problem for the Goldstein-Taylor model in diffusive regime ($\epsilon = 10^{-5}$). Solution of the forward problem obtained with a standard PINN (top, left) or with an APNN (top, right) with respect to the ground truth (bottom, right) in terms of total density ρ; comparison of the PINN and APNN solution at $t = 0.01$ (bottom, left). White crosses in the right plot identify the 80 sparse training data points used to train both the neural networks

$$\rho(x, 0) = 6 + 3\cos(3\pi x), \qquad j(x, 0) = 9\pi\sigma^{-1}\sin(3\pi x),$$

with $\sigma = 4$, model parameter that is assumed to be unknown and need to be estimated by the neural network. We consider again periodic boundary conditions and we fix $\epsilon = 10^{-4}$ to simulate the diffusive, parabolic regime of the model until $t_{end} = 0.1$.

For both standard PINN and APNN formulations, we train the network model on measurements composed of $N_d = 24000$ equally spaced samples (taken from the synthetic solution) in the domain $(x, t) \in [-1, 1] \times [0, 0.1]$, from which 20% (4800) points are randomly selected for validation purpose. For the APNN model, to account for an even poorer partially available dataset, we consider measurements only for the density ρ, hence assuming to have no information on the flux j, whereas for the standard PINN formulation we employ data samples for both the densities f^+ and f^- (therefore, in the latter case we assume we have more information on the system). Finally, $N_r = 24000$ residual points are employed to enforce the physics, with the same 20% data split for the validation set.

We show the convergence of the target parameter σ in Fig. 6 for both the neural networks. Despite the partial availability of information on the dynamic (of which

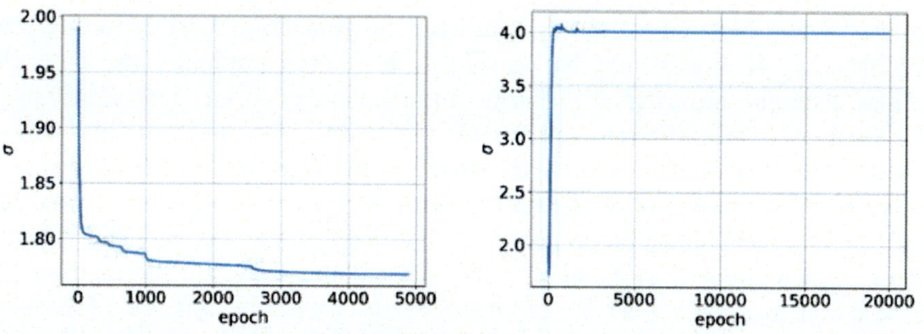

Fig. 6 Inverse problem for the Goldstein-Taylor model in diffusive regime ($\epsilon = 10^{-4}$). Convergence of the target parameter $\sigma = 4$ with respect to epochs using the standard PINN (left) and the APNN (right). The correct inference of the parameter is obtained only when recurring to the APNN

flux data are completely missing), a very fast convergence of the APNN towards the expected value of σ is observed, obtaining a final relative error $O(10^{-3})$. On the other hand, it can be observed that the standard PINN, although trained with a full dataset of $f^{\pm}$, fails to retrieve the correct value of the scattering coefficient σ (early stopping the iterations almost at epoch 5000 to avoid further training).

6 Application to Epidemic Dynamics

A particularly interesting area in which the use of the APNN framework discussed here can play a key role is that of the study of epidemic dynamics. In this context, also driven by the COVID-19 pandemic that has hit the world, a number of mathematical models have recently been proposed that require the estimation of several parameters from data to provide predictive scenarios and to test their reliability. Examples are [1, 5, 6, 37–43]. Here we focus on a new class of epidemic models defined by multiscale PDEs capable of describing both hyperbolic-type phenomena, characteristic of epidemic propagation over long distances and main lines of communication between cities, and parabolic-type phenomena, in which classical diffusion prevails at the urban level [5, 7, 10].

6.1 The Multiscale Hyperbolic SIR Model

Let us consider for simplicity the space dependent epidemiological modeling in the case of a classic SIR compartmental dynamic. Thus, we start subdividing the population in susceptible S (individuals who may be infected by the disease), infectious I (individuals who may transmit the disease) and removed R (individuals healed and immune or died due to the disease), following the pioneering work of Kermack and McKendrick [44]. We assume that no subjects have prior immunity and we neglect the vital dynamic represented by births and deaths due to the time scale considered. By analogy with discrete-velocity kinetic theory [2], and in particular with the Goldstein-Taylor model (1)–(2), we consider now that individuals are moving in a 1D domain $\mathcal{D} \subseteq \mathbb{R}$ in two opposite directions, with velocities $\pm\lambda_{S,I,R} = \pm\lambda_{S,I,R}(x)$, distinguished for each epidemic compartment. Notice that the characteristic velocities reflect the heterogeneity of geographical areas, and, therefore, are chosen dependent on the spatial location $x \in \mathcal{D}$. We can describe the space-time dynamic of this population for $t > 0$ through the following two-velocity SIR epidemic transport model [7, 37]:

$$\frac{\partial S^{\pm}}{\partial t} + \lambda_S \frac{\partial S^{\pm}}{\partial x} = -\beta S^{\pm} I \mp \frac{1}{2\tau_S}\left(S^{+} - S^{-}\right), \tag{17}$$

$$\frac{\partial I^{\pm}}{\partial t} + \lambda_I \frac{\partial I^{\pm}}{\partial x} = \beta S^{\pm} I - \gamma I^{\pm} \mp \frac{1}{2\tau_I}\left(I^{+} - I^{-}\right), \tag{18}$$

$$\frac{\partial R^{\pm}}{\partial t} + \lambda_R \frac{\partial R^{\pm}}{\partial x} = \gamma I^{\pm} \mp \frac{1}{2\tau_R}\left(R^{+} - R^{-}\right), \tag{19}$$

with the total densities of each compartment, $S(x,t)$, $I(x,t)$, and $R(x,t)$, given by

$$S = S^{+} + S^{-}, \quad I = I^{+} + I^{-}, \quad R = R^{+} + R^{-},$$

and $S(t) + I(t) + R(t) = P$, total population size, which remains constant over time.

The transport dynamic of the population is governed by the scaling parameters $\lambda_{S,I,R}$ as well as the relaxation times $\tau_{S,I,R} = \tau_{S,I,R}(x)$. The quantity $\gamma = \gamma(x,t)$ is the recovery rate of infected, which corresponds to the inverse of the infectious period. The transmission of the infection is defined by an incidence function $\beta S I$ modeling the transmission of the disease [44–46], where the transmission rate $\beta = \beta(x,t)$ characterizes the average number of contacts per person per time, multiplied by the probability of disease transmission in a contact between a susceptible and an infectious subject.

When investigating real epidemic scenarios, the above-mentioned parameters are, in general, unknown. While the recovery rate might be fixed based on clinical data, the transmission rate must always be estimated through a delicate calibration process in order to match observations. It is also well-known that this process is highly heterogeneous, which makes the inverse problem even more challenging [8, 47].

If we look at system (17)–(19), it is worth observing that, for each compartment, if we exclude the presence of the epidemic source term, we are essentially prescribing a dynamic that directly recalls that presented for the Goldstein-Taylor model (1)–(2) in Sect. 2. Furthermore, defining now the flux for each compartment as

$$J_S = \lambda_S\left(S^{+} - S^{-}\right), \quad J_I = \lambda_I\left(I^{+} - I^{-}\right), \quad J_R = \lambda_R\left(R^{+} - R^{-}\right),$$

in analogy with the Goldstein-Taylor model, it is possible to derive the following macroscopic formulation of system (17)–(19):

$$\frac{\partial S}{\partial t}+\frac{\partial J_S}{\partial x}=-\beta SI\,, \tag{20}$$

$$\frac{\partial I}{\partial t}+\frac{\partial J_I}{\partial x}=\beta SI-\gamma I\,, \tag{21}$$

$$\frac{\partial R}{\partial t}+\frac{\partial J_R}{\partial x}=\gamma I\,, \tag{22}$$

$$\frac{\partial J_S}{\partial t}+\lambda_S^2\frac{\partial S}{\partial x}=-\beta J_S I-\frac{J_S}{\tau_S}\,, \tag{23}$$

$$\frac{\partial J_I}{\partial t}+\lambda_I^2\frac{\partial I}{\partial x}=\frac{\lambda_I}{\lambda_S}\beta J_S I-\gamma J_I-\frac{J_I}{\tau_I}\,, \tag{24}$$

$$\frac{\partial J_R}{\partial t}+\lambda_R^2\frac{\partial R}{\partial x}=\frac{\lambda_R}{\lambda_I}\gamma J_I-\frac{J_R}{\tau_R}\,. \tag{25}$$

Let us now consider the behavior of this model in the zero-relaxation limit [29]. To this aim, we introduce the space dependent diffusion coefficients,

$$D_S=\lambda_S^2\tau_S\,,\quad D_I=\lambda_I^2\tau_I\,,\quad D_R=\lambda_R^2\tau_R\,, \tag{26}$$

which characterize the diffusive transport mechanism of susceptible, infectious and removed, respectively. Keeping the above quantities fixed while letting the relaxation times $\tau_{S,I,R}\to 0$ (and so the characteristic velocities $\lambda_{S,I,R}\to\infty$), from Eqs. (23)–(25) we obtain three proportionality relations, one for each epidemic compartment, between the flux and the spatial derivative of the corresponding density (the so-called Fick's law):

$$J_S=-D_S\frac{\partial S}{\partial x}\,,\quad J_I=-D_I\frac{\partial I}{\partial x}\,,\quad J_R=-D_R\frac{\partial R}{\partial x}\,. \tag{27}$$

Substituting (27) into Eqs. (20)–(22), we recover the following parabolic reaction-diffusion model, widely used in literature to study the spread of infectious diseases [48–51]:

$$\frac{\partial S}{\partial t}=-\beta SI+\frac{\partial}{\partial x}\left(D_S\frac{\partial S}{\partial x}\right)\,, \tag{28}$$

$$\frac{\partial I}{\partial t}=\beta SI-\gamma I+\frac{\partial}{\partial x}\left(D_I\frac{\partial I}{\partial x}\right)\,, \tag{29}$$

$$\frac{\partial R}{\partial t}=\gamma I+\frac{\partial}{\partial x}\left(D_R\frac{\partial R}{\partial x}\right)\,. \tag{30}$$

This model is therefore capable to account for different regimes, ranging from hyperbolic to parabolic, according to the space dependent values $\tau_{S,I,R}$ and $\lambda_{S,I,R}$. This feature makes it particularly suitable for describing the multiscale dynamics of human beings [1, 7, 37].

For rigorous results on the diffusion limit of kinetic models of the type of system (17)–(19) we refer to [29, 52]. Finally, we underline that, although the model here discussed, for simplicity of presentation, is based on a simple SIR structure, the approach can be extended naturally to more realistic compartmental models designed to take into account specific features of the infectious disease of interest, such as those specifically developed to deal with the COVID-19 pandemic [39, 40].

6.2 APNN for the Hyperbolic SIR Model

The multiscale nature of the problem poses a concrete challenge to the PINN construction, and preservation of the AP property is essential in order to obtain reliable results. To satisfy the AP property, we follow the approach presented in Sect. 4.1, starting from the system written in macroscopic form, defined by Eqs. (20)–(25). After multiplying both members of each Eq. by the corresponding scaling parameter $\tau_{S,I,R}$, we can rewrite the system in the following compact form:

$$\tau(x)\frac{\partial U(x,t)}{\partial t} + D(x)\frac{\partial F(U)}{\partial x} = G(U)\,, \quad (x,t) \in \Omega\,,$$

where

$$U=\begin{bmatrix} S \\ I \\ R \\ J_S \\ J_I \\ J_R \end{bmatrix}, \tau = \begin{bmatrix} 1 \\ 1 \\ 1 \\ \tau_S \\ \tau_I \\ \tau_R \end{bmatrix}, D = \begin{bmatrix} 1 \\ 1 \\ 1 \\ D_S \\ D_I \\ D_R \end{bmatrix}, F(U) = \begin{bmatrix} J_S \\ J_I \\ J_R \\ S \\ I \\ R \end{bmatrix}, G(U) = \begin{bmatrix} -\beta SI \\ \beta SI - \gamma I \\ \gamma I \\ -\tau_S \beta J_S I - J_S \\ \tau_I \frac{\lambda_I}{\lambda_S} \beta J_S I - \tau_I \gamma J_I - J_I \\ \tau_R \frac{\lambda_R}{\lambda_I} \gamma J_I - J_R \end{bmatrix}.$$

We consider $U_{NN}(x, t; \theta)$ to be the output vector of a DNN with inputs x and t and trainable parameters θ, which approximate the solution of our system: $U(x, t) \approx U_{NN}(x, t; \theta)$. Then, we define the residual vector term

$$\mathcal{R}_{NN}^{\tau} = \tau \frac{\partial U_{NN}}{\partial t} + D \frac{\partial F(U_{NN})}{\partial x} - G(U_{NN}) \tag{31}$$

and embed it into the loss function of the neural network to obtain the sought APNN [22]. We omit for brevity the detailed analysis of the AP property. Anyway, in the limit $\tau_{S,I,R} \to 0$, $\lambda_{S,I,R} \to \infty$, under conditions (26), such analysis follows the same steps presented in Sect. 4.1 for the prototype model, and $\mathcal{R}_{NN}^{0}$ results in agreement with the equilibrium system (28)–(30).

If we consider both the epidemic parameters β and γ to be unknown, in the training process of the APNN we minimize the following AP-loss function, which recalls Eq. (10):

$$\mathcal{L}(\theta, \beta, \gamma) = \omega_d^T \mathcal{L}_d(\theta, \beta, \gamma) + \omega_b^T \mathcal{L}_b(\theta, \beta, \gamma) + \omega_r^T \mathcal{L}_r(\theta, \beta, \gamma), \tag{32}$$

where $\mathcal{L}_r$ enforces the minimization of the residual (31):

$$\omega_r^T \mathcal{L}_r(\theta, \beta, \gamma) = \frac{\omega_r^T}{N_r} \sum_{n=1}^{N_r} \left|\mathcal{R}_{NN}^{\tau}(x_r^n, t_r^n; \theta, \beta, \gamma)\right|^2 .$$

The reader is referred to [22] for a detailed expansion of each term in this AP-loss function.

6.3 APNN Performance with Epidemic Dynamics

Let us now examine through a numerical example the APNN performance in inferring unknown epidemic parameters, reconstructing the spatio-temporal dynamic and forecasting the spread of an infectious disease. We design a numerical test with an initial condition that simulates the presence of two epidemic hot-spots, aligned in the spatial domain $\mathcal{D} = [0, 20]$, with different numbers of infected individuals, distributed according to a Gaussian function,

$$I(x, 0) = \alpha_1 e^{-(x-x_1)^2} + \alpha_2 e^{-(x-x_2)^2},$$

where $x_1 = 5$ and $x_2 = 15$ are the coordinates of the hot-spots, while $\alpha_1 = 0.01$ and $\alpha_2 = 0.0001$ define the different initial epidemic concentration in the two cities. Assuming that there are no immune individuals at $t = 0$ and that the total population is uniformly distributed in the domain, we impose

$$S(x, 0) = 1 - I(x, 0), \qquad R(x, 0) = 0 .$$

Initial fluxes are set to be in equilibrium, following (27), and we consider periodic boundary conditions to allow both directions of connection for the two cities.

As with the numerical test with the Goldstein-Taylor model, the numerical solution obtained with a second-order AP-IMEX Runge-Kutta finite volume method [7, 37] is used as synthetic data for the training of the APNN when solving the inverse problem. Nevertheless, since data of fluxes J_S, J_I, J_R are not accessible in real-world applications, we only enforce measurements of the densities S, I, R in $\mathcal{L}_d$.

We run two test cases for this problem:

- in the first case, a hyperbolic configuration of speeds and relaxation parameters is considered, with $\lambda_{S,I,R}^2 = 1$ and $\tau_{S,I,R} = 1$;

- in the second case, we simulate a parabolic regime setting $\lambda^2_{S,I,R} = 10^3$ and $\tau_{S,I,R} = 10^{-3}$.

For each scenario, both an inverse and a forward problem are evaluated. PDEs residuals are enforced on $N_r = 40000$ and $N_r = 23600$ in the spatio-temporal domain respectively when simulating the parabolic and the hyperbolic regime, always using 20% of training and residual points for validation purposes only. To strictly impose also the periodic boundary conditions in the neural network (accounted in $\mathcal{L}_b$ in the loss function), we employ the periodic mapping technique taken from [53] in the input layer:

$$U_{NN}(x,t) = U_{NN}\left(\cos(\alpha x), \sin(\alpha x), t\right), \tag{33}$$

where α is a hyperparameter controlling the frequency of the solution, here chosen equal to 3.

To solve the problem, we adopt again a feed-forward neural network, this time with depth 8 and width 32, choosing `tanh` as activation function and fixing a learning rate LR=10^{-2}. Finally, the Adam method [31] is employed for the optimization process. Weights associated to each loss term are listed in Table 1.

Inverse Problems

We initially suppose that the epidemic parameters $\beta = 12$ and $\gamma = 6$ are unknown, so that the APNN is used to infer both of them as well as to reconstruct the spatio-temporal solution. To mimic the availability of data close to reality, we use a sparse dataset of only $N_d = 20$ measurements of densities S, I, R for the training process, sampling the spatio-temporal points from the available dataset with probability proportional to the magnitude of I. Indeed, in real epidemic scenarios, data on the evolution of the infectious disease are only available in regions where the virus has already started to spread. Specifically, the probability of each spatio-temporal location (x,t) chosen for the training dataset is given by

$$p(x,t) = \frac{I(x,t)}{\int_\Omega I(x,t)}.$$

Moreover, in this inverse problem, both initial and boundary conditions are regarded as unknown.

Results of the parameter inference task based on the sparse measurement dataset are reported in Table 2 for the test in both regimes. An excellent estimation of

Table 1 Weights used in the loss function of the APNN for the SIR test cases

Test regime	$(\omega_d^S, \omega_d^I, \omega_d^R)$	$(\omega_b^S, \omega_b^I, \omega_b^R)$	$(\omega_r^S, \omega_r^I, \omega_r^R, \omega_r^{J_S}, \omega_r^{J_I}, \omega_r^{J_R})$
Hyperbolic	(1, 100, 10)	(1, 10, 1)	(1, 100, 10, 1, 100, 10)
Parabolic	(1, 100, 10)	(1, 10, 1)	(1, 10, 1, 1, 10, 1)

Table 2 Inverse problem with the SIR transport model considering a partially observed dynamic (sparse density data only and no initial conditions available). Inferred results for transmission rate β and recovery rate γ from a sparse measurement dataset of $N_d = 20$ samples, and the relative error with respect to the ground truth values

Test regime	Parameter	Ground truth	Initial guess	APNN estimation	Relative error
Hyperbolic	β	12	8	12.0126	1.05×10^{-3}
	γ	6	3	6.0447	7.45×10^{-3}
Parabolic	β	12	8	11.9428	4.76×10^{-3}
	γ	6	3	5.9772	3.80×10^{-3}

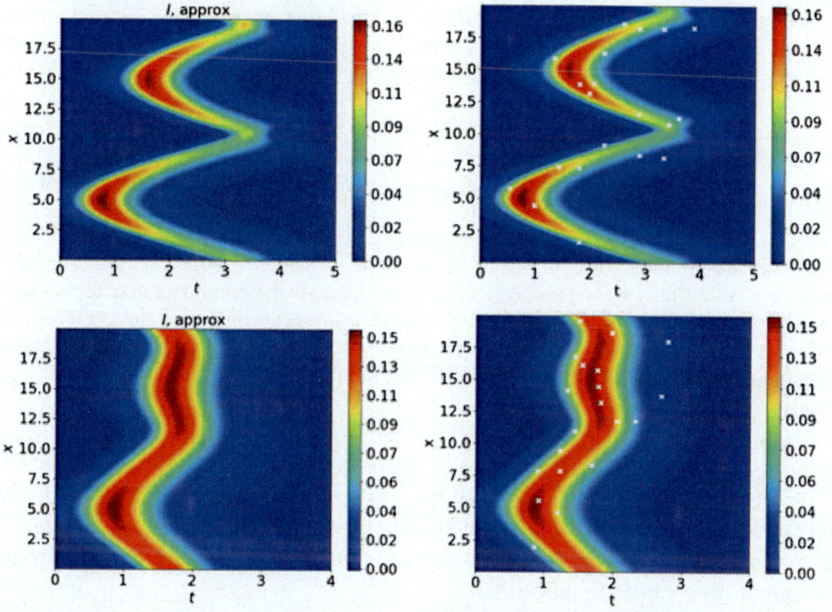

Fig. 7 Inverse problem with the SIR transport model considering a partially observed dynamic (sparse density data only and no initial conditions available) in hyperbolic regime (top) or parabolic regime (bottom). Approximation obtained with the APNN (left) and ground truth of the densities of infected I (right) with the identification of the sparse data samples used for the training ($N_d = 20$) marked with white crosses

parameters β and γ can be observed with respect to the ground truth. At the same time, the APNN is also capable of reconstructing the correct dynamic of the epidemic spread in the whole domain besides the consistent sparsity and incompleteness of data, as shown in Fig. 7 for both the test cases.

Forward Problems

As a second task, we seek to investigate the forecasting capability of the APNN. In contrast to the previous inference test, in which we sampled the available mea-

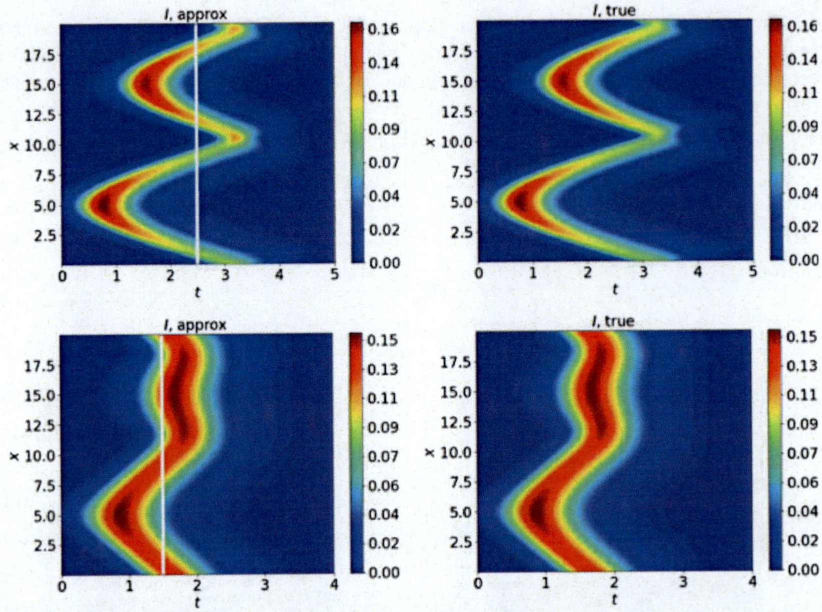

Fig. 8 Forward problem with the SIR transport model considering a partially observed dynamic (sparse density data only and no initial conditions available) in hyperbolic regime (top) or parabolic regime (bottom). Approximation and forecast with measurements on a short time, with $t \in [0, 2.5]$ in the hyperbolic test and $t \in [0, 1.5]$ in the parabolic case (left), and ground truth of the densities of infected I (right). The continuous white line on the left plots identifies the end of the data training domain

surements over the entire spatio-temporal domain, in this test we generate a training dataset over a shorter time domain and evaluate not only the correctness of the APNN approximations in this restricted time period, but also the prediction performance in subsequent time steps. In particular,

- for the test in hyperbolic regime, we train the APNN with $N_d = 8500$ measurements of densities S, I and R generated from $t \in [0, 2.5]$, and then evaluate the APNN performance over the complete time domain $t \in [0, 5]$, with predictions in $t \in [2.5, 5]$;
- for the test in parabolic regime, we consider a training dataset of $N_d = 5300$ points for densities S, I and R uniformly distributed in $t \in [0, 1.5]$ and we assess the correctness of APNN approximations in $t \in [0, 1.5]$ and the forecasting performance in $t \in [1.5, 4]$.

As shown in Fig. 8, the approximated and predicted dynamics perfectly match the ground truth in the entire domain in both the test cases. These results further emphasize the great capabilities of the APNN, which is also able to predict the spread of an infectious disease in multiscale regimes due to the physical knowledge of the PDEs system that has been incorporated into the loss function of the neural network in such a way as to ensure the preservation of the AP property.

Nevertheless, it is worth to remark that these predictions could only be plausible on the assumption that, at least in the short term analysis, there are no significant changes in population mobility, in the contact rate between individuals and in the contagiousness of the virus. In other words, assuming that there are no restrictions or removal of restrictions by governments and no viral variants with markedly different characteristics from the virus already in the field begin to spread.

7 Conclusion

In this Chapter, we discussed a new class of Physics-Informed Neural Networks (PINNs) that are able to appropriately deal with problems characterized by multi-scale dynamics that may have several orders of magnitude difference, in particular hyperbolic problems that exhibit diffusive scaling. If we consider the residual term relating to the physical knowledge of the phenomenon under study inserted in the loss function of the PINN optimization process as a numerical approximation of the original equations, we can say that these neural networks satisfy the asymptotic-preservation (AP) property, and are therefore called Asymptotic-Preserving Neural Networks (APNNs) [22, 35].

Several numerical tests were presented to illustrate the performance of this new class of neural networks, addressing the problems of inverse learning and dynamics prediction. We have shown, both analytically and numerically, how APNNs provide considerably better results with respect to the different scales of the problem when compared with standard PINNs using the Goldstein-Taylor model as prototype system of equations.

One of the application areas of particular interest, also used in this work as an example case study, concerns the investigation of the spread of infectious diseases. Indeed, it is well known that epidemic models require a delicate calibration phase of the parameters involved for their validation and subsequent use for forecasting purposes. It is perhaps less obvious to think that even the dynamic of a population, and the associated propagation of a virus, is actually defined by a system of multiscale hyperbolic equations, with individuals moving in suburban areas following a convective-transport mechanism and individuals interacting in high-density urban areas with a purely diffusive dynamic. The results obtained confirm the great usefulness and effectiveness of APNNs, even and especially when analyzing scenarios in which only little and scattered information is available.

Acknowledgements The author acknowledges the support by INdAM-GNCS and by MIUR-PRIN Project 2017, No. 2017KKJP4X, *Innovative numerical methods for evolutionary partial differential equations and applications.*

References

1. Albi, G., Bertaglia, G., Boscheri, W., Dimarco, G., Pareschi, L., Toscani, G., Zanella, M.: Kinetic modelling of epidemic dynamics: social contacts, control with uncertain data, and multiscale spatial dynamics. In: Bellomo, N., Chaplain, M.A.J., (eds.), Predicting Pandemics in a Globally Connected World, Volume 1. Toward a Multiscale, Multidisciplinary Framework through Modeling and Simulation, pp. 43–108. Birkhauser-Springer Series: Modeling and Simulation in Science, Engineering and Technology (2022)
2. Pareschi, L., Toscani, G.: Interacting Multiagent Systems. Oxford University Press, Kinetic Equations And Monte Carlo Methods (2013)
3. Carrillo, J.A., Fornasier, M., Toscani, G., Vecil, F.: Particle, kinetic, and hydrodynamic models of swarming. In: Mathematical Modeling of Collective Behavior in Socio-Economic and Life Sciences, pp. 297–336. Birkhäuser Boston, Boston (2010)
4. Bellomo, N., Bingham, R., Chaplain, M.A.J., Dosi, G., Forni, G., Knopoff, D.A., Lowengrub, J., Twarock, R., Virgillito, M.E.: A multiscale model of virus pandemic: Heterogeneous interactive entities in a globally connected world. Math. Models Methods Appl. Sci. **30**(08), 1591–1651 (2020)
5. Boscheri, W., Dimarco, G., Pareschi, L.: Modeling and simulating the spatial spread of an epidemic through multiscale kinetic transport equations. Math. Models Methods Appl. Sci. **31**(06), 1059–1097 (2021)
6. Albi, G., Pareschi, L., Zanella, M.: Modelling lockdown measures in epidemic outbreaks using selective socio-economic containment with uncertainty. Math. Biosci. Eng. **18**(6), 7161–7190 (2021)
7. Bertaglia, G., Boscheri, W., Dimarco, G., Pareschi, L.: Spatial spread of COVID-19 outbreak in Italy using multiscale kinetic transport equations with uncertainty. Math. Biosci. Eng. **18**(5), 7028–7059 (2021)
8. Bertaglia, G., Liu, L., Pareschi, L., Zhu, X.: Bi-fidelity stochastic collocation methods for epidemic transport models with uncertainties. Netw. Heterogen. Media **17**(3), 401–425 (2022)
9. Bertaglia, G., Caleffi, V., Pareschi, L., Valiani, A.: Uncertainty quantification of viscoelastic parameters in arterial hemodynamics with the a-FSI blood flow model. J. Comput. Phys. **430**, 110102 (2021)
10. Bertaglia, G., Pareschi, L.: Hyperbolic compartmental models for epidemic spread on networks with uncertain data: application to the emergence of COVID-19 in Italy. Math. Models Methods Appl. Sci. **31**(12), 2495–2531 (2021)
11. Higham, C.F., Higham, D.J.: Deep Learning: An Introduction for Applied Mathematicians. SIAM Rev. **61**(3), 860–891 (2019)
12. Weinan, E.: The dawning of a new era in applied mathematics. Notices Am. Math. Soc. **68**(04), 565–571 (2021)
13. Peng, G.C., Alber, M., Buganza Tepole, A., Cannon, W.R., De, S., Dura-Bernal, S., Garikipati, K., Karniadakis, G., Lytton, W.W., Perdikaris, P., Petzold, L., Kuhl, E.: Multiscale modeling meets machine learning: what can we learn? Arch. Comput. Methods Eng. **28**(3), 1017–1037 (2021)
14. Baker, N., Alexander, F., Bremer, T., Hagberg, A., Kevrekidis, Y., Najm, H., Parashar, M., Patra, A., Sethian, J., Wild, S. and Willcox, K., Lee, S.: Workshop Report on Basic Research Needs for Scientific Machine Learning: Core Technologies for Artificial Intelligence. Technical Report 1, USDOE Office of Science (SC) (United States) (2019)
15. Fabiani, G., Calabrò, F., Russo, L., Siettos, C.: Numerical solution and bifurcation analysis of nonlinear partial differential equations with extreme learning machines. J. Sci. Comput. **89**(2), 44 (2021)
16. Lou, Q., Meng, X., Karniadakis, G.E.: Physics-informed neural networks for solving forward and inverse flow problems via the Boltzmann-BGK formulation. J. Comput. Phys. **447**, 110676 (2021)

17. Raissi, M., Perdikaris, P., Karniadakis, G.: Physics-informed neural networks: A deep learning framework for solving forward and inverse problems involving nonlinear partial differential equations. J. Comput. Phys. **378**, 686–707 (2019)
18. Karniadakis, G.E., Kevrekidis, I.G., Lu, L., Perdikaris, P., Wang, S., Yang, L.: Physics-informed machine learning. Nat. Rev. Phys. **3**(6), 422–440 (2021)
19. Coutinho, E.J.R., Dall'Aqua, M., McClenny, L., Zhong, M., Braga-Neto, U., Gildin, E.: Physics-informed neural networks with adaptive localized artificial viscosity. ArXiv:2203.08802, 2022
20. Yu, J., Lu, L., Meng, X., Karniadakis, G.E.: Gradient-enhanced physics-informed neural networks for forward and inverse PDE problems. Comput. Methods Appl. Mech. Eng. **393**, 114823 (2022)
21. Moseley, B., Markham, A. and Nissen-Meyer, T.: Solving the wave equation with physics-informed deep learning (2020). ArXiv:2006.11894
22. Bertaglia, G., Lu, C., Pareschi, L., Zhu, X.: Asymptotic-Preserving Neural Networks for multiscale hyperbolic models of epidemic spread. Math. Models Methods Appl, Sci (2022)
23. Jin, S., Ma, Z., Wu, K.: Asymptotic-preserving neural networks for multiscale time-dependent linear transport equations (2022). ArXiv:2111.02541v4
24. Jin, S., Pareschi, L., Toscani, G.: Uniformly accurate diffusive relaxation schemes for multiscale transport equations. SIAM J. Numer. Anal. **38**(3), 913–936 (2000)
25. Jin, S., Xiu, D., Zhu, X.: Asymptotic-preserving methods for hyperbolic and transport equations with random inputs and diffusive scalings. J. Comput. Phys. **289**, 35–52 (2015)
26. Albi, G., Dimarco, G., Pareschi, L.: Implicit-Explicit multistep methods for hyperbolic systems with multiscale relaxation. SIAM J. Sci. Comput. **42**(4), A2402–A2435 (2020)
27. Boscarino, S., Pareschi, L., Russo, G.: A unified IMEX Runge-Kutta approach for hyperbolic systems with multiscale relaxation. SIAM J. Numer. Anal. **55**(4), 2085–2109 (2017)
28. Cercignani, C., Illner, R., Pulvirenti, M.: Hydrodynamical Limits, pp. 312–335. Springer, New York (1994)
29. Lions, P.L., Toscani, G.: Diffusive limit for finite velocity Boltzmann kinetic models. Revista Matematica Iberoamericana **13**(3), 473–513 (1997)
30. Goodfellow, I., Bengio, Y., Courville, A.: Deep Learning. The MIT Press, Adaptive Computation and Machine Learning Series (2016)
31. Kingma, D., Ba, J.: Adam: a method for stochastic optimization (2014). ArXiv:1412.6980
32. Ma, C., Wojtowytsch, S., Wu, L.: Towards a mathematical understanding of neural network-based machine learning: what we know and what we don't. CSIAM Trans. Appl. Math. **1**(4), 561–615 (2020)
33. Kharazmi, E., Cai, M., Zheng, X., Zhang, Z., Lin, G., Karniadakis, G.E.: Identifiability and predictability of integer- and fractional-order epidemiological models using physics-informed neural networks. Nat. Comput. Sci. **1**(11), 744–753 (2021)
34. Baydin, A.G., Pearlmutter, B.A., Radul, A.A., Siskind, J.M.: Automatic differentiation in machine learning: a survey. J. Mach. Learn. Res. **18**(1), 5595–5637 (2017)
35. Jin, S.: Asymptotic-preserving schemes for multiscale physical problems. Acta Numer. **31**, 415–489 (2022)
36. Dimarco, G., Pareschi, L.: Numerical methods for kinetic equations. Acta Numer. **23**, 369–520 (2014)
37. Bertaglia, G., Pareschi, L.: Hyperbolic models for the spread of epidemics on networks: kinetic description and numerical methods. ESAIM: Math. Model. Numer. Analys. **55**(2), 381–407 (2021)
38. Buonomo, B., Della Marca, R.: Effects of information-induced behavioural changes during the COVID-19 lockdowns: the case of Italy. R. Soc. Open Sci. **7**(10), 201635 (2020)
39. Gatto, M., Bertuzzo, E., Mari, L., Miccoli, S., Carraro, L., Casagrandi, R., Rinaldo, A.: Spread and dynamics of the COVID-19 epidemic in Italy: Effects of emergency containment measures. Proc. Natl. Acad. Sci. **117**(19), 10484–10491 (2020)
40. Giordano, G., Colaneri, M., Di Filippo, A., Blanchini, F., Bolzern, P., De Nicolao, G., Sacchi, P., Colaneri, P., Bruno, R.: Modeling vaccination rollouts, SARS-CoV-2 variants and the requirement for non-pharmaceutical interventions in Italy. Nat. Med. **27**(6), 993–998 (2021)

41. Marca, R.D., Loy, N., Tosin, A.: An SIR-like kinetic model tracking individuals' viral load. Netw. Heterogen. Media **17**(3), 467 (2022)
42. Scarabel, F., Pellis, L., Ogden, N.H., Wu, J.: A renewal equation model to assess roles and limitations of contact tracing for disease outbreak control. R. Soc. Open Sci. **8**, 202091 (2021)
43. Guglielmi, N., Iacomini, E., Viguerie, A.: Delay differential equations for the spatially resolved simulation of epidemics with specific application to COVID-19. Math. Methods Appl. Sci. **45**(8), 4752–4771 (2022)
44. Kermack, W.O., McKendrick, A.G.: A contribution to the mathematical theory of epidemics. Proc. R. Soc. London. Ser. A, Contain. Papers Math. Phys. Charact. **115**(772), 700–721 (1927)
45. Hethcote, H.W.: The mathematics of infectious diseases. SIAM Rev. **42**(4), 599–653 (2000)
46. Capasso, V., Serio, G.: A generalization of the Kermack-McKendrick deterministic epidemic model. Math. Biosci. **42**(1–2), 43–61 (1978)
47. Dimarco, G., Liu, L., Pareschi, L., Zhu, X.: Multi-fidelity methods for uncertainty propagation in kinetic equations (2021)
48. Magal, P., Webb, G.F., Wu, Y.: On the basic reproduction number of reaction-diffusion epidemic models. SIAM J. Appl. Math. **79**(1), 284–304 (2019)
49. Sun, G.-Q.: Pattern formation of an epidemic model with diffusion. Nonlinear Dyn. **69**(3), 1097–1104 (2012)
50. Berestycki, H., Roquejoffre, J.-M., Rossi, L.: Propagation of epidemics along lines with fast diffusion. Bull. Math. Biol. **83**(1), 2 (2021)
51. Viguerie, A., Veneziani, A., Lorenzo, G., Baroli, D., Aretz-Nellesen, N., Patton, A., Yankeelov, T.E., Reali, A., Hughes, T.J., Auricchio, F.: Diffusion-reaction compartmental models formulated in a continuum mechanics framework: application to COVID-19, mathematical analysis, and numerical study. Comput. Mech. **66**(5), 1131–1152 (2020)
52. Salvarani, F., Vázquez, J.L.: The diffusive limit for Carleman-type kinetic models. Nonlinearity **18**(3), 1223–1248 (2005)
53. Zhang, D., Guo, L., Karniadakis, G.E.: Learning in modal space: solving time-dependent stochastic PDEs using physics-informed neural networks. SIAM J. Sci. Comput. **42**(2), A639–A665 (2020)

A Non-local System Modeling Bi-directional Traffic Flows

Felisia Angela Chiarello and Paola Goatin

Abstract We present a non-local model describing the dynamics of two groups of agents moving in opposite directions. The model consists of a 2×2 system of conservation laws with non-local fluxes, coupled in the speed function. We prove local in time existence of weak solutions and present some numerical tests illustrating their behaviour.

Keywords System of conservation laws · Non-local flux · Macroscopic traffic flow models · Finite volume schemes · Bi-directional model

1 Introduction

Conservation laws with non-local flux are suitable to describe several phenomena arising in many fields of application. In this paper, we are interested in describing two groups of agents moving in opposite directions using this class of equations.

The first macroscopic traffic flow model based on fluid-dynamics equations was introduced in the transportation literature in the mid-fifties of last century, with the Lighthill, Whitham and Richards (LWR) model [11, 12]. In recent years, "non-local" versions of the LWR model have been proposed in [2, 3, 7, 9]. In most of these models, the speed depends on a weighted mean of the downstream traffic density, describing the behaviour of agents that adapt their velocity with respect to what happens in front of them. Therefore, the flux function depends on a "downstream" convolution term of the density of agents and with kernel function. In [4, 6], the

F. A. Chiarello (✉)
Dipartimento di Ingegneria e Scienze dell'Informazione e Matematica (DISIM), University of L'Aquila, Via Vetoio, 67100 L'Aquila, Italy
e-mail: felisiaangela.chiarello@univaq.it

P. Goatin
Université Côte d'Azur, Inria, CNRS, LJAD, 2004 route des Lucioles - BP 93, 06902 Sophia Antipolis Cedex, France
e-mail: paola.goatin@inria.fr

G. Albi et al. (eds.), *Advances in Numerical Methods for Hyperbolic Balance Laws and Related Problems*, SEMA SIMAI Springer Series 32,
https://doi.org/10.1007/978-3-031-29875-2_3

authors consider a multi-class traffic model expressed by a system of conservation laws with non-local fluxes obtained generalizing the $n-$populations model for traffic flow introduced in [1], where each equation of the system describes the evolution of the density ρ_i of the vehicles belonging to the $i-$th class. In particular, the non-local multi-class model takes into account the distribution of heterogeneous agents characterized by their maximal speeds and look-ahead visibility in a traffic stream.

In this paper, we consider two non-local conservation laws describing two classes of agents moving in opposite directions. The resulting system is a non-local version of the model presented in [8], where the authors study a mixed type system of conservation laws describing two populations moving in opposite directions. The latter model is not hyperbolic for certain density values, because the Jacobian matrix of the flux exhibits complex eigenvalues in a subset of the phase space, and oscillations arise in the elliptic region. In particular, existence and uniqueness of solutions are still open problems. On the contrary, we will show that introducing a non-local dependence in the speed function allows to prove existence of solutions through the convergence of a suitable finite volume scheme, at least for sufficiently small times.

The paper is organized as follows. In Sect. 2 we present a non-local version of the mixed system studied in [8] and we discretize it with an upwind scheme; in Sect. 3 we recover uniform $\mathbf{L}^\infty$ and **BV** bounds on the sequence of approximate solutions, which allow to prove the existence of weak solutions locally in time. Finally, in Sect. 4, we exhibit some numerical simulations showing the behaviour of weak solutions of our system considering different kernel and velocity functions. In particular, we compare non-local solutions to their local counterpart presented in [8].

2 A Non-local Bi-directional Traffic Flow Model

We consider the following system

$$\begin{cases} \partial_t \rho_1 + \partial_x \left(\rho_1 v_1(r * \omega_1)\right) = 0, \\ \partial_t \rho_2 - \partial_x \left(\rho_2 v_2(r * \omega_2)\right) = 0, \end{cases} \tag{1}$$

where

$$r = \rho_1 + \rho_2,$$

$$r * \omega_1(t, x) = \int_x^{x+\eta_1} \omega_1(y - x) r(t, y) dy,$$

$$r * \omega_2(t, x) = \int_{x-\eta_2}^{x} \omega_2(y - x) r(t, y) dy.$$

In (1), $\rho_1 = \rho_1(t, x)$ is the density of the population moving in the direction of increasing space coordinate with speed $v_1 = v_1(r * \omega_1)$ depending on the downstream weighed mean of the total density r, while $\rho_2 = \rho_2(t, x)$ is the density of the population moving in the opposite direction, with non-local speed v_2.

In the following, we will assume

(H1) $\omega_1 \in \mathbf{C}^1([0, \eta_1]; \mathbb{R}^+)$, $\omega_2 \in \mathbf{C}^1([-\eta_2, 0]; \mathbb{R}^+)$, $\eta_i > 0$ for $i \in \{1, 2\}$, $\omega_1' \leq 0$ and $\omega_2' \geq 0$, $\int_0^{\eta_1} \omega_1(x)dx = \int_{-\eta_2}^{0} \omega_2(x)dx = 1$; we set $\omega_{\max} := \max\{\omega_1(0), \omega_2(0)\}$;

(H2) $v_i : \mathbb{R}^+ \to \mathbb{R}^+$, $i \in \{1, 2\}$, are smooth non-increasing functions such that $v_i(0) = V_i^{\max} > 0$ and $v_i(r) = 0$ for $r \geq 1$.

We couple (1) with an initial datum

$$\rho_i(0, x) = \rho_i^0(x), \qquad i = 1, 2, \tag{2}$$

and we construct a sequence of finite volume approximate solutions as follows. We fix a space step Δx and a time step Δt subject to a CFL condition that will be specified later. Let $x_{j+1/2} = j\Delta x$ for $j \in \mathbb{Z}$ and $n \in \mathbb{N}$ be the cells interfaces, $x_j = (j - 1/2)\Delta x$ the cells centers and $t^n = n\Delta t$ the time mesh. We aim at constructing finite volume approximate solutions of (1) of the form $\boldsymbol{\rho}^{\Delta x} = (\rho_1^{\Delta x}, \rho_2^{\Delta x})$ with $\rho_i^{\Delta x}(t, x) = \rho_{i,j}^n$ for $i = 1, 2$, $(t, x) \in C_j^n = [t^n, t^{n+1}[\times [x_{j-1/2}, x_{j+1/2}[$. To this end, we approximate the initial data with piece-wise constant functions

$$\rho_{i,j}^0 = \frac{1}{\Delta x} \int_{x_{j-1/2}}^{x_{j+1/2}} \rho_i^0(x)dx, \quad \forall j \in \mathbb{Z}, \ i = 1, 2. \tag{3}$$

Similarly, for the kernels, we set

$$\omega_{i,k} := \frac{1}{\Delta x} \int_{k\Delta x}^{(k+1)\Delta x} \omega_i(x)dx, \quad \forall k \in \mathbb{Z}, \ i = 1, 2, \tag{4}$$

so that $\Delta x \sum_{k=-\infty}^{+\infty} \omega_{i,k} = \int_{-\infty}^{+\infty} \omega_i(x)dx = 1$ (extending $\omega_i(x) = 0$ outside their domain).

We use the following upwind scheme, see [4, 7]:

$$\rho_{1,j}^{n+1} = \rho_{1,j}^n - \frac{\Delta t}{\Delta x} \left[\rho_{1,j}^n v_1(R_{1,j+1}^n) - \rho_{1,j-1}^n v_1(R_{1,j}^n)\right], \tag{5a}$$

$$\rho_{2,j}^{n+1} = \rho_{2,j}^n + \frac{\Delta t}{\Delta x} \left[\rho_{2,j+1}^n v_2(R_{2,j}^n) - \rho_{2,j}^n v_2(R_{2,j-1}^n)\right], \tag{5b}$$

where

$$R_{1,j}^n = \Delta x \sum_{k=0}^{+\infty} \omega_{1,k} r_{j+k}^n, \qquad R_{2,j}^n = \Delta x \sum_{k=-\infty}^{0} \omega_{2,k} r_{j+k}^n, \tag{6}$$

where the sums are indeed finite due to the compact supports of ω_1 and ω_2.

3 Existence of Weak Solutions

In this section, we prove existence of weak solutions for sufficiently small times.

Theorem 1 *Let* $\rho_i^0(x) \in (\mathbf{BV} \cap \mathbf{L}^\infty)(\mathbb{R}; \mathbb{R}^+)$, *for* $i = 1, 2$, *and assumptions* **(H1)**–**(H2)** *hold. Then the Cauchy problem* (1), (2) *admits a weak solution on* $[0, T[\times\mathbb{R}$, *for some* $T > 0$ *sufficiently small.*

The proof relies on the following estimates, which ensure the convergence of the numerical scheme (5).

Lemma 1 (Positivity) *For any* $T > 0$, *under the CFL stability condition*

$$\frac{\Delta t}{\Delta x} \leq \frac{1}{\max\{||v_1||_\infty, ||v_2||_\infty\}}, \tag{7}$$

the scheme (5) *is positivity preserving on* $[0, T] \times \mathbb{R}$.

Proof Let us assume that $\rho_{i,j}^n \geq 0$ for all $j \in \mathbb{Z}$ and $i \in \{1, 2\}$. Let us show that the same holds for $\rho_{i,j}^{n+1}$. Let us compute

$$\rho_{1,j}^{n+1} = \left(1 - \frac{\Delta t}{\Delta x} v_1(R_{1,j+1}^n)\right) \rho_{1,j}^n + \rho_{1,j-1}^n \frac{\Delta t}{\Delta x} v_1(R_{1,j}^n),$$

$$\rho_{2,j}^{n+1} = \left(1 - \frac{\Delta t}{\Delta x} v_2(R_{2,j-1}^n)\right) \rho_{2,j}^n + \rho_{2,j+1}^n \frac{\Delta t}{\Delta x} v_2(R_{2,j}^n).$$

By assumption (7), we ensure $\rho_{i,j}^{n+1} \geq 0$ for $i \in \{1, 2\}$ and for all $j \in Z$. □

Corollary 1 ($\mathbf{L}^1$-bound) *If* $\rho_{i,j}^0 \geq 0$ *for all* $j \in \mathbb{Z}$ *and* $i \in \{1, 2\}$ *and the CFL condition* (7) *holds, the approximate solutions constructed via the scheme* (5) *satisfy*

$$\left\|\rho_i^n\right\|_1 = \left\|\rho_i^0\right\|_1, \qquad i = 1, \ldots, M, \tag{8}$$

for any $n \in \mathbb{N}$, *where* $\left\|\rho_i^n\right\|_1 := \Delta x \sum_j \left|\rho_{i,j}^n\right|$ *denotes the* $\mathbf{L}^1$ *norm of the i-th component of* $\rho^{\Delta x}$.

Proof Thanks to Lemma 1, we have

$$\left\|\rho_1^{n+1}\right\|_1 = \Delta x \sum_j \rho_{1,j}^{n+1} = \Delta x \sum_j \rho_{1,j}^n - \frac{\Delta t}{\Delta x}\left[\rho_{1,j}^n v_1(R_{1,j+1}^n) - \rho_{1,j-1}^n v_1(R_{1,j}^n)\right] = \Delta x \sum_j \rho_{1,j}^n,$$

$$\left\|\rho_2^{n+1}\right\|_1 = \Delta x \sum_j \rho_{2,j}^{n+1} = \Delta x \sum_j \rho_{2,j}^n + \frac{\Delta t}{\Delta x}\left[\rho_{2,j+1}^n v_2(R_{2,j}^n) - \rho_{2,j}^n v_2(R_{2,j-1}^n)\right] = \Delta x \sum_j \rho_{2,j}^n,$$

proving (8). □

Lemma 2 ($\mathbf{L}^\infty$-bound) *If $\rho^0_{i,j} \geq 0$ for all $j \in \mathbb{Z}$ and $i \in \{1, 2\}$ and (7) holds, then the approximate solution $\boldsymbol{\rho}^{\boldsymbol{\Delta x}}$ constructed by the algorithm (5) is uniformly bounded on $[0, T] \times \mathbb{R}$ for any T such that*

$$T \leq \frac{1}{4 \max\{\|v_1'\|_\infty, \|v_2'\|_\infty\} w_{\max} \|\boldsymbol{\rho}^0\|_\infty}.$$

Proof Let us define $\bar{\rho} = \max\{\rho^n_{1,j-1}, \rho^n_{1,j}, \rho^n_{2,j}, \rho^n_{2,j+1}\}$. Then we get

$$\begin{aligned}
\rho^{n+1}_{1,j} &= \left(1 - \frac{\Delta t}{\Delta x} v_1(R^n_{1,j+1})\right) \rho^n_{1,j} + \frac{\Delta t}{\Delta x} v_1(R^n_{1,j}) \rho^n_{1,j-1} \\
&\leq \left(1 - \frac{\Delta t}{\Delta x} \left(v_1(R^n_{1,j+1}) - v_1(R^n_{1,j})\right)\right) \bar{\rho} \\
\rho^{n+1}_{2,j} &= \left(1 - \frac{\Delta t}{\Delta x} v_2(R^n_{2,j-1})\right) \rho^n_{2,j} + \frac{\Delta t}{\Delta x} v_2(R^n_{2,j}) \rho^n_{2,j+1} \\
&\leq \left(1 - \frac{\Delta t}{\Delta x} \left(v_2(R^n_{2,j-1}) - v_2(R^n_{2,j})\right)\right) \bar{\rho},
\end{aligned}$$

and

$$\begin{aligned}
\left|v_1(R^n_{1,j}) - v_1(R^n_{1,j+1})\right| &= \left| v_1'(\xi_{j+1/2}) \left(\Delta x \sum_{k=0}^{+\infty} \omega_{1,k} r^n_{j+k+1} - \Delta x \sum_{k=0}^{+\infty} \omega_{1,k} r^n_{j+k} \right) \right| \\
&= \Delta x \left| v_1'(\xi_{j+1/2}) \right| \left| \sum_{k=1}^{+\infty} (\omega_{1,k-1} - \omega_{1,k}) r^n_{j+k} - \omega_{1,0}\, r^n_j \right| \\
&\leq 4\, \Delta x\, \omega_1(0) \|v_1'\|_\infty \|\boldsymbol{\rho}^n\|_\infty, \qquad (9)
\end{aligned}$$

$$\begin{aligned}
\left|v_2(R^n_{2,j-1}) - v_2(R^n_{2,j})\right| &= \left| v_2'(\xi_{j-1/2}) \left(\Delta x \sum_{k=-\infty}^{0} \omega_{2,k} r^n_{j+k-1} - \Delta x \sum_{k=-\infty}^{0} \omega_{2,k} r^n_{j+k} \right) \right| \\
&= \Delta x \left| v_2'(\xi_{j-1/2}) \right| \left| \sum_{k=-\infty}^{-1} (\omega_{2,k+1} - \omega_{2,k}) r^n_{j+k} - \omega_{2,0}\, r^n_j \right| \\
&\leq 4\, \Delta x\, \omega_2(0) \|v_2'\|_\infty \|\boldsymbol{\rho}^n\|_\infty, \qquad (10)
\end{aligned}$$

where $\xi_{j+1/2} \in \mathrm{i}(R^n_{1,j}, R^n_{1,j+1})$, with $\mathcal{I}(a, b) = [\min\{a, b\}, \max\{a, b\}]$, and $\|\boldsymbol{\rho}^n\|_\infty = \|(\rho^n_1, \rho^n_2)\|_\infty = \max_{i,j} \left|\rho^n_{i,j}\right|$. So, until $\|\boldsymbol{\rho}^n\|_\infty \leq K$ for some $K \geq \|\boldsymbol{\rho}^0\|_\infty$, we get

$$\|\boldsymbol{\rho}^{n+1}\|_\infty \leq \|\boldsymbol{\rho}^n\|_\infty \left(1 + 4K \max\left\{\|v_1'\|_\infty, \|v_2'\|_\infty\right\} \Delta t\, w_{\max}\right)$$

with $w_{\max} = \max\{\omega_1(0), \omega_2(0)\}$. This implies

$$\|\boldsymbol{\rho}^n\|_\infty \leq \|\boldsymbol{\rho}^0\|_\infty e^{Cn\Delta t},$$

where $C = 4K \max\{\|v_1'\|_\infty, \|v_2'\|_\infty\} w_{\max}$. Therefore we get that $\|\boldsymbol{\rho}(t,\cdot)\|_\infty \le K$ for

$$t \le \frac{1}{4K \max\{\|v_1'\|_\infty, \|v_2'\|_\infty\} w_{\max}} \ln\left(\frac{K}{\|\boldsymbol{\rho}^0\|_\infty}\right) \le \frac{1}{4e \max\{\|v_1'\|_\infty, \|v_2'\|_\infty\} w_{\max} \|\boldsymbol{\rho}^0\|_\infty},$$

where the maximum is attained for $K = e\|\boldsymbol{\rho}^0\|_\infty$. Let us iterate the procedure: at time t^m, $m \ge 1$, we set $K = e^m \|\boldsymbol{\rho}^0\|_\infty$ and we get that the solution is bounded by K until t^{m+1} such that

$$t^{m+1} \le t^m + \frac{m}{4e^m \max\{\|v_1'\|_\infty, \|v_2'\|_\infty\} w_{\max} \|\boldsymbol{\rho}^0\|_\infty}.$$

Therefore, the approximate solution remains bounded, uniformly in Δx, at least for $t \le T$ with

$$T \le \frac{1}{4 \max\{\|v_1'\|_\infty, \|v_2'\|_\infty\} w_{\max} \|\boldsymbol{\rho}^0\|_\infty} \sum_{m=1}^{+\infty} \frac{m}{e^m} \le \frac{1}{4 \max\{\|v_1'\|_\infty, \|v_2'\|_\infty\} w_{\max} \|\boldsymbol{\rho}^0\|_\infty}.$$

□

Lemma 3 (Spatial **BV**-bound) *Let $\rho_i^0 \in (\mathbf{BV} \cap \mathbf{L}^\infty)(\mathbb{R}; \mathbb{R}^+)$ for $i \in \{1, 2\}$. If (7) holds, then the approximate solution $\boldsymbol{\rho}^{\Delta x}(t, \cdot)$ constructed by the algorithm (5) has uniformly bounded total variation for $t \in [0, T]$ for any T such that*

$$T \le \min_{i=1,2} \frac{1}{\mathcal{H}\left(TV(\rho_i^0) + 1\right)}, \tag{11}$$

where $\mathcal{H} = 2\|\boldsymbol{\rho}^n\|_\infty w_{\max} \left(12\|\boldsymbol{\rho}^n\|_\infty \max\{\|v_1''\|_\infty, \|v_2''\|_\infty\} + \max\{\|v_1'\|_\infty, \|v_2'\|_\infty\}\right)$.

Proof Let us consider the component ρ_1 and subtract the identities

$$\begin{aligned}
\rho_{1,j+1}^{n+1} &= \rho_{1,j+1}^n - \frac{\Delta t}{\Delta x}\left[\rho_{1,j+1}^n v_1(R_{1,j+2}^n) - \rho_{1,j}^n v_1(R_{1,j+1}^n)\right] \\
\rho_{1,j}^{n+1} &= \rho_{1,j}^n - \frac{\Delta t}{\Delta x}\left[\rho_{1,j}^n v_1(R_{1,j+1}^n) - \rho_{1,j-1}^n v_1(R_{1,j}^n)\right],
\end{aligned}$$

analogously for the second component ρ_2,

$$\begin{aligned}
\rho_{2,j+1}^{n+1} &= \rho_{2,j+1}^n + \frac{\Delta t}{\Delta x}\left[\rho_{2,j+2}^n v_2(R_{2,j+1}^n) - \rho_{2,j+1}^n v_2(R_{2,j}^n)\right] \\
\rho_{2,j}^{n+1} &= \rho_{2,j}^n + \frac{\Delta t}{\Delta x}\left[\rho_{2,j+1}^n v_2(R_{2,j}^n) - \rho_{2,j}^n v_2(R_{2,j-1}^n)\right].
\end{aligned}$$

Setting $\Delta_{i,j+1/2}^n = \rho_{i,j+1}^n - \rho_{i,j}^n$ for $i \in \{1, 2\}$, we get

$$\Delta^{n+1}_{1,j+1/2} = \Delta^n_{1,j+1/2} - \frac{\Delta t}{\Delta x}\left(\rho^n_{1,j+1} v_1(R^n_{1,j+2}) - 2\rho^n_{1,j} v_1(R^n_{1,j+1}) + \rho^n_{1,j-1} v_1(R^n_{1,j})\right),$$

$$\Delta^{n+1}_{2,j+1/2} = \Delta^n_{2,j+1/2} + \frac{\Delta t}{\Delta x}\left(\rho^n_{2,j+2} v_2(R^n_{2,j+1}) - 2\rho^n_{2,j+1} v_2(R^n_{2,j}) + \rho^n_{2,j} v_2(R^n_{2,j-1})\right).$$

We can write

$$\Delta^{n+1}_{1,j+1/2} = \left(1 - \frac{\Delta t}{\Delta x} v_1(R^n_{1,j+2})\right)\Delta^n_{1,j+1/2} \tag{12}$$

$$- \frac{\Delta t}{\Delta x}\rho^n_{1,j}\left(v_1(R^n_{1,j+2}) - 2v_1(R^n_{1,j+1}) + v_1(R^n_{1,j})\right) \tag{13}$$

$$+ \frac{\Delta t}{\Delta x}\Delta^n_{1,j-1/2} v_1(R^n_{1,j}),$$

and

$$\Delta^{n+1}_{2,j+1/2} = \left(1 - \frac{\Delta t}{\Delta x} v_2(R^n_{2,j-1})\right)\Delta^n_{2,j+1/2} \tag{14}$$

$$+ \frac{\Delta t}{\Delta x}\rho^n_{2,j+1}\left(v_2(R^n_{2,j+1}) - 2v_2(R^n_{2,j}) + v_2(R^n_{2,j-1})\right) \tag{15}$$

$$+ \frac{\Delta t}{\Delta x}\Delta^n_{2,j+3/2} v_2(R^n_{2,j+1}).$$

Observe that assumption (7) guarantees the positivity of (12) and (14). The term (13) can be estimated as

$$\begin{aligned}
& v_1(R^n_{1,j+2}) - 2v_1(R^n_{1,j+1}) + v_1(R^n_{1,j}) \\
&= v_1\left(\Delta x \sum_{k=0}^{+\infty} \omega_{1,k} r^n_{j+k+2}\right) - 2v_1\left(\Delta x \sum_{k=0}^{+\infty} \omega_{1,k} r^n_{j+k+1}\right) + v_1\left(\Delta x \sum_{k=0}^{+\infty} \omega_{1,k} r^n_{j+k}\right) \\
&= v_1'(\xi^n_{1,j+3/2})\Delta x\left(\sum_{k=0}^{+\infty} \omega_{1,k} r^n_{j+k+2} - \sum_{k=0}^{+\infty} \omega_{1,k} r^n_{j+k+1}\right) \\
&\quad - v_1'(\xi^n_{1,j+1/2})\Delta x\left(\sum_{k=0}^{+\infty} \omega_{1,k} r^n_{j+k+1} - \sum_{k=0}^{+\infty} \omega_{1,k} r^n_{j+k}\right) \\
&= N v_1'(\xi^n_{1,j+3/2})\Delta x\left(\sum_{k=1}^{+\infty} (\omega_{1,k-1} - \omega_{1,k}) r^n_{j+k+1} - \omega_{1,0} r^n_{j+1}\right) \\
&\quad - v_1'(\xi^n_{1,j+1/2})\Delta x\left(\sum_{k=1}^{+\infty} (\omega_{1,k-1} - \omega_{1,k}) r^n_{j+k} - \omega_{1,0} r^n_{j}\right) \\
&= v_1''(\tilde{\xi}^n_{1,j+1})(\xi^n_{1,j+3/2} - \xi^n_{1,j+1/2})\Delta x\left(\sum_{k=0}^{+\infty} \omega_{1,k} \sum_{i=1}^{2} \Delta^n_{i,j+k+3/2}\right)
\end{aligned}$$

$$+ v_1'(\xi^n_{1,j+1/2})\Delta x \left(\sum_{k=1}^{+\infty} \sum_{i=1}^{2} (\omega_{1,k-1} - \omega_{1,k}) \Delta^n_{i,j+k+1/2} - \omega_{1,0} \Delta^n_{i,j+1/2} \right),$$

with $\xi^n_{1,j+1/2} \in \mathcal{I}(R^n_{1,j}, R^n_{1,j+1})$ and $\tilde{\xi}^n_{1,j+1} \in \mathcal{I}(\xi^n_{1,j+1/2}, \xi^n_{1,j+3/2})$. For some $\theta, \mu \in [0, 1]$, we compute

$$\begin{aligned}
\xi^n_{1,j+3/2} - \xi^n_{1,j+1/2} &= \theta \Delta x \sum_{k=0}^{+\infty} \omega_{1,k} \sum_{i=1}^{2} \rho^n_{i,j+k+2} + (1-\theta)\Delta x \sum_{k=0}^{+\infty} \omega_{1,k} \sum_{i=1}^{2} \rho^n_{i,j+k+1} \\
&\quad - \mu \Delta x \sum_{k=0}^{+\infty} \omega_{1,k} \sum_{i=1}^{2} \rho^n_{i,j+k+1} - (1-\mu)\Delta x \sum_{k=0}^{+\infty} \omega_{1,k} \sum_{i=1}^{2} \rho^n_{i,j+k} \\
&= \theta \Delta x \sum_{k=1}^{+\infty} \omega_{1,k-1} \sum_{i=1}^{2} \rho^n_{i,j+k+1} + (1-\theta)\Delta x \sum_{k=0}^{+\infty} \omega_{1,k} \sum_{i=1}^{2} \rho^n_{i,j+k+1} \\
&\quad - \mu \Delta x \sum_{k=0}^{+\infty} \omega_{1,k} \sum_{i=1}^{2} \rho^n_{i,j+k+1} - (1-\mu)\Delta x \sum_{k=-1}^{+\infty} \omega_{1,k+1} \sum_{i=1}^{2} \rho^n_{i,j+k+1} \\
&= \Delta x \sum_{k=1}^{+\infty} \left[\theta \omega_{1,k-1} + (1-\theta)\omega_{1,k} - \mu \omega_{1,k} - (1-\mu)\omega_{1,k+1} \right] \sum_{i=1}^{2} \rho^n_{i,j+k+1} \\
&\quad + \Delta x (1-\theta)\omega_{1,0} \sum_{i=1}^{2} \rho^n_{i,j+1} - \Delta x \mu\, \omega_{1,0} \sum_{i=1}^{2} \rho^n_{i,j+1} \\
&\quad - \Delta x (1-\mu) \left(\omega_{1,0} \sum_{i=1}^{2} \rho^n_{i,j} + \omega_{1,1} \sum_{i=1}^{2} \rho^n_{i,j+1} \right).
\end{aligned}$$

By monotonicity we have

$$\theta \omega_{1,k-1} + (1-\theta)\omega_{1,k} - \mu \omega_{1,k} - (1-\mu)\omega_{1,k+1} \geq 0.$$

Taking the absolute values, we get

$$\begin{aligned}
\left| \xi^n_{1,j+3/2} - \xi^n_{1,j+1/2} \right| &\leq \Delta x \left\{ \sum_{k=1}^{+\infty} \left[\theta \omega_{1,k-1} + (1-\theta)\omega_{1,k} \right. \right. \\
&\qquad \left. \left. - \mu \omega_{1,k} - (1-\mu)\omega_{1,k+1} \right] + 4\, \omega_{1,0} \right\} 2 \|\boldsymbol{\rho}^n\|_\infty \\
&\leq 12\, \Delta x\, \omega_1(0) \|\boldsymbol{\rho}^n\|_\infty.
\end{aligned}$$

Applying the same argument to the second component, we compute

$$\begin{aligned}
& v_2(R^n_{2,j+1}) - 2 v_2(R^n_{2,j}) + v_2(R^n_{2,j-1}) \\
&= v_2''(\tilde{\xi}^n_{2,j})(\xi^n_{2,j+1/2} - \xi^n_{2,j-1/2}) \Delta x \left(\sum_{k=-\infty}^{0} \omega_{2,k} \sum_{i=1}^{2} \Delta^n_{i,j+k+1/2} \right)
\end{aligned}$$

$$+ v_2'(\xi^n_{2,j-1/2})\Delta x \left(\sum_{k=-\infty}^{0} \sum_{i=1}^{2} (\omega_{2,k-1} - \omega_{2,k})\Delta^n_{i,j+k-1/2} - \omega_{2,0}\Delta^n_{i,j-1/2} \right),$$

with $\xi^n_{2,j+1/2} \in \mathcal{I}(R^n_{2,j}, R^n_{2,j+1})$ and $\tilde{\xi}^n_{2,j} \in \mathcal{I}(\xi^n_{2,j-1/2}, \xi^n_{2,j+1/2})$, and

$$\left|\xi^n_{2,j+1/2} - \xi^n_{2,j-1/2}\right| \leq 12\,\Delta x\,\omega_2(0)\|\boldsymbol{\rho}^n\|_\infty.$$

Let now $K_1 > 0$ be such that $\sum_j \left|\Delta^\ell_{i,j}\right| \leq K_1$ for $i \in \{1, 2\}, \ell = 0, \dots, n$. Taking the absolute values and rearranging the indexes, we have

$$\sum_j \left|\Delta^{n+1}_{1,j+1/2}\right| \leq \sum_j \left|\Delta^n_{1,j+1/2}\right| \left(1 - \lambda \left(v_1(R^n_{1,j+2}) - v_1(R^n_{1,j+1})\right)\right) + \Delta t\,\mathcal{H}K_1,$$

$$\sum_j \left|\Delta^{n+1}_{2,j+1/2}\right| \leq \sum_j \left|\Delta^n_{2,j+1/2}\right| \left(1 - \lambda \left(v_2(R^n_{2,j-1}) - v_2(R^n_{2,j})\right)\right) + \Delta t\,\mathcal{H}K_1,$$

where $\mathcal{H} = 2\|\boldsymbol{\rho}^n\|_\infty\, w_{\max} \left(12\|\boldsymbol{\rho}^n\|_\infty \max\{\|v_1''\|_\infty, \|v_2''\|_\infty\} + \max\{\|v_1'\|_\infty, \|v_2'\|_\infty\}\right)$. Therefore, by (9)–(10) we get

$$\sum_j \left|\Delta^{n+1}_{i,j+1/2}\right| \leq \sum_j \left|\Delta^n_{i,j+1/2}\right| (1 + \Delta t\,\mathcal{G}) + \Delta t\,\mathcal{H}K_1,$$

with $\mathcal{G} = 4\max\{\|v_1'\|_\infty, \|v_2'\|_\infty\} w_{\max}\|\boldsymbol{\rho}^n\|_\infty$. We thus obtain

$$\sum_j \left|\Delta^n_{i,j+1/2}\right| \leq e^{\mathcal{G}n\Delta t} \sum_j \left|\Delta^0_{i,j+1/2}\right| + e^{\mathcal{H}K_1 n\Delta t} - 1,$$

that we can rewrite as

$$\begin{aligned} TV(\rho_i^{\Delta x})(n\Delta t, \cdot) &\leq e^{\mathcal{G}n\Delta t} TV(\rho_i^0) + e^{\mathcal{H}K_1 n\Delta t} - 1 \\ &\leq e^{\mathcal{H}K_1 n\Delta t}\left(TV(\rho_i^0) + 1\right) - 1, \end{aligned}$$

since $\mathcal{H} \geq \mathcal{G}$ and it is not restrictive to assume $K_1 \geq 1$.

Therefore, we have that $TV(\rho_i^{\Delta x}) \leq K_1$ for

$$t \leq \frac{1}{\mathcal{H}K_1} \ln\left(\frac{K_1 + 1}{TV(\rho_i^0) + 1}\right),$$

where the maximum is attained for some $K_1 < e\left(TV(\rho_i^0) + 1\right) - 1$ such that

$$\ln\left(\frac{K_1 + 1}{TV(\rho_i^0) + 1}\right) = \frac{K_1}{K_1 + 1}.$$

Therefore the total variation is uniformly bounded for

$$t \leq \frac{1}{\mathcal{H}e\left(TV(\rho_i^0)+1\right)}.$$

Iterating the procedure, at time t^m, $m \geq 1$ we set $K_1 = e^m\left(TV(\rho_i^0)+1\right)-1$ and we get that the solution is bounded by K_1 until t^{m+1} such that

$$t^{m+1} \leq t^m + \frac{m}{\mathcal{H}e^m\left(TV(\rho_i^0)+1\right)}. \tag{16}$$

Therefore, the approximate solution has bounded total variation for $t \leq T$ with

$$T \leq \frac{1}{\mathcal{H}\left(TV(\rho_i^0)+1\right)}.$$

□

Corollary 2 *Let $\rho_i^0 \in (\mathbf{BV} \cap \mathbf{L}^\infty)(\mathbb{R};\mathbb{R}^+)$. If (7) holds, then the approximate solution $\rho^{\Delta x}$ constructed by the algorithm (5) has uniformly bounded total variation on $[0,T]\times\mathbb{R}$, for any T satisfying (11).*

Proof If $T \leq \Delta t$, then $TV(\rho_i^{\Delta x};[0,T]\times\mathbb{R}) \leq T\,TV(\rho_i^0)$. Let us assume now that $T > \Delta t$. Let $n_T \in \mathbb{N}\backslash\{0\}$ such that $n_T\Delta t < T \leq (n_T+1)\Delta t$. Then, for $i \in \{1,2\}$,

$$\begin{aligned}&TV(\rho_i^{\Delta x};[0,T]\times\mathbb{R})\\ &= \underbrace{\sum_{n=0}^{n_T-1}\sum_{j\in\mathbb{Z}}\Delta t\left|\rho_{i,j+1}^n-\rho_{i,j}^n\right| + (T-n_T\Delta t)\sum_{j\in\mathbb{Z}}\left|\rho_{i,j+1}^{n_T}-\rho_{i,j}^{n_T}\right|}_{\leq T\,\sup_{t\in[0,T]} TV(\rho_i^{\Delta x})(t,\cdot)} + \sum_{n=0}^{n_T-1}\sum_{j\in\mathbb{Z}}\Delta x\left|\rho_{i,j}^{n+1}-\rho_{i,j}^n\right|.\end{aligned}$$

We then need to bound the term

$$\sum_{n=0}^{n_T-1}\sum_{j\in\mathbb{Z}}\Delta x\left|\rho_{i,j}^{n+1}-\rho_{i,j}^n\right|.$$

From the definition of the numerical scheme (5a)–(5b), we obtain

$$\begin{aligned}\rho_{1,j}^{n+1}-\rho_{1,j}^n &= -\frac{\Delta t}{\Delta x}\left[\rho_{1,j}^n v_1(R_{1,j+1}^n)-\rho_{1,j-1}^n v_1(R_{1,j}^n)\right],\\ &= \frac{\Delta t}{\Delta x}\left[\rho_{1,j-1}^n\left(v_1(R_{1,j}^n)-v_1(R_{1,j+1}^n)\right)+v_1(R_{1,j+1}^n)\left(\rho_{1,j-1}^n-\rho_{1,j}^n\right)\right],\\ \rho_{2,j}^{n+1}-\rho_{2,j}^n &= \frac{\Delta t}{\Delta x}\left[\rho_{2,j+1}^n v_2(R_{2,j}^n)-\rho_{2,j}^n v_2(R_{2,j-1}^n)\right]\\ &= \frac{\Delta t}{\Delta x}\left[\rho_{2,j+1}^n\left(v_2(R_{2,j}^n)-v_2(R_{2,j-1}^n)\right)+v_2(R_{2,j-1}^n)\left(\rho_{2,j+1}^n-\rho_{2,j}^n\right)\right].\end{aligned}$$

Taking the absolute values and using (9)–(10) we get

$$\left|\rho_{1,j}^{n+1} - \rho_{1,j}^{n}\right| \le \frac{\Delta t}{\Delta x}\left(4\,\Delta x\,\omega_1(0)\|v_1'\|_\infty\|\boldsymbol{\rho}^n\|_\infty\left|\rho_{1,j-1}^{n}\right| + \|v_1\|_\infty\left|\rho_{1,j-1}^{n} - \rho_{1,j}^{n}\right|\right)$$
$$\left|\rho_{2,j}^{n+1} - \rho_{2,j}^{n}\right| \le \frac{\Delta t}{\Delta x}\left(4\,\Delta x\,\omega_2(0)\|v_2'\|_\infty\|\boldsymbol{\rho}^n\|_\infty\left|\rho_{2,j+1}^{n}\right| + \|v_2\|_\infty\left|\rho_{2,j+1}^{n} - \rho_{2,j}^{n}\right|\right).$$

Summing on j, we get

$$\begin{aligned}\sum_{j\in\mathbb{Z}}\Delta x\left|\rho_{1,j}^{n+1} - \rho_{1,j}^{n}\right| &\le 4\,\omega_1(0)\|v_1'\|_\infty\|\boldsymbol{\rho}^n\|_\infty\Delta t\sum_{j\in\mathbb{Z}}\Delta x\left|\rho_{1,j-1}^{n}\right|\\ &\quad + \|v_1\|_\infty\Delta t\sum_{j\in\mathbb{Z}}\left|\rho_{1,j-1}^{n} - \rho_{1,j}^{n}\right|\\ \sum_{j\in\mathbb{Z}}\Delta x\left|\rho_{2,j}^{n+1} - \rho_{2,j}^{n}\right| &\le 4\,\omega_2(0)\|v_2'\|_\infty\|\boldsymbol{\rho}^n\|_\infty\Delta t\sum_{j\in\mathbb{Z}}\Delta x\left|\rho_{2,j+1}^{n}\right|\\ &\quad + \|v_2\|_\infty\Delta t\sum_{j\in\mathbb{Z}}\left|\rho_{2,j+1}^{n} - \rho_{2,j}^{n}\right|.\end{aligned}$$

which yields, for $i \in \{1, 2\}$,

$$\begin{aligned}&\sum_{n=0}^{n_T-1}\sum_{j\in\mathbb{Z}}\Delta x\left|\rho_{i,j}^{n+1} - \rho_{i,j}^{n}\right|\\ &\le \max\{\|v_1\|_\infty, \|v_2\|_\infty\}T\sup_{t\in[0,T]}TV(\rho_i^{\Delta x})(t,\cdot)\\ &\quad + 4\max\{\omega_1(0), \omega_2(0)\}\max\{\|v_1'\|_\infty, \|v_2'\|_\infty\}\|\boldsymbol{\rho}^n\|_\infty T\sup_{t\in[0,T]}\left\|\rho_i^{\Delta x}(t,\cdot)\right\|_1,\end{aligned}$$

that is bounded by Corollary 1, Lemmas 2 and 3. □

Proof (Proof of Theorem 1) To complete the proof of the existence of solutions to the Cauchy problem (1)–(2), we apply Helly's theorem and follow a Lax-Wendroff type argument as in [2–4], see also [10], to show that the approximate solutions constructed by scheme (5) converge to a weak solution of (1)–(2). □

4 Numerical Tests

In this section, we perform some numerical tests implementing the scheme (5)–(6). For high-order of accuracy, the Finite Volume WENO (FV-WENO) schemes in [5] can be adapted.

4.1 Kernel Support Tending to Zero

In this subsection, we observe the solution behaviour as the length of kernel support diminishes. We consider the space domain given by the interval $[-1, 1]$ and the space discretization mesh size is set to $\Delta x = 0.001$.

We impose absorbing conditions at the boundaries, adding $N = \eta_i/\Delta x$ ghost cells at the right boundary for the first population ρ_1 and at the left boundary for the second population ρ_2, where we extend the solution constantly equal to the last value inside the domain.

We take the following linear decreasing kernel and velocity functions:

$$\omega_1(x) = \frac{2}{\eta_1}\left(1 - \frac{x}{\eta_1}\right), \qquad \omega_2(x) = \frac{2}{\eta_2}\left(1 + \frac{x}{\eta_2}\right); \tag{17a}$$

$$v_1(\rho) = 1 - \rho\,, \qquad v_2(\rho) = 1 - \rho\,. \tag{17b}$$

In Figs. 1 and 2, we consider the same initial data of [8, Figs. 5 and 6], observing that the solutions have a similar behaviour, especially when the kernel supports $\eta_1 = \eta_2$ are taken sufficiently small. Moreover, comparing Figs. 3 and 4 with the numerical tests in [8, Figs. 14 and 15], we note that when we consider initial data in the elliptic region of the local system, oscillations increase as the kernel support diminish.

4.2 Asymptotic Behaviour in a Periodic Setting

We consider now sinusoidal initial data with periodic boundary conditions, see Fig. 5. In the following we set $\Delta x = 0.001$ and $\rho_{\max} = 1$, the CFL condition is that one in (7). We are interested in the behaviour of solutions with different types of kernel functions and velocities.

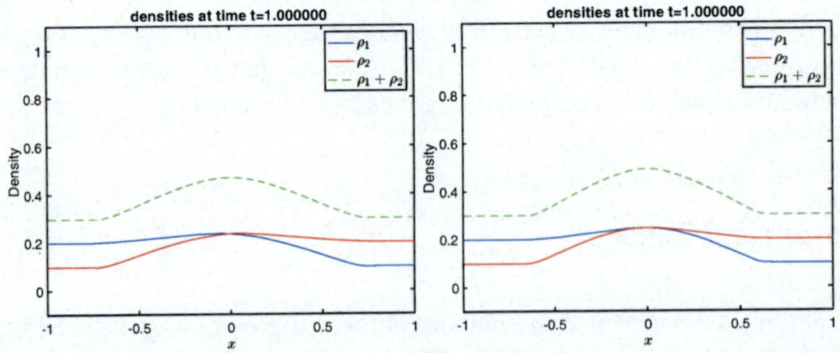

Fig. 1 Test 1. Solution to (1) with (13) and $\rho_1^0 = 0.2 * (x < 0) + 0.1 * (x >= 0)$ and $\rho_2^0 = 0.1 * (x < 0) + 0.2 * (x >= 0)$ with $\eta_1 = \eta_2 = 0.1$ on the left and $\eta_1 = \eta_2 = 0.01$ on the right

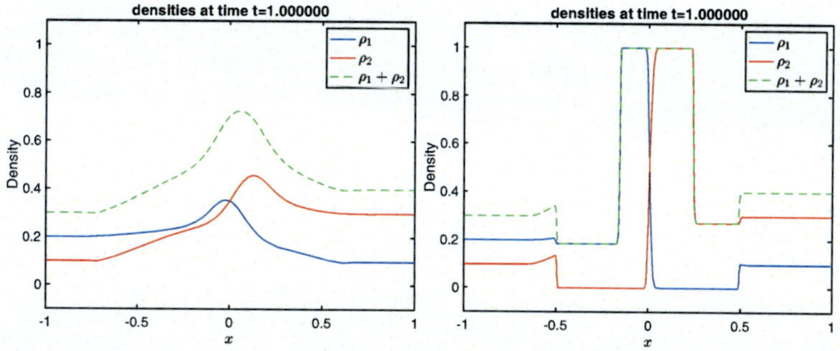

Fig. 2 Test 2. Solution to (1) with (13) and $\rho_1 = 0.2 * (x < 0) + 0.1 * (x >= 0)$ and $\rho_2 = 0.1 * (x < 0) + 0.3 * (x >= 0)$ with $\eta_1 = \eta_2 = 0.1$ on the left and $\eta_1 = \eta_2 = 0.01$ on the right

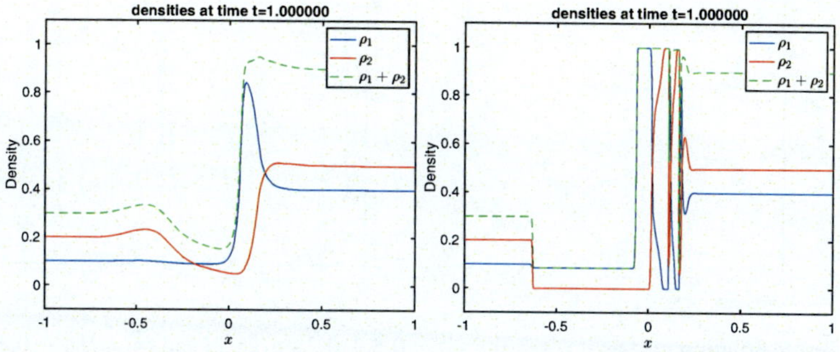

Fig. 3 Test 3. Solution to (1) with (13) and $\rho_1^0 = 0.1 * (x < 0) + 0.4 * (x >= 0)$ and $\rho_2^0 = 0.2 * (x < 0) + 0.5 * (x >= 0)$. With $\eta_1 = \eta_2 = 0.1$ at left, $\eta_1 = \eta_2 = 0.01$ at right

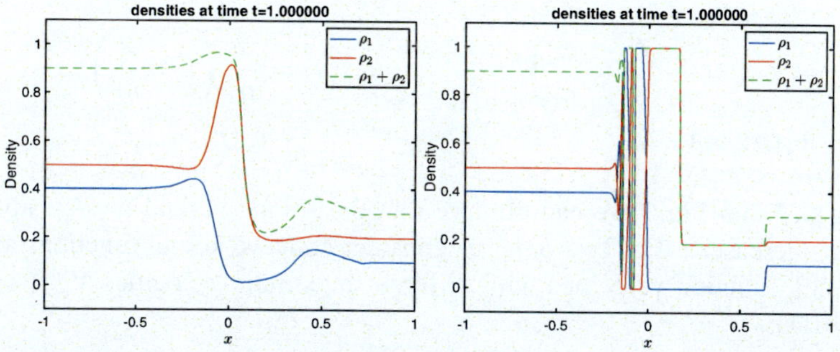

Fig. 4 Test 4. Solution to (1) with (13) and $\rho_1^0 = 0.4 * (x < 0) + 0.1 * (x >= 0)$ and $\rho_2^0 = 0.5 * (x < 0) + 0.2 * (x >= 0)$. With $\eta_1 = \eta_2 = 0.1$ at left, $\eta_1 = \eta_2 = 0.01$ at right

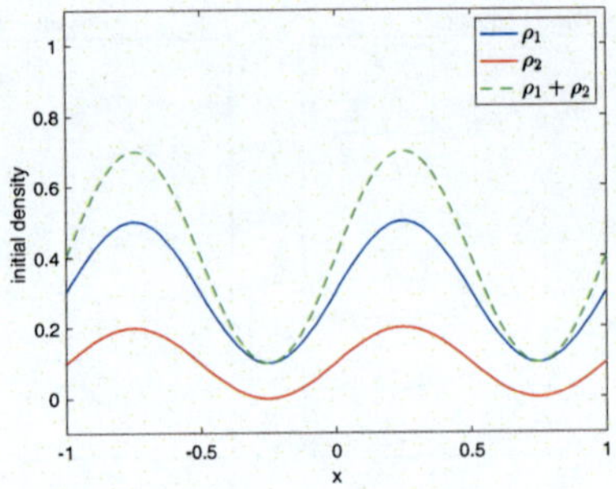

Fig. 5 Initial condition for tests 5, 6 and 7 with periodic boundary conditions. $\rho_1^0 = 0.3 + 0.2\sin(2\pi x)$ and $\rho_2^0 = 0.1 + 0.1\sin(2\pi x)$

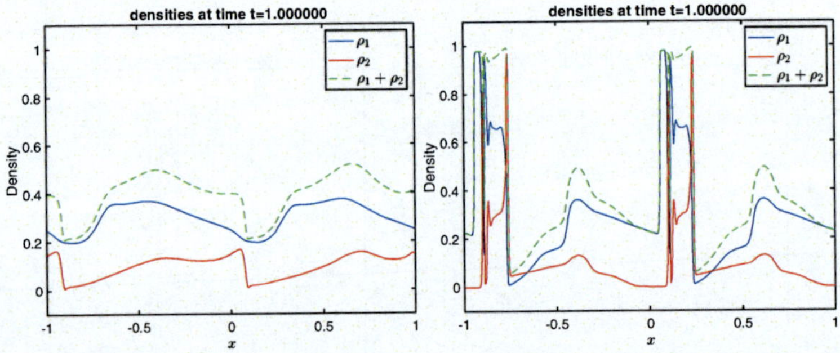

Fig. 6 Test 5. Solution to (1) with (14) and $\rho_1^0 = 0.3 + 0.2\sin(2\pi x)$ and $\rho_2^0 = 0.1 + 0.1\sin(2\pi x)$ with $\eta_1 = \eta_2 = 0.1$ at left, $\eta_1 = \eta_2 = 0.01$ at right

For Test 5, let us consider the linear kernels and velocity functions

$$\omega_1(x) = \frac{2}{\eta_1}\left(1 - \frac{x}{\eta_1}\right), \qquad \omega_2(x) = \frac{2}{\eta_2}\left(1 + \frac{x}{\eta_2}\right); \tag{18a}$$

$$v_1(\rho) = 1 - \rho\,, \qquad v_2(\rho) = 1 - \rho\,, \tag{18b}$$

see Fig. 6. From Fig. 7 we can observe that the solutions tend to a steady-state as the time increases. For Test 6, let us consider concave kernel functions and linear decreasing function velocities with different maximum velocities $V_1^{\max} = 0.8$ and $V_2^{\max} = 1.3$:

$$\omega_1(x) = \frac{3}{2\eta_1^3}\left(\eta_1^2 - x^2\right), \qquad \omega_2(x) = \frac{3}{2\eta_2^3}\left(\eta_2^2 - x^2\right); \tag{19a}$$

$$v_1(\rho) = V_1^{\max}(1 - \rho)\,, \qquad v_2(\rho) = V_2^{\max}(1 - \rho)\,; \tag{19b}$$

In Fig. 8 we can see the profile of the solutions where the agents of the second class are faster than the agents belonging to the first class. Also in this case, reducing the

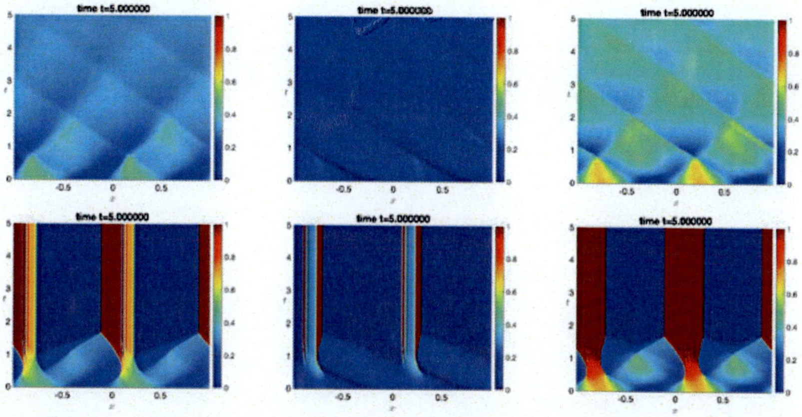

Fig. 7 (t, x)−plot solutions of (1) with (14) and initial conditions $\rho_1^0 = 0.3 + 0.2\sin(2\pi x)$ and $\rho_2^0 = 0.1 + 0.1\sin(2\pi x)$, see Fig. 5; from top to bottom: $\eta_1 = \eta_2 = 0.1,\ 0.01$; from left to right: ρ_1, ρ_2 and $\rho_1 + \rho_2$

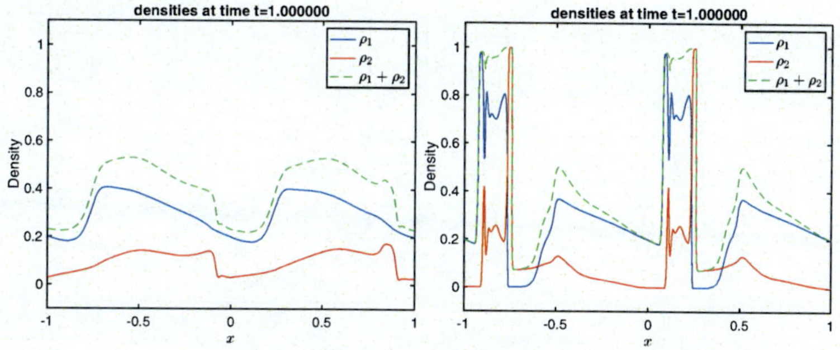

Fig. 8 Test 6. Solution to (1) with (15), $V_1^{\max} = 0.8$ and $V_2^{\max} = 1.3$, and $\rho_1^0 = 0.3 + 0.2\sin(2\pi x)$ and $\rho_2^0 = 0.1 + 0.1\sin(2\pi x)$; $\eta_1 = \eta_2 = 0.1$ on the left, $\eta_1 = \eta_2 = 0.01$ on right

support of the kernels oscillations are increasing. Figure 9 is very similar to Fig. 7; here we observe that solutions tend to a steady-state after time 2.

For Test 7, let us consider the concave kernel functions (19a) with different supports and same maximum velocities $V_1^{\max} = V_2^{\max} = 1$ in (19b). In Fig. 10 we can observe the behaviour of the solutions when the agents of the two classes have different look-ahead distances, i.e. supports of the kernels. We can notice that when one of the agents has a sufficiently large look-ahead distance oscillations are not evident. From Fig. 11, we see that the solutions reach the steady-state as the time increases.

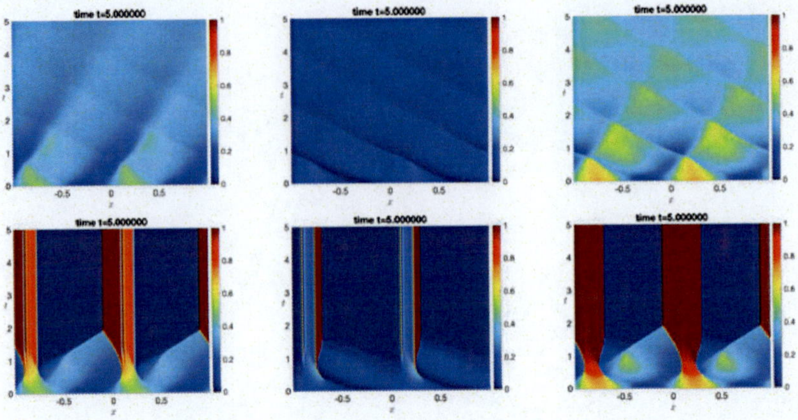

Fig. 9 $(t, x)-$plot solutions of (1) with (15) and initial conditions $\rho_1^0 = 0.3 + 0.2\sin(2\pi x)$ and $\rho_2^0 = 0.1 + 0.1\sin(2\pi x)$, see Fig. 5; from top to bottom: $\eta_1 = \eta_2 = 0.1,\ 0.01$; from left to right: ρ_1, ρ_2 and $\rho_1 + \rho_2$

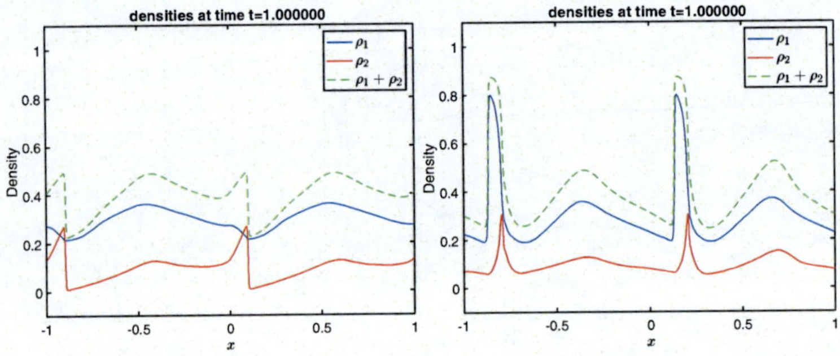

Fig. 10 Test 7. Solution to (1) with (15), $V_1^{\max} = V_2^{\max} = 1$, and $\rho_1^0 = 0.3 + 0.2\sin(2\pi x)$ and $\rho_2^0 = 0.1 + 0.1\sin(2\pi x)$; $\eta_1 = 0.1,\ \eta_2 = 0.01$ on the left, $\eta_1 = 0.01,\ \eta_2 = 0.1$ on the right

4.3 Maximum Principle

The simplex

$$\mathcal{S} := \{(\rho_1, \rho_2) \in \mathbb{R}^2 : \rho_1 + \rho_2 \leq 1,\ \rho_i \geq 0 \quad \text{for } i = 1, 2\},$$

is not an invariant domain, as we can see in Fig. 12, as we already noticed for the multi-class system presented in [4]. In particular, in Fig. 12 we observe that the sum $r(t, x)$ can exceed 1 even if the initial condition is such that $\rho_1(0, x) + \rho_2(0, x) \leq 1$.

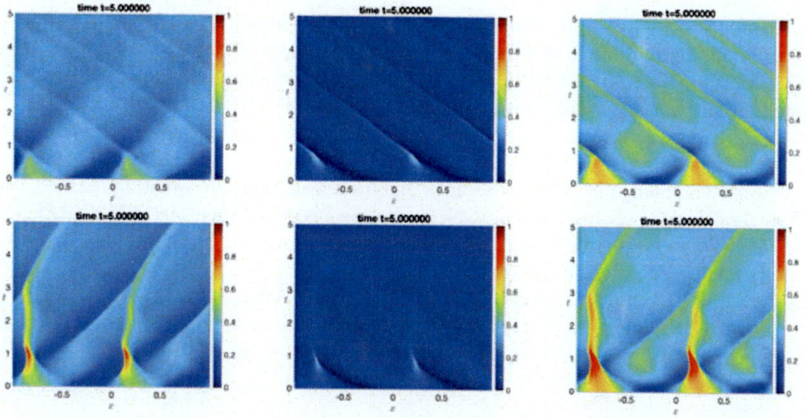

Fig. 11 $(t, x)-$plot solutions of (1) with (14) and initial conditions $\rho_1^0 = 0.3 + 0.2\sin(2\pi x)$ and $\rho_2^0 = 0.1 + 0.1\sin(2\pi x)$, see Fig. 5; from top to bottom: $\eta_1 = 0.1\ \eta_2 = 0.01$ and $\eta_1 = 0.01\ \eta_2 = 0.1$; from left to right: ρ_1, ρ_2 and $\rho_1 + \rho_2$

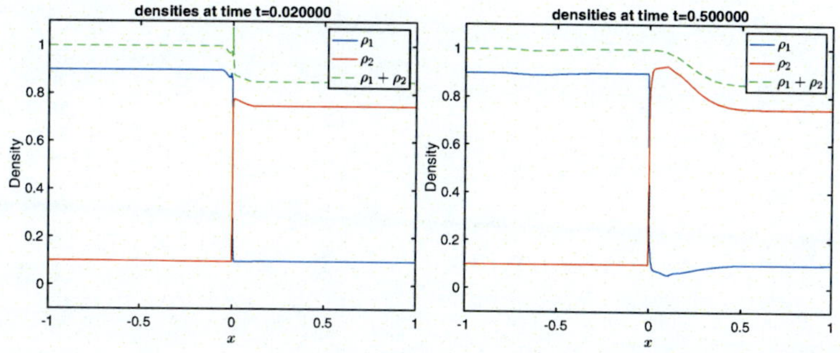

Fig. 12 Solution to (1) obtained with initial data $\rho_1^0 = 0.9 * (x < 0) + 0.1 * (x >= 0)$ and $\rho_2^0 = 0.1 * (x < 0) + 0.75 * (x >= 0)$, linear kernels, $\eta_1 = 0.01$ and $\eta_2 = 0.1$, $V_1^{\max} = 1.5$ and $V_2^{\max} = 0.8$ and absorbing boundary conditions

References

1. Benzoni-Gavage, S., Colombo, R.M.: An n-populations model for traffic flow. European J. Appl. Math. **14**(5), 587–612 (2003)
2. Blandin, S., Goatin, P.: Well-posedness of a conservation law with non-local flux arising in traffic flow modeling. Numer. Math. **132**(2), 217–241 (2016)
3. Chiarello, F.A., Goatin, P.: Global entropy weak solutions for general non-local traffic flow models with anisotropic kernel. ESAIM: M2AN **52**(1), 163–180 (2018)
4. Chiarello, F.A., Goatin, P.: Non-local multi-class traffic flow models. Netw. Heterog. Media **14**(2), 371–387 (2019)
5. Chiarello, F.A., Goatin, P., Villada, L.M.: High-order finite volume WENO schemes for non-local multi-class traffic flow models. In: AIMS on Applied Mathematics. Proceedings of the XVII International Conference on Hyperbolic Problems Theory, Numerics, Applications in Penn State, June 2018, vol. 10, pp. 353–560 (2020)

6. Chiarello, F.A., Goatin, P., Villada, L.M.: Lagrangian-antidiffusive remap schemes for non-local multi-class traffic flow models. Comput. Appl. Math. **39**, 60 (2020)
7. Friedrich, J., Kolb, O., Göttlich, S.: A Godunov type scheme for a class of LWR traffic flow models with non-local flux. Netw. Heterog. Media **13**(4), 531–547 (2018)
8. Goatin, P., Mimault, M.: A mixed system modeling two-directional pedestrian flows. Math. Biosci. Eng. **12**(2), 375–392 (2015)
9. Goatin, P., Scialanga, S.: Well-posedness and finite volume approximations of the LWR traffic flow model with non-local velocity. Netw. Heterog. Media **11**(1), 107–121 (2016)
10. LeVeque, R.J.: Finite volume methods for hyperbolic problems. In: Cambridge Texts in Applied Mathematics, pp. xx+558 (2002)
11. Lighthill, M.J., Whitham, G.B.: On kinematic waves. II. A theory of traffic flow on long crowded roads. In: Proceedings of the Royal Society London Series A, vol. 229, pp. 317–345 (1955)
12. Richards, P.I.: Shock waves on the highway. Oper. Res. **4**, 42–51 (1956)

Semi-implicit Finite-Difference Methods for Compressible Gas Dynamics with Curved Boundaries: A Ghost-Point Approach

Armando Coco and Santina Chiara Stissi

Abstract We present a finite-difference numerical method to solve systems of hyperbolic conservation laws in domains with arbitrary shapes. The curved boundary is immersed in a uniform Cartesian grid and implicitly defined by a level-set function. The method is based on a semi-implicit discretization of the differential equations coupled with a ghost-point approach to impose the boundary conditions. The method is designed to be straightforwardly extended to higher order accuracy. The semi-implicit approach alleviates the stability restriction on the time step that is associated with acoustic waves in explicit methods, while preventing the numerical dissipation introduced in fully implicit methods. Several numerical tests to solve the Euler equations of gas dynamics past steady obstacles with arbitrary shapes are presented to show the efficiency of the semi-implicit method and the efficacy of the ghost-point approach.

Keywords Immersed boundary method · Ghost-point extrapolation · Unfitted boundary method · Uniform Cartesian grid · Finite-difference method · Boundary conditions for gas dynamics

1 Introduction

Hyperbolic systems of conservation laws embrace various fields of application, ranging from meteorology, aerodynamics, traffic flow models, astrophysics and physics of the plasmas [19, 21, 22]. A correct evaluation of these phenomena has made the

A. Coco (✉)
Dipartimento di Matematica e Informatica, Università degli Studi di Catania, Viale Andrea Doria 6, 95125 Catania, Italy
e-mail: armando.coco@unict.it

S. C. Stissi
Istituto Nazionale di Geofisica e Vulcanologia, Osservatorio Etneo, Piazza Roma 2, 95125 Catania, Italy
e-mail: santina.stissi@ingv.it

G. Albi et al. (eds.), *Advances in Numerical Methods for Hyperbolic Balance Laws and Related Problems*, SEMA SIMAI Springer Series 32,
https://doi.org/10.1007/978-3-031-29875-2_4

development of numerical methods for this class of problems a very active research field [3, 6, 7, 9, 15, 16, 23, 25].

Even with smooth initial/boundary conditions and source terms, these models may develop shocks that must be correctly captured by the numerical methods for a better assessment and prediction/interpretation of phenomena. Shock-capturing approaches are usually based on Discontinuous Galerkin (DG), Finite-Volume (FV) and Finite-Difference (FD) methods.

DG and FV methods have the advantage of being built on non-uniform and unstructured meshes, extending their application on models with arbitrary domains. On the other hand, FD methods are designed on Cartesian grids and the source is evaluated pointwise, allowing higher flexibility when implicit solvers are implemented.

Several real-world applications ask for accurate and efficient methods for 2D/3D models in non-rectangular domains, for example to simulate compressible gas dynamics past complex-shaped objects. To this purpose, FD methods are usually preferred due to the easier implementation strategies for higher accurate schemes and for flux evaluation procedures, that are performed dimension-by-dimension as tensor products of 1D interpolations. The domain is embedded in a fixed Cartesian grid and a suitable extrapolation procedure will define the physical quantities outside the domain on those external grid points that do not necessarily lie on the boundary (ghost-points), with the aim of fulfilling the boundary conditions by highly accurate techniques. Computational efforts to generate a mesh that conforms to the boundary is therefore not needed. This advantage is more evident when moving objects are modelled.

Various studies have already been carried out for the numerical solution of equations of conservation laws on domains with arbitrary shape. In [18] the authors propose a very high-order FD scheme on Cartesian meshes for conservation laws with curved boundaries, based on a ghost-point technique designed through a least square method and matched at the boundary with a fifth-order WENO [26–28] reconstruction scheme. The scheme can be extended to arbitrary high-order accuracy, although the stencil adopted near the boundary is not very compact and the possible parallelization of the method to solve large 3D problems in a High Performance Computing framework is not straightforward.

In [4] the authors have implemented a finite-difference method by discretizing the equations on a uniform Cartesian grid and in which Lagrange interpolations have been applied with a filter capable of detecting any discontnuities to apply non-oscillatory properties, and smooth regions to apply higher order interpolations.

In [9] the authors have developed a second-order explicit conservative finite-difference shock-capturing method for the numerical solution of the compressible Euler equation of gas dynamics discretized on a regular Cartesian grid and defined in a rectangular domain with a circular obstacle extending the boundary approach successfully employed in [14] for elliptic problems on arbitrary domains.

Time advancement is usually based on explicit or implicit methods according to the stiffness of the problem. Hyperbolic systems develop waves that propagate at finite but different speeds, so an accurate method would require the solution of all spatial and temporal scales [7]. However, some systems develop waves that are not very relevant and therefore there is no interest in solving them. For example, in the low Mach number regimes of compressible Euler equations in gas dynamics, the acoustic waves carry a negligible amount of energy. If the equations are solved with an explicit time discretization, stiffness problems, related to the CFL condition that the time step must satisfy in order to guarantee the stability of the numerical method, may occur [3, 7]. The CFL condition in fact imposes a limit on the time step, which is limited by the space step divided by the maximum speed of the waves. This restriction on the time step is not a problem if the order of accuracy in space and time are the same, since the stability and accuracy restrictions are similar [3]. For the Euler equations of gas dynamics, on the other hand, the CFL condition is determined by the acoustic waves since they move at a greater speed than the material waves (that move at the speed of the fluid), imposing a fairly rigid restriction on the time step. This restriction on the time step results in an increasingly large computational time. The system becomes *stiff*.

Solving the system with a fully implicit time discretization, for example the classic upwind methods, would avoid the restriction on the time step imposed by the acoustic CFL, allowing a wider time step. However, this discretization does not represent an optimal solution to the stiffness problem for two reasons: (i) it could introduce an excessive amount of numerical dissipation on slow waves causing a loss in accuracy; (ii) it is complicated to solve, as it is highly non-linear.

To solve the issues of both explicit and implicit techniques, *semi-implicit methods* have been proposed in literature. In [3] the authors solve implicitly only the linear operators related to acoustic waves, eliminating the restriction on the time step imposed by the acoustic CFL, while the non-linear operators related to convective or material speeds are solved explicitly, maintaining their accuracy. The former are discretized with central finite-differences, while the latter are treated with local Lax-Friedrichs fluxes. The authors show that the method does not introduce an excessive numerical dissipation for low Mach numbers and captures the solutions well enough for not very small Mach numbers.

In this chapter we couple the semi-implicit approach with the ghost-point technique that was proposed in [9] for explicit schemes, providing a unified framework to numerically solve 2D conservation laws on non-rectangular domains. The method is designed to be straightforwardly extended to 3D problems. Unlike the explicit approach, where ghost values are updated after the internal values, in the semi-implicit approach the equations for ghost and internal points are coupled each other, resulting in an implicit problem to solve. We do not address the issue of proposing a specific solver for the arising linear systems and use the built-in solvers of MATLAB [24].

The chapter is structured as follows. In Sect. 2 the finite-difference method for systems of conservation laws is presented, with particular attention to the system of compressible Euler equations of the gas dynamics. The explicit and semi-implicit finite-difference schemes for the spatial and temporal discretizations of conservation law systems are described. The semi-implicit schemes are then applied to the 2D compressible Euler equations. In Sect. 3 we focus on the Euler equations on a two-dimensional domain with an obstacle of arbitrary shape and we describe the ghost-point technique to impose the boundary conditions. The ghost-point technique is coupled with the semi-implicit approach and numerical results for square and circular obstacles are presented in Sect. 4.

2 Finite-Difference Methods for Conservation Laws

A system of *conservation laws* takes the following form (in 1D):

$$\frac{\partial}{\partial t}u(x,t) + \frac{\partial}{\partial x}f(u(x,t)) = 0, \tag{1}$$

in which $u : \mathbb{R} \times \mathbb{R} \longrightarrow \mathbb{R}^m$ represents the m-dimensional vector of conserved quantities or state variables and $f : \mathbb{R}^m \longrightarrow \mathbb{R}^m$ is the *flux function* or *flux*, which is supposed smooth and known.

Classical space discretizations are based on finite volume schemes, where the unknowns are represented by cell averages of physical quantities. An alternative approach is the finite-difference method in which the unknown is the pointwise value of the function u.

We follow [25] to describe the finite-difference method. Let us consider a uniform spatial discretization of the domain with spatial step Δx. Let $x_i = i\ \Delta x$ be the grid points at the center of each cell (where the numerical solution u is defined) and $x_{i\pm\frac{1}{2}} = i\ \Delta x \pm \frac{1}{2}$ be the grid points at the edges of the cell at which the flux function is evaluated.

From (1), we write the exact expression:

$$\frac{\partial f}{\partial x}(u(x,t)) = \frac{\hat{f}\left(u\left(x + \frac{\Delta x}{2}\right), t\right) - \hat{f}\left(u\left(x - \frac{\Delta x}{2}\right), t\right)}{\Delta x}. \tag{2}$$

To convert this expression into a numerical scheme we approximate the right-hand side of (2). To this purpose, we first find the relation between f and $\hat{f}$. We consider the average operator

$$\bar{u}(x,t) = \frac{1}{\Delta x}\int_{x-\frac{\Delta x}{2}}^{x+\frac{\Delta x}{2}} u(\xi,t)\mathrm{d}\xi$$

and we differentiate it with respect to x:

$$\frac{\partial \bar{u}}{\partial x} = \frac{1}{\Delta x}\left(\left(u\left(x + \frac{\Delta x}{2}\right), t\right) - \left(u\left(x - \frac{\Delta x}{2}\right), t\right)\right). \tag{3}$$

The expression (3) shows that the relation between f and $\hat{f}$ is the same as the one between $\bar{u}(x,t)$ and $u(x,t)$. Thus, the flux function f is the cell average of the function $\hat{f}$. It is therefore possible to compute the values of $\hat{f}$ at the cell edges $x_{i\pm\frac{1}{2}}$

from the values of f in x_i using the techniques employed, for example, in finite volume methods, to reconstruct pointwise values $u_{i\pm\frac{1}{2}}$ from cell averages $\bar{u}_i$.

In detail, a piecewise polynomial reconstruction $\mathcal{R}(x)$ of $\hat{f}_i$ is computed from the values of f_i and the value of $\hat{f}$ at the border of the cell $x_{i+\frac{1}{2}}$ is computed from the reconstruction: $\hat{f}^{\pm}_{i+\frac{1}{2}} = \lim_{x\to x^{\pm}_{i+\frac{1}{2}}} \mathcal{R}(x)$. For example, a second-order reconstruction is computed as $\hat{f}_{x^-_{i+\frac{1}{2}}} = f_i + \frac{\Delta x}{2}\hat{f}'_i$, where $\hat{f}'_i$, computed with a slop limiter, is an approximation of the first-order of the derivative of $\hat{f}$.

Since the reconstruction at $x_{i\pm\frac{1}{2}}$ can be discontinuous, we consider flux functions that can be split as follow:

$$f(u) = f^+(u) + f^-(u),$$

where f^+ and f^- respectively satisfy the conditions to ensure that the flux is monotone consistent:

$$\frac{\mathrm{d}f^+(u)}{\mathrm{d}u} \geq 0, \quad \frac{\mathrm{d}f^-(u)}{\mathrm{d}u} \leq 0.$$

For example, the Local Lax-Friedrichs flux, which we will talk about in Sect. 2.2, satisfies these conditions.

A finite-difference scheme is therefore given by:

$$\frac{\mathrm{d}}{\mathrm{d}t}u_i = -\frac{\hat{f}_{i+\frac{1}{2}} - \hat{f}_{i-\frac{1}{2}}}{\Delta x},$$

where the numerical flux function at the edge of the cell is:

$$\hat{f}_{i+\frac{1}{2}} = \hat{f}^+(u^-_{i+\frac{1}{2}}) + \hat{f}^-(u^+_{i+\frac{1}{2}})$$

and the quantities $\hat{f}^{\pm}(u^-_{i+\frac{1}{2}})$ are computed as follow: firstly, $f^{\pm}(u_i)$ are computed and interpreted as cell average of $\hat{f}^{\pm}$; secondly, a pointwise reconstruction of $\hat{f}^{\pm}$ in cell i is performed and it is evaluated at the edge $x_{i+\frac{1}{2}}$ of the cell.

This technique will be detailed in Sect. 2.2, where first and second-order spatial reconstructions are employed to compute the numerical flux at the cell edge.

Furthermore, the method discussed so far for the scalar equation can be extended to systems of conservation laws.

In the following sections we describe the finite-difference approach for the *compressible Euler equations of gas dynamics* in 2D.

2.1 Compressible Euler Equations of Gas Dynamics

A hyperbolic system of particular importance in the field of systems of conservation laws is represented by the *compressible Euler equations of gas dynamics*, concerning the equations of the conservation of mass, energy and momentum.

The Euler equations for an ideal gas are given by [7]:

$$\begin{cases} \rho_t + \nabla \cdot (\rho \mathbf{u}) = 0 \\ (\rho \mathbf{u})_t + \nabla \cdot (\rho \mathbf{u} \otimes \mathbf{u}) + \frac{\nabla p}{\varepsilon^2} = 0 \\ E_t + \nabla \cdot [(E + p)\mathbf{u}] = 0 \end{cases}, \tag{4}$$

where ρ is the density of the fluid, $\mathbf{u}$ is the velocity of the fluid, p is the pressure and E is the energy density per unit of volume. Here, ε is the global Mach number, whose relation to the reference Mach number M is $M = \frac{\varepsilon}{\sqrt{\gamma}}$, where $\gamma = \frac{c_p}{c_v} \geq 1$ is the ratio of the specific heats of the gas. When ε vanishes the compressible flow converges to the incompressible one.

A detailed description can be found in [7], in which the authors describe the importance of developing all Mach number methods. For example, to describe homogeneous media with different mechanical properties such as air-water flows, it can occur that the speed of sound is lower than both the speed of sound in water and the speed of sound in air.

To close the system (4) we consider the equation of state (EOS), which for a polytropic gas is:

$$E = \frac{p}{\gamma - 1} + \frac{\varepsilon^2}{2} \rho |\mathbf{u}|^2. \tag{5}$$

Thus, the energy or the pressure can be computed in terms of the other variables, according to (5).

Denoting by $\mathbf{m} = \rho \mathbf{u}$ the momentum and by $h = \frac{E+p}{\rho}$ the total enthalpy, the system (4) can be written as:

$$\begin{cases} \rho_t + \nabla \cdot \mathbf{m} = 0 \\ \mathbf{m}_t + \nabla \cdot (\mathbf{m} \otimes \mathbf{u}) + \frac{\nabla p}{\varepsilon^2} = 0 \\ E_t + \nabla \cdot (h\mathbf{m}) = 0 \end{cases}, \tag{6}$$

which in 2D takes the following form:

$$\begin{cases} \rho_t + m^1_x + m^2_y = 0 \\ m^1_t + \left(\frac{m^{1^2}}{\rho} + \frac{p}{\varepsilon^2}\right)_x + \left(m^1 \frac{m^2}{\rho}\right)_y = 0 \\ m^2_t + \left(m^2 \frac{m^1}{\rho}\right)_x + \left(\frac{m^{2^2}}{\rho} + \frac{p}{\varepsilon^2}\right)_y = 0 \\ E_t + (hm^1)_x + (hm^2)_y = 0 \end{cases}, \tag{7}$$

where m^1 and m^2 are the momentum along x and y axes, respectively. The system is closed by the equation of state (EOS) $p = (\gamma - 1)\left(E - \frac{1}{2}\varepsilon^2\left(\frac{{m^1}^2}{\rho} + \frac{{m^2}^2}{\rho}\right)\right)$.

In the isentropic case, the energy E is constant over time and space and the last equation of (6) disappears. Therefore, the first two equations of (6) represent the *Isentropic Euler equations* and the EOS is given by $p = p(\rho) = \rho^\gamma$ [7].

Discretization in time can be performed by explicit or implicit solvers. Despite being easy to implement, explicit methods suffer from a restrictive CFL condition on the time step. While for high Mach numbers the explicit methods are able to capture the shocks with a reasonable time step, for low Mach numbers the flow speed is slow compared to the speed of sound and therefore material waves are much slower than acoustic ones and the time step must satisfy a CFL stability condition:

$$\mathrm{CFL} = \frac{\lambda_{\max}\,\Delta t}{\Delta x} < 1,$$

which sets a limit on the time step:

$$\Delta t < \frac{\Delta x}{\lambda_{\max}} \approx \mathrm{O}(\varepsilon \Delta x),$$

where $\lambda_{\max} = \max_\Omega\left(|\mathbf{u}| + \frac{c_s}{\varepsilon}\right)$, $\mathbf{u}$ is the wave velocity, $c_s = \sqrt{\frac{\gamma p}{\rho}}$ is the speed of sound, ρ and p are the density and pressure, respectively. Therefore, Δt is of order of ε [7].

On the other hand, implicit solvers are usually more difficult to solve because of the non linearity features, as well as they can lose some accuracy due to the numerical dissipation introduced on the slow waves.

For this reason, a semi-implicit approach is preferred in order to relax the CFL condition while maintaining a good overall accuracy and preserve stability also for larger time steps [3].

Semi-implicit schemes involve differences in both temporal and space discretizations compared to classical explicit methods. Temporal discretization implies that stiff terms, related to acoustic waves, are discretized implicitly, while non-stiff terms, related to material waves, are discretized explicitly. Non-stiff terms are discretized via local Lax-Friedrichs fluxes, while stiff terms are treated with a central finite-difference scheme.

From a numerical point of view, while for explicit schemes it is necessary that CFL < 1, in semi-implicit schemes CFL can be larger than 1. However, since the material waves are treated explicitly, there is a restriction on the material CFL, that is, $\mathrm{CFL}_\mathrm{u} < 1$, where

$$\mathrm{CFL}_\mathrm{u} = \frac{\max|\mathbf{u}|\,\Delta t}{\Delta x}.$$

In the next section we describe the discretizations in space and in time of the compressible Euler equations of gas dynamics in 2D (Eq. 7) with both explicit and semi-implicit methods. The 1D and 3D methods can be easily obtained from it.

2.2 *Explicit and Semi-implicit Spatial Discretization*

The 2D hyperbolic system of conservation laws (7) can be written in the following vector form:

$$\mathbf{U}_t + \mathbf{F}(\mathbf{U})_x + \mathbf{G}(\mathbf{U})_y = 0, \tag{8}$$

with

$$\mathbf{U} = \begin{pmatrix} \rho \\ m^1 \\ m^2 \\ E \end{pmatrix}, \ \mathbf{F}(U) = \begin{pmatrix} m^1 \\ \frac{{m^1}^2}{\rho} + \frac{p}{\varepsilon^2} \\ m^2 \frac{m^1}{\rho} \\ hm^1 \end{pmatrix} \text{ and } \mathbf{G}(U) = \begin{pmatrix} m^2 \\ m^1 \frac{m^2}{\rho} \\ \frac{{m^2}^2}{\rho} + \frac{p}{\varepsilon^2} \\ hm^2 \end{pmatrix}.$$

The rectangular domain $\Omega = [0, 1]^2$ is discretized through a uniform Cartesian grid with N^2 cells. Spatial steps are Δx and Δy along x and y axes, respectively. We consider the case of a regular grid, in which $\Delta x = \Delta y = 1/N$, but all numerical schemes proposed in this chapter can be easily extended to the case of $\Delta x \neq \Delta y$. The numerical solution $\mathbf{U}_{i,j}(t) \approx \mathbf{U}(x_i, y_j, t)$ is computed at the centre of each cell, with $x_i = \left(i - \frac{1}{2}\right)\Delta x$ and $y_j = \left(j - \frac{1}{2}\right)\Delta y$, $i, j = 1, \ldots, N$.

2.2.1 Explicit Method

In the explicit approach, the values of the numerical solution on the internal points progress over time following a semi-discrete scheme [9]:

$$\frac{d\mathbf{U}_{i,j}}{dt} = -\frac{\hat{f}_{i+\frac{1}{2},j} - \hat{f}_{i-\frac{1}{2},j}}{\Delta x} - \frac{\hat{g}_{i,j+\frac{1}{2}} - \hat{g}_{i,j-\frac{1}{2}}}{\Delta y}, \ i, j = 1, 2, \ldots, N,$$

in which $\{\hat{f}_{i\pm\frac{1}{2},j}\}$ and $\{\hat{g}_{i,j\pm\frac{1}{2}}\}$ are the numerical fluxes at the edges of each cell centered at (x_i, y_j).

An approximation of the numerical fluxes $\{\hat{f}_{i\pm\frac{1}{2},j}\}$ and $\{\hat{g}_{i,j\pm\frac{1}{2}}\}$ is obtained using the local Lax-Friedrichs fluxes. We first split the flux functions $\mathbf{F}$ and $\mathbf{G}$ as follows:

$$\mathbf{F}(\mathbf{U}_{i,j}) = \mathbf{F}^+_{i,j} + \mathbf{F}^-_{i,j}, \ \mathbf{F}^\pm_{i,j} = \frac{1}{2}\left(\mathbf{F}(\mathbf{U}_{i,j}) \pm \alpha_{i,j}\mathbf{U}_{i,j}\right), \tag{9}$$

$$\mathbf{G}(\mathbf{U}_{i,j}) = \mathbf{G}^+_{i,j} + \mathbf{G}^-_{i,j}, \ \mathbf{G}^\pm_{i,j} = \frac{1}{2}\left(\mathbf{G}(\mathbf{U}_{i,j}) \pm \beta_{i,j}\mathbf{U}_{i,j}\right).$$

with $\alpha_{i,j} = |u_{ij}| + \frac{c_{ij}}{\varepsilon}$ and $\beta_{i,j} = |v_{ij}| + \frac{c_{ij}}{\varepsilon}$ the spectral radii of the Jacobian matrices $\frac{\partial \mathbf{F}(\mathbf{U}_{i,j})}{\partial \mathbf{U}}$ and $\frac{\partial \mathbf{G}(\mathbf{U}_{i,j})}{\partial \mathbf{U}}$, u and v the velocities of the fluid along x and y axes, respectively, and $c_{ij} = \sqrt{\frac{\gamma p_{ij}}{\rho_{ij}}}$ the speed of sound.

A first-order approximation of the numerical fluxes at the cell interfaces $(x_{i+\frac{1}{2}}, y_j)$ and $(x_i, y_{j+\frac{1}{2}})$ is given by:

$$\hat{f}_{i+\frac{1}{2},j} = \mathbf{F}^+_{i,j} + \mathbf{F}^-_{i+1,j} = \frac{\mathbf{F}(\mathbf{U}_{i+1,j}) + \mathbf{F}(\mathbf{U}_{i,j})}{2} - \alpha_{i+\frac{1}{2},j}\frac{\mathbf{U}_{i+1,j} - \mathbf{U}_{i,j}}{2},$$

$$\hat{g}_{i,j+\frac{1}{2}} = \mathbf{G}^+_{i,j} + \mathbf{G}^-_{i,j+1} = \frac{\mathbf{G}(\mathbf{U}_{i,j+1}) + \mathbf{G}(\mathbf{U}_{i,j})}{2} - \beta_{i,j+\frac{1}{2}}\frac{\mathbf{U}_{i,j+1} - \mathbf{U}_{i,j}}{2},$$

where

$$\alpha_{i+\frac{1}{2},j} = \max\left\{\alpha_{ij}, \alpha_{i+1,j}\right\} \text{ and } \beta_{i,j+\frac{1}{2}} = \max\left\{\beta_{ij}, \beta_{i,j+1}\right\}.$$

In the same way, we can compute the numerical fluxes at the cell interfaces $(x_{i-\frac{1}{2}}, y_j)$ and $(x_i, y_{j-\frac{1}{2}})$.

To obtain a second-order accurate discretization, we use a non-oscillatory generalized minmod reconstruction to compute $\mathbf{F}^{\pm}$ and $\mathbf{G}^{\pm}$ of (9) at the cell interfaces $\left(x_{i+\frac{1}{2}}, y_j\right)$ and $\left(x_i, y_{j+\frac{1}{2}}\right)$ [3, 9].

In particular, we first compute:

$$\mathbf{F}^{\mathrm{E}}_{i,j} := \mathbf{F}^+_{i,j} + \frac{\Delta x}{2}(\mathbf{F}_x)^+_{i,j}, \quad \mathbf{F}^{\mathrm{W}}_{i,j} := \mathbf{F}^-_{i,j} - \frac{\Delta x}{2}(\mathbf{F}_x)^-_{i,j},$$

and

$$\mathbf{G}^{\mathrm{N}}_{i,j} := \mathbf{G}^+_{i,j} + \frac{\Delta y}{2}(\mathbf{G}_y)^+_{i,j}, \quad \mathbf{G}^{\mathrm{S}}_{i,j} := \mathbf{G}^-_{i,j} - \frac{\Delta y}{2}(\mathbf{G}_y)^-_{i,j}.$$

Having defined the minimod function as follows:

$$\operatorname{minmod}(t_1, t_2, \ldots) := \begin{cases} \min(t_1, t_2, \ldots), & \text{if } t_i > 0\ \forall i \\ \max(t_1, t_2, \ldots), & \text{if } t_i < 0\ \forall i \\ 0 & \text{otherwise} \end{cases},$$

the slopes $(\mathbf{F}_x)^{\pm}_{i,j}$ and $(\mathbf{G}_y)^{\pm}_{i,j}$ are evaluated with the generalized minmod limiter:

$$(\mathbf{F}_x)^{\pm}_{i,j} = \operatorname{minmod}\left(\theta\frac{\mathbf{F}^{\pm}_{i,j} - \mathbf{F}^{\pm}_{i-1,j}}{\Delta x}, \frac{\mathbf{F}^{\pm}_{i+1,j} - \mathbf{F}^{\pm}_{i-1,j}}{2\Delta x}, \theta\frac{\mathbf{F}^{\pm}_{i+1,j} - \mathbf{F}^{\pm}_{i,j}}{\Delta x}\right),$$

$$(\mathbf{G}_y)^{\pm}_{i,j} = \operatorname{minmod}\left(\theta\frac{\mathbf{G}^{\pm}_{i,j} - \mathbf{G}^{\pm}_{i,j-1}}{\Delta y}, \frac{\mathbf{G}^{\pm}_{i,j+1} - \mathbf{G}^{\pm}_{i,j-1}}{2\Delta y}, \theta\frac{\mathbf{G}^{\pm}_{i,j+1} - \mathbf{G}^{\pm}_{i,j}}{\Delta y}\right),$$

where $\theta \in [1, 2]$ is a parameter related to the numerical dissipation.

The numerical fluxes are therefore computed as follow:

$$\hat{f}_{i+\frac{1}{2},j} = \mathbf{F}^{\mathrm{E}}_{i,j} + \mathbf{F}^{\mathrm{W}}_{i+1,j}, \ \hat{g}_{i,j+\frac{1}{2}} = \mathbf{G}^{\mathrm{N}}_{i,j} + \mathbf{G}^{\mathrm{S}}_{i,j+1}.$$

In [9] the authors explain how the choice of the values of θ and $\alpha_{i,j}$ is fundamental to prevent oscillations. It is convenient not to use high values of θ, as they favor sharper resolution of discontinuities, but, at the same time, they can be responsible for oscillations. In the same way, it is convenient to replace $\alpha_{i,j}$ with $\max_{i,j} \alpha_{i,j}$ to increase the numerical diffusion and prevent oscillations.

Another approach to avoiding the onset of oscillations is to impose constraints on the reconstruction process to satisfy, for example, the TVD condition. More recent schemes are represented by ENO (Essentially Non-Oscillatory) [21, 26] and WENO (Weighted Essentially Non-Oscillatory) reconstructions from cell averages to pointwise values [26].

2.2.2 Semi-implicit Method

In semi-implicit methods, non-stiff terms are treated explicitly as in the previous section. We denote by $\hat{D}_x f = \frac{\hat{f}_{i+\frac{1}{2},j} - \hat{f}_{i-\frac{1}{2},j}}{\Delta x}$ and $\hat{D}_y g = \frac{\hat{g}_{i,j+\frac{1}{2}} - \hat{g}_{i,j-\frac{1}{2}}}{\Delta y}$ the numerical flux derivatives.

The choice of $\alpha_{i\pm\frac{1}{2},j}$ and $\beta_{i,j\pm\frac{1}{2}}$ for the formulation of numerical fluxes requires some considerations in semi-implicit methods [3].

In the explicit schemes, they represent a limit on the maximum speed of the waves and they are given by:

$$\alpha_{i,j} = |u_{i,j}| + \frac{c_{i,j}}{\varepsilon}, \quad \beta_{i,j} = |v_{i,j}| + \frac{c_{i,j}}{\varepsilon}. \tag{10}$$

In semi-implicit methods, acoustic waves are treated implicitly, consequently α is proportional to the material speed $\mathbf{u}$. For low Mach numbers, we therefore choose:

$$\alpha_{i,j} = |u_{i,j}|, \ \beta_{i,j} = |v_{i,j}|, \tag{11}$$

For Mach numbers larger than one, instead, the sound speed is bounded by the speed of fluid, therefore:

$$\alpha_{i,j} = |u_{i,j}| + \frac{c_{i,j}}{\varepsilon}, \quad \beta_{i,j} = |v_{i,j}| + \frac{c_{i,j}}{\varepsilon}.$$

For stiff terms, on the other hand, the spatial discretization is performed with the standard central finite-difference method:

$$\frac{d\mathbf{U}_{i,j}}{dt} = -\frac{\mathbf{F}_{i+\frac{1}{2},j} - \mathbf{F}_{i-\frac{1}{2},j}}{\Delta x} - \frac{\mathbf{G}_{i,j+\frac{1}{2}} - \mathbf{G}_{i,j-\frac{1}{2}}}{\Delta y} =$$

$$= -\frac{\mathbf{F}_{i+1,j} - \mathbf{F}_{i-1,j}}{2\Delta x} - \frac{\mathbf{G}_{i,j+1} - \mathbf{G}_{i,j-1}}{2\Delta y}, \; i, j = 1, 2, \ldots, N.$$

We denote by $D_x\mathbf{F} = \frac{\mathbf{F}_{i+\frac{1}{2},j} - \mathbf{F}_{i-\frac{1}{2},j}}{\Delta x} = \frac{\mathbf{F}_{i+1,j} - \mathbf{F}_{i-1,j}}{2\Delta x}$ and $D_y\mathbf{G} = \frac{\mathbf{G}_{i,j+\frac{1}{2}} - \mathbf{G}_{i,j-\frac{1}{2}}}{\Delta y} = \frac{\mathbf{G}_{i,j+1} - \mathbf{G}_{i,j-1}}{2\Delta y}$.

2.3 *Explicit and Semi-implicit Time Discretization*

We now proceed in the description of the temporal discretization of hyperbolic systems of conservation laws.

In the explicit methods, the numerical solution at time t^{n+1} with a first-order scheme is obtained by a first order Euler method:

$$\mathbf{U}^{n+1} = \mathbf{U}^n - \Delta t\mathbf{F}(\mathbf{U}^n)_x - \Delta t\mathbf{G}(\mathbf{U}^n)_y.$$

In the semi-implicit method, stiff terms $\mathbf{U}_{\mathrm{I}}$ are evaluated at time t^{n+1}, while no-stiff terms $\mathbf{U}_{\mathrm{E}}$ are evaluated at time t^n, thus:

$$\mathbf{U}^{n+1} = \mathbf{U}^n - \Delta t\mathbf{F}(\mathbf{U}_{\mathrm{E}}^n, \mathbf{U}_{\mathrm{I}}^{n+1})_x - \Delta t\mathbf{G}(\mathbf{U}_{\mathrm{E}}^n, \mathbf{U}_{\mathrm{I}}^{n+1})_y.$$

High order schemes in time are obtained following the approach proposed in [3, 6], in which, defined with $\mathbf{U}_{\mathrm{I}} = \mathcal{S}(\mathbf{U}_*, \mathbf{U}_{\mathrm{E}}, \Delta t)$ the solution of the problem

$$\mathbf{U}_{\mathrm{I}} = \mathbf{U}_* - \Delta t\mathbf{F}(\mathbf{U}_{\mathrm{E}}, \mathbf{U}_{\mathrm{I}})_x - \Delta t\mathbf{G}(\mathbf{U}_{\mathrm{E}}, \mathbf{U}_{\mathrm{I}})_y,$$

the numerical solution $\mathbf{U}^{n+1}$ at time t^{n+1} is computed using the following algorithm:

$$\mathbf{U}_{\mathrm{I}}^{(1)} = \mathcal{S}(\mathbf{U}^n, \mathbf{U}^n, \beta\Delta t), \qquad \mathbf{U}_{\mathrm{E}}^{(2)} = \left(1 - \frac{\hat{c}}{\beta}\right)\mathbf{U}^n + \frac{\hat{c}}{\beta}\mathbf{U}_{\mathrm{I}}^{(1)}$$

$$\mathbf{U}_*^{(2)} = \frac{2\beta - 1}{\beta}\mathbf{U}^n + \frac{1-\beta}{\beta}\mathbf{U}_{\mathrm{I}}^{(1)}, \qquad \mathbf{U}_{\mathrm{I}}^{(2)} = \mathcal{S}(\mathbf{U}_*^{(2)}, \mathbf{U}_{\mathrm{E}}^{(2)}, \beta\Delta t).$$

where $\beta = 1 - \frac{1}{\sqrt{2}}$ and $\hat{c} = \frac{1}{2\beta}$.
The numerical solution is calculated by setting $\mathbf{U}^{n+1} = \mathbf{U}_{\mathrm{I}}^{(2)}$.

In explicit methods, all terms are treated explicitly:

$$\mathbf{U}^{(1)} = \mathcal{S}(\mathbf{U}^n, \mathbf{U}^n, \beta\Delta t), \qquad \mathbf{U}^{(2)} = \left(1 - \frac{\hat{c}}{\beta}\right)\mathbf{U}^n + \frac{\hat{c}}{\beta}\mathbf{U}^{(1)}$$

$$\mathbf{U}_*^{(2)} = \frac{2\beta - 1}{\beta}\mathbf{U}^n + \frac{1-\beta}{\beta}\mathbf{U}^{(1)}, \qquad \mathbf{U}^{n+1} = \mathcal{S}(\mathbf{U}_*^{(2)}, \mathbf{U}^{(2)}, \beta\Delta t).$$

Observe that, in the explicit schemes, the first-order method is simply $\mathbf{U}^{n+1} = \mathcal{S}(\mathbf{U}^n, \mathbf{U}^n, \Delta t)$.

2.4 *Discretization of Compressible Euler Equations in 2D*

We now apply the discretizations to the Full Euler equations in 2D (the isentropic case will be obtained by neglecting the equation for the energy E).

The approach followed for time and spatial discretization, called *Pressure Splitting*, is proposed in [3], where the pressure terms are treated implicitly, while the density is computed explicitly:

$$\begin{cases} \rho^{n+1} = \rho^n - \Delta t \hat{D}_x m1^n - \Delta t \hat{D}_y m2^n & (a) \\ m1^{n+1} = m1^n - \Delta t \tilde{D}_x \left(\left(\frac{m1^2}{\rho} \right)^n + \frac{p^{n+1}}{\varepsilon^2} \right) - \Delta t \hat{D}_y \left(m1 \frac{m2}{\rho} \right)^n & (b) \\ m2^{n+1} = m2^n - \Delta t \hat{D}_x \left(m2 \frac{m1}{\rho} \right)^n - \Delta t \tilde{D}_y \left(\left(\frac{m2^2}{\rho} \right)^n + \frac{p^{n+1}}{\varepsilon^2} \right) & (c) \\ E^{n+1} = E^n - \Delta t \tilde{D}_x \left(h^n m1^{n+1} \right) - \Delta t \tilde{D}_y \left(h^n m2^{n+1} \right) & (d) \end{cases},$$

$$p^{n+1} = (\gamma - 1)\left(E^{n+1} - \frac{1}{2}\varepsilon^2 \left(\frac{m1^2}{\rho} + \frac{m2^2}{\rho} \right)^n \right).$$

We observe that in [3] the authors solve the EOS for the pressure p^{n+1} in terms of the energy E^{n+1} and plug it in into the equations of the momentum. We do not follow this approach here and leave the pressure term in the momentum equations.

We can write the equations for the density:

$$\rho_{\mathrm{I}}^{(1)} = \rho^n - \beta\Delta t \hat{D}_x m1^n - \beta\Delta t \hat{D}_y m2^n,$$

$$\rho_{\mathrm{E}}^{(2)} = \left(1 - \frac{\hat{c}}{\beta}\right)\rho^n + \frac{\hat{c}}{\beta}\rho_{\mathrm{I}}^{(1)},$$

$$\rho_*^{(2)} = \frac{2\beta - 1}{\beta}\rho^n + \frac{1-\beta}{\beta}\rho_{\mathrm{I}}^{(1)},$$

that can be computed explicitly.

We split now the equations (b) and (c) for the momentums in this way:

$$\begin{aligned} m_{\mathrm{I}}^{\mathbf{1}(1)} &= m^{\mathbf{1}^n} - \beta\Delta t\hat{D}_x\left(\frac{m^{\mathbf{1}^2}}{\rho}\right)^n - \beta\Delta t\hat{D}_y\left(m^{\mathbf{1}}\frac{m^{\mathbf{2}}}{\rho}\right)^n - \frac{\beta\Delta t}{\varepsilon^2}D_x p_{\mathrm{I}}^{(1)} \\ &= M_{\mathrm{E}}^{\mathbf{1}(1)} - \frac{\beta\Delta t}{\varepsilon^2}D_x p_{\mathrm{I}}^{(1)}, \\ m_{\mathrm{I}}^{\mathbf{2}(1)} &= m^{\mathbf{2}^n} - \beta\Delta t\hat{D}_x\left(m^{\mathbf{2}}\frac{m^{\mathbf{1}}}{\rho}\right)^n - \beta\Delta t\hat{D}_y\left(\frac{m^{\mathbf{2}^2}}{\rho}\right)^n - \frac{\beta\Delta t}{\varepsilon^2}D_y p_{\mathrm{I}}^{(1)} \\ &= M_{\mathrm{E}}^{\mathbf{2}(1)} - \frac{\beta\Delta t}{\varepsilon^2}D_y p_{\mathrm{I}}^{(1)}, \end{aligned}$$

with $M_{\mathrm{E}}^{\mathbf{1}(1)} = m^{\mathbf{1}^n} - \beta\Delta t\hat{D}_x\left(\frac{m^{\mathbf{1}^2}}{\rho}\right)^n - \beta\Delta t\hat{D}_y\left(m^{\mathbf{1}}\frac{m^{\mathbf{2}}}{\rho}\right)^n$ and $M_{\mathrm{E}}^{\mathbf{2}(1)} = m^{\mathbf{2}^n} - \beta\Delta t\hat{D}_x\left(m^{\mathbf{2}}\frac{m^{\mathbf{1}}}{\rho}\right)^n - \beta\Delta t\hat{D}_y\left(\frac{m^{\mathbf{2}^2}}{\rho}\right)^n$.

The quantities $M_{\mathrm{E}}^{\mathbf{1}(1)}$ and $M_{\mathrm{E}}^{\mathbf{2}(1)}$ are computed explicitly.

Substituting the expressions of $M_{\mathrm{E}}^{\mathbf{1}(1)}$ and $M_{\mathrm{E}}^{\mathbf{2}(1)}$ in the equation (d) on the energy, we obtain:

$$\begin{aligned} E_{\mathrm{I}}^{(1)} &= E^n - \beta\Delta t\hat{D}_x\left(h^n M_{\mathrm{E}}^{\mathbf{1}(1)}\right) - \beta\Delta t\hat{D}_y\left(h^n M_{\mathrm{E}}^{\mathbf{2}(1)}\right) + \\ &\quad \frac{\beta^2\Delta t^2}{\varepsilon^2}\left(D_x\left(h^n\left(D_x p_{\mathrm{I}}^{(1)}\right)\right) + D_y\left(h^n\left(D_y p_{\mathrm{I}}^{(1)}\right)\right)\right) \\ &= E^* + \frac{\beta^2\Delta t^2}{\varepsilon^2}\left(D_x\left(h^n\left(D_x p_{\mathrm{I}}^{(1)}\right)\right) + D_y\left(h^n\left(D_y p_{\mathrm{I}}^{(1)}\right)\right)\right), \end{aligned}$$

with $E^* = E^n - \beta\Delta t\hat{D}_x\left(h^n M_{\mathrm{E}}^{\mathbf{1}(1)}\right) - \beta\Delta t\hat{D}_y\left(h^n M_{\mathrm{E}}^{\mathbf{2}(1)}\right)$.

Now, being

$$\begin{aligned} p_{\mathrm{I}}^{(1)} &= (\gamma - 1)\left(E_{\mathrm{I}}^{(1)} - \frac{1}{2}\varepsilon^2\frac{\left(m^{\mathbf{1}^n}\right)^2}{\rho^n} - \frac{1}{2}\varepsilon^2\frac{\left(m^{\mathbf{2}^n}\right)^2}{\rho^n}\right) \\ &= (\gamma - 1)\left(E^* - \frac{1}{2}\varepsilon^2\frac{\left(m^{\mathbf{1}^n}\right)^2}{\rho^n} - \frac{1}{2}\varepsilon^2\frac{\left(m^{\mathbf{2}^n}\right)^2}{\rho^n}\right) + \\ &\quad (\gamma - 1)\frac{\beta^2\Delta t^2}{\varepsilon^2}\left(D_x\left(h^n\left(D_x p_{\mathrm{I}}^{(1)}\right)\right) + D_y\left(h^n\left(D_y p_{\mathrm{I}}^{(1)}\right)\right)\right) \\ &= (\gamma - 1)p^* + (\gamma - 1)\frac{\beta^2\Delta t^2}{\varepsilon^2}\left(D_x\left(h^n\left(D_x p_{\mathrm{I}}^{(1)}\right)\right) + D_y\left(h^n\left(D_y p_{\mathrm{I}}^{(1)}\right)\right)\right), \end{aligned}$$

with $p^* = E^* - \frac{1}{2}\varepsilon^2\frac{\left(m^{\mathbf{1}^n}\right)^2}{\rho^n} - \frac{1}{2}\varepsilon^2\frac{\left(m^{\mathbf{2}^n}\right)^2}{\rho^n}$ computed explicitly, we obtain:

$$\frac{p_{\mathrm{I}}^{(1)}}{\gamma-1}-\frac{\beta^2\Delta t^2}{\varepsilon^2}\left(D_x\left(h^n\left(D_x p_{\mathrm{I}}^{(1)}\right)\right)+D_y\left(h^n\left(D_y p_{\mathrm{I}}^{(1)}\right)\right)\right)=p^*. \tag{12}$$

$p_{\mathrm{I}}^{(1)}$ can then be calculated by solving (12):

$$\frac{p_{\mathrm{I}i,j}^{(1)}}{\gamma-1}-\frac{\beta^2\Delta t^2}{\varepsilon^2\Delta y^2}\left(\frac{h_{i,j}^n+h_{i+1,j}^n}{2}\left(p_{\mathrm{I}i+1,j}^{(1)}-p_{\mathrm{I}i,j}^{(1)}\right)-\frac{h_{i-1,j}^n+h_{i,j}^n}{2}\left(p_{\mathrm{I}i,j}^{(1)}-p_{\mathrm{I}i-1,j}^{(1)}\right)\right)+$$
$$-\frac{\beta^2\Delta t^2}{\varepsilon^2\Delta x^2}\left(\frac{h_{i,j}^n+h_{i,j+1}^n}{2}\left(p_{\mathrm{I}i,j+1}^{(1)}-p_{\mathrm{I}i,j}^{(1)}\right)-\frac{h_{i,j-1}^n+h_{i,j}^n}{2}\left(p_{\mathrm{I}i,j}^{(1)}-p_{\mathrm{I}i,j-1}^{(1)}\right)\right)=p_{i,j}^*.$$

Substituting $p_{\mathrm{I}}^{(1)}$ in the momentums equations (b) and (c), we find $m_{\mathrm{I}}^{\mathbf{1}(1)}$ and $m_{\mathrm{I}}^{\mathbf{2}(1)}$. Thus, we can compute:

$$m_{\mathrm{E}}^{\mathbf{1}(2)}=\left(1-\frac{\hat{c}}{\beta}\right)m^{\mathbf{1}^n}+\frac{\hat{c}}{\beta}m_{\mathrm{I}}^{\mathbf{1}(1)},\quad m_{\mathrm{E}}^{\mathbf{2}(2)}=\left(1-\frac{\hat{c}}{\beta}\right)m^{\mathbf{2}^n}+\frac{\hat{c}}{\beta}m_{\mathrm{I}}^{\mathbf{2}(1)},$$

$$m_*^{\mathbf{1}(2)}=\frac{2\beta-1}{\beta}m^{\mathbf{1}^n}+\frac{1-\beta}{\beta}m_{\mathrm{I}}^{\mathbf{1}(1)},\quad m_*^{\mathbf{2}(2)}=\frac{2\beta-1}{\beta}m^{\mathbf{2}^n}+\frac{1-\beta}{\beta}m_{\mathrm{I}}^{\mathbf{2}(1)},$$

$$\rho_{\mathrm{I}}^{(2)}=\rho_*^{(2)}-\beta\Delta t\hat{D}_x m_{\mathrm{E}}^{\mathbf{1}(2)}-\beta\Delta t\hat{D}_y m_{\mathrm{E}}^{\mathbf{2}(2)}\ \text{(computed explicitly)},$$

$$p_{\mathrm{E}}^{(2)}=\left(1-\frac{\hat{c}}{\beta}\right)p^n+\frac{\hat{c}}{\beta}p_{\mathrm{I}}^{(1)},\quad p_*^{(2)}=\frac{2\beta-1}{\beta}p^n+\frac{1-\beta}{\beta}p_{\mathrm{I}}^{(1)}.$$

To compute $m_{\mathrm{I}}^{\mathbf{1}(2)}$, $m_{\mathrm{I}}^{\mathbf{2}(2)}$ and $p_{\mathrm{I}}^{(2)}$ we use the same procedure used to compute $m_{\mathrm{I}}^{\mathbf{1}(1)}$, $m_{\mathrm{I}}^{\mathbf{2}(1)}$ and $p_{\mathrm{I}}^{(1)}$. Therefore, we have:

$$m_{\mathrm{I}}^{\mathbf{1}(2)}=M_{\mathrm{E}}^{\mathbf{1}(2)}-\frac{\beta\Delta t}{\varepsilon^2}D_x p_{\mathrm{I}}^{(2)},\quad m_{\mathrm{I}}^{\mathbf{2}(2)}=M_{\mathrm{E}}^{\mathbf{2}(2)}-\frac{\beta\Delta t}{\varepsilon^2}D_y p_{\mathrm{I}}^{(2)},$$

with $M_{\mathrm{E}}^{\mathbf{1}(2)}=m_*^{\mathbf{1}(2)}-\beta\Delta t\hat{D}_x\left(\frac{\left(m_{\mathrm{E}}^{\mathbf{1}(2)}\right)^2}{\rho_{\mathrm{E}}^{(2)}}\right)-\beta\Delta t\hat{D}_y\left(m_{\mathrm{E}}^{\mathbf{1}(2)}\frac{m_{\mathrm{E}}^{\mathbf{2}(2)}}{\rho_{\mathrm{E}}^{(2)}}\right)$,

$M_{\mathrm{E}}^{\mathbf{2}(2)}=m_*^{\mathbf{2}(2)}-\beta\Delta t\hat{D}_x\left(m_{\mathrm{E}}^{\mathbf{2}(2)}\frac{m_{\mathrm{E}}^{\mathbf{1}(2)}}{\rho_{\mathrm{E}}^{(2)}}\right)-\beta\Delta t\hat{D}_y\left(\frac{\left(m_{\mathrm{E}}^{\mathbf{2}(2)}\right)^2}{\rho_{\mathrm{E}}^{(2)}}\right)$, and

$$\frac{p_{\mathrm{I}}^{(2)}}{\gamma-1}-\frac{\beta^2\Delta t^2}{\varepsilon^2}\left(D_x\left(h_{\mathrm{E}}^{(2)}\left(D_x p_{\mathrm{I}}^{(2)}\right)\right)+D_y\left(h_{\mathrm{E}}^{(2)}\left(D_y p_{\mathrm{I}}^{(2)}\right)\right)\right)=p^{**},$$

with $p^{**} = E_*^{(2)} - \beta \Delta t \hat{D}_x \left(h_{\mathrm{E}}^{(2)} M_{\mathrm{E}}^{\mathbf{1}(2)} \right) - \beta \Delta t \hat{D}_y \left(h_{\mathrm{E}}^{(2)} M_{\mathrm{E}}^{\mathbf{2}(2)} \right) - \frac{1}{2}\varepsilon^2 \frac{\left(m_{\mathrm{E}}^{\mathbf{1}(2)}\right)^2}{\rho_{\mathrm{E}}^{(2)}} + - \frac{1}{2}\varepsilon^2 \frac{\left(m_{\mathrm{E}}^{\mathbf{2}(2)}\right)^2}{\rho_{\mathrm{E}}^{(2)}}$.

In summary, the algorithm is:

- compute $\rho_{\mathrm{I}}^{(1)}, \rho_{\mathrm{E}}^{(2)}, \rho_*^{(2)}, M_{\mathrm{E}}^{\mathbf{1}(1)}, M_{\mathrm{E}}^{\mathbf{2}(1)}, p^*$;
- compute $p_{\mathrm{I}}^{(1)}, p_{\mathrm{E}}^{(2)}, p_*^{(2)}, E_*^{(2)}, m_{\mathrm{I}}^{\mathbf{1}(1)}, m_{\mathrm{E}}^{\mathbf{1}(2)}, m_*^{\mathbf{1}(2)}, m_{\mathrm{I}}^{\mathbf{2}(1)}, m_{\mathrm{E}}^{\mathbf{2}(2)}, m_*^{\mathbf{2}(2)}$;
- compute $\rho_{\mathrm{I}}^{(2)}, M_{\mathrm{E}}^{\mathbf{1}(2)}, M_{\mathrm{E}}^{\mathbf{2}(2)}, p^{**}$;
- compute $p_{\mathrm{I}}^{(2)}, m_{\mathrm{I}}^{\mathbf{1}(2)}, m_{\mathrm{I}}^{\mathbf{2}(2)}$;
- compute $p^{n+1} = p_{\mathrm{I}}^{(2)}$, E^{n+1} from EOS, $\rho^{n+1} = \rho_{\mathrm{I}}^{(2)}$, $m^{\mathbf{1}^{n+1}} = m_{\mathrm{I}}^{\mathbf{1}(2)}$, $m^{\mathbf{2}^{n+1}} = m_{\mathrm{I}}^{\mathbf{2}(2)}$.

We observe that more specific schemes can be applied in order to improve numerical stability. For example, constant pressure and a constant velocity fields should be preserved through a density discontinuity, as pointed out in [5]. This is achieved by taking a suitable implicit representation of the kinetic energy and enthalpy as in [8]. However, the focus of this chapter is the ghost point technique to implement boundary conditions and the coupling strategy with internal equations. The technique does not rely on the specific scheme adopted to solve internal equations and then more sophisticated schemes for internal discretizations can be adopted.

3 Ghost-Point Method for Boundary Conditions

To model the motion of a compressible gas past a non-rectangular obstacle, suitable boundary conditions must be set up and discretized on the curved boundary of the object. In this chapter the obstacle is embedded in a fixed uniform Cartesian grid and the boundary conditions are enforced by adopting a ghost-point technique as follows. Grid points inside the gas domain are called *internal points*. Euler equations are discretized at each internal point as in Sect. 2.2.2. For each internal point (x_i, y_j) the value at the four (six in 3D) neighbour points $\left\{(x_{i\pm1}, y_j), (x_i, y_{j\pm1})\right\}$ is needed. If the internal point (x_i, y_j) is close to the boundary, some of the four neighbour points might lie outside the gas domain. We call those external points *ghost points*. We indicate with N_{I} and N_{G} the number of internal points and of ghost points respectively. Figure 1, on the left, shows the discretization of a square domain with a circular obstacle.

Let us suppose that the gas domain Ω is defined by a level set function ϕ, namely as the set $\left\{\mathbf{x} \in \mathbb{R}^2 : \phi(\mathbf{x}) < 0\right\}$, while the obstacle is identified by $\left\{\mathbf{x} \in \mathbb{R}^2 : \phi(\mathbf{x}) > 0\right\}$ and the boundary by $\left\{\mathbf{x} \in \mathbb{R}^2 : \phi(\mathbf{x}) = 0\right\}$. Internal and ghost points can be identified by the level-set function.

The definition of physical quantities at ghost cells requires the solution of equations that arise from the discretization of the boundary conditions. In the following, we describe how the boundary conditions are derived from physical principles [13].

We denote by $\mathbf{n}$ and $\boldsymbol{\tau}$ respectively the normal and the tangential unit vector to the boundary, while κ is the signed curvature (see Fig. 2). We assume that the

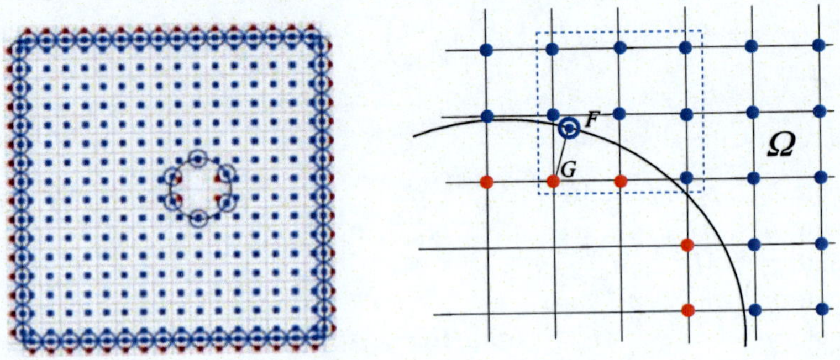

Fig. 1 Left: discretization of a domain $\Omega \subset [0, 1]^2$ with $N = 16$ and a circular obstacle: the blue dots are the internal points, the red dots are the ghost points. We impose the boundary conditions in the orthogonal projections of the ghost points to the boundary (blue circles). Right: Example of St_G stencil enclosed within the dashed rectangular line

Fig. 2 Locally convex boundary $\kappa < 0$ (left) and locally concave boundary $\kappa > 0$ (right)

unit normal vector points outside the fluid domain. Let us denote by $u_n = \mathbf{u} \cdot \mathbf{n}$ and $u_\tau = \mathbf{u} \cdot \boldsymbol{\tau}$ respectively the normal and tangential velocity. The boundary condition on the normal velocity is derived from the impenetrability assumption:

$$u_n = 0 \text{ on } \partial\Omega, \tag{13}$$

The equation of motion for a fluid particle (balance of momentum) for Euler equations is:

$$\rho \frac{D\mathbf{u}}{Dt} + \nabla p = 0 \tag{14}$$

where $D/Dt = \partial/\partial t + \mathbf{u} \cdot \nabla$ denotes the Lagrangian derivative.

Condition (13) implies that, along the boundary of the domain, the velocity vector is $\mathbf{u} = u_\tau \boldsymbol{\tau}$. It is therefore:

$$\frac{D\mathbf{u}}{Dt} = \frac{Du_\tau}{Dt}\boldsymbol{\tau} + u_\tau \frac{D\boldsymbol{\tau}}{Dt} = a_\tau \boldsymbol{\tau} + u_\tau^2 \, \kappa \, \mathbf{n}. \tag{15}$$

With the notation of Fig. 2, the sign of κ is negative for locally convex regions, and positive for locally concave regions, and a_τ denotes the tangential acceleration of the fluid. By projecting Eq. (14) on the normal direction, and making use of (15), one obtain the boundary condition on the pressure:

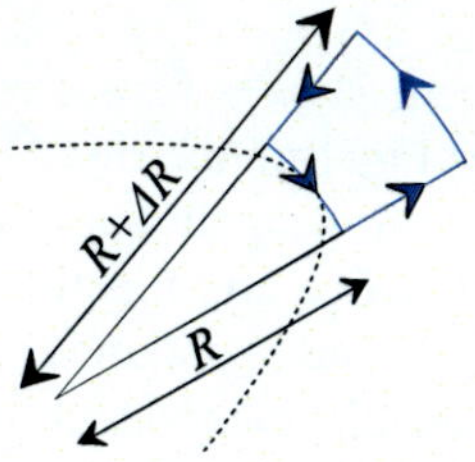

Fig. 3 To obtain the boundary condition for the tangential velocity, we impose that the vorticity along the blue curve is zero. The dashed line is the boundary of the domain, that has the same tangent and curvature of the smaller arc

$$\frac{\partial p}{\partial \mathbf{n}} = -\rho\, u_\tau^2\, \kappa. \tag{16}$$

Boundary condition for u_τ is:

$$\frac{\partial u_\tau}{\partial \mathbf{n}} = u_\tau\, \kappa. \tag{17}$$

Such condition can be obtained by imposing that the vorticity is zero, namely $\oint_\Gamma \mathbf{u} \cdot d\mathbf{l} = 0$ for each closed circuit Γ. Choosing the circuit of Fig. 3, where the arc with the smaller radius and the boundary of the domain are tangent and have the same curvature, we obtain (assuming that u_τ is constant along each of the two arcs):

$$\oint_\Gamma \mathbf{u} \cdot d\mathbf{l} = (R + \Delta R)\, u_\tau|_{R+\Delta R} - R\, u_\tau|_R = 0, \tag{18}$$

where $|\kappa| = 1/R$. By Taylor expansion and taking the limit for $\Delta R \to 0$, we obtain (17).

Lastly, the boundary condition on the density is obtained by imposing that the boundary is adiabatic, i.e. that the entropy s is locally flat, meaning $\partial s/\partial n = 0$, from which:

$$\frac{\partial p}{\partial \mathbf{n}} = c_s^2\, \frac{\partial \rho}{\partial \mathbf{n}}. \tag{19}$$

This set of boundary conditions refers to the case of a steady obstacle. A detailed treatment of moving objects can be found in [9].

Boundary conditions (13), (17), (16), (19) are used to extrapolate $\mathbf{u}$, p, ρ in the ghost points, as follows.

Let G be a ghost point. First, we compute the projection point F to the interface, making use of the level-set function ϕ by solving the nonlinear equation $\phi(\text{G} - \delta\, h\, \mathbf{n}_G) = 0$ for $\delta \in [0, 1)$ (using a bisection method, for example). The normal vector is approximated using $\mathbf{n}_G = \nabla\phi/|\nabla\phi|$ and discretizing it by standard central difference methods.

Then, we identify the 3×3 stencil St_G in the upwind direction containing G (see Fig. 1 on the right). In detail:

$$St_G = \left\{(x_G + s_x\, k_1\, h,\, y_G + s_y\, k_2\, h) \colon (k_1, k_2) \in \{0, 1, 2\}^2\right\},$$

where $\quad s_x = \lceil |x_F - x_G| \,/\, h \rceil \mathrm{sgn}(x_F - x_G), \quad s_y = \lceil |y_F - y_G| \,/\, h \rceil \mathrm{sgn}(y_F - y_G)$. See Fig. 1 on the right.

Following the same approach of [9], we denote by $Q[\psi; S]$ the biquadratic interpolant of a grid function $\psi_{i,j}$ in a 3×3 stencil $\mathcal{S}$. The discretization of the boundary conditions is obtained by approximating the values of pressure, density, normal and tangential components of the velocity at point F using the corresponding biquadratic interpolants:

$$Q[u_n; \mathcal{S}_{\mathrm{G}}](\mathrm{F}) = 0 \tag{20}$$

$$\frac{\partial Q[p; \mathcal{S}_{\mathrm{G}}]}{\partial \mathbf{n}}(\mathrm{F}) = -\kappa Q[\rho; \mathcal{S}_{\mathrm{G}}](\mathrm{F}) \left(Q[u_\tau; \mathcal{S}_{\mathrm{G}}](\mathrm{F})\right)^2 \tag{21}$$

$$\frac{\partial Q[\rho; \mathcal{S}_{\mathrm{G}}]}{\partial \mathbf{n}}(\mathrm{F}) = \frac{Q[\rho; \mathcal{S}_{\mathrm{G}}](\mathrm{F})}{\gamma Q[p; \mathcal{S}_{\mathrm{G}}](\mathrm{F})} \frac{\partial Q[p; \mathcal{S}_{\mathrm{G}}]}{\partial \mathbf{n}}(\mathrm{F}) \tag{22}$$

$$\frac{\partial Q[u_\tau; \mathcal{S}_{\mathrm{G}}]}{\partial \mathbf{n}}(\mathrm{F}) = \kappa Q[u_\tau; \mathcal{S}_{\mathrm{G}}](\mathrm{F}) \tag{23}$$

We observe that the stencil associated with a ghost point G can contain other ghost points, therefore each 4×4 system, obtained from the Eqs. (20), (21), (22) and (23), can be coupled with the corresponding 4×4 systems obtained at other ghost points of the stencil. Consequentially, it is not possible to solve separately the 4×4 system for each ghost point G, but a coupled $4N_g \times 4N_g$ system is solved, where N_g is the number of ghost points.

To solve the system of boundary conditions, in [9] the authors use an iterative scheme, which is built first, by transforming the system in a fictitious time-dependent problem with time σ:

$$\begin{cases} \frac{\partial u_n}{\partial \sigma} = -u_n \\ \frac{\partial u_\tau}{\partial \sigma} = -\mu_2 \left(\frac{\partial u_\tau}{\partial \mathbf{n}} - \kappa u_\tau\right) \\ \frac{\partial p}{\partial \sigma} = -\mu_1 \left(\frac{\partial p}{\partial \mathbf{n}} + \kappa \rho u_\tau^2\right) \\ \frac{\partial \rho}{\partial \sigma} = -\mu_3 \left(\frac{\partial \rho}{\partial \mathbf{n}} - \frac{1}{c_s^2} \frac{\partial p}{\partial \mathbf{n}}\right) \end{cases}, \tag{24}$$

where μ_1, μ_2 and μ_3 are constants chosen appropriately to guarantee convergence and stability, and then, discretizing the system (24) in space and time. Ghost values are then obtained from the steady state solution of (24).

In [9], the authors discretize the partial derivatives with respect to σ at the ghost point G using the first-order forward Euler method. Therefore, the iterative scheme is given by:

$$u_n^{(m+1)}(\mathrm{G}) = u_n^{(m)}(\mathrm{G}) - \Delta\sigma Q[u_n^{(m)}; \mathcal{S}_\mathrm{G}](\mathrm{F}),$$

$$u_\tau^{(m+1)}(\mathrm{G}) = u_\tau^{(m)}(\mathrm{G}) - \mu_2 \Delta\sigma \left(\frac{\partial Q[u_\tau^{(m)}; \mathcal{S}_\mathrm{G}](\mathrm{F})}{\partial \mathbf{n}} - \kappa Q[u_\tau^{(m)}; \mathcal{S}_\mathrm{G}](\mathrm{F}) \right),$$

$$p^{(m+1)}(\mathrm{G}) = p^{(m)}(\mathrm{G}) - \mu_1 \Delta\sigma \left(\frac{\partial Q[p^{(m)}; \mathcal{S}_\mathrm{G}]}{\partial \mathbf{n}}(\mathrm{F}) + \kappa Q[\rho^{(m)}; \mathcal{S}_\mathrm{G}](\mathrm{F}) \left(Q[u_\tau^{(m)}; \mathcal{S}_\mathrm{G}](\mathrm{F}) \right)^2 \right),$$

$$\rho^{(m+1)}(\mathrm{G}) = \rho^{(m)}(\mathrm{G}) - \mu_3 \Delta\sigma \left(\frac{\partial Q[\rho^{(m)}; \mathcal{S}_\mathrm{G}]}{\partial \mathbf{n}}(\mathrm{F}) - \frac{Q[\rho^{(m)}; \mathcal{S}_\mathrm{G}](\mathrm{F})}{\gamma Q[p^{(m)}; \mathcal{S}_\mathrm{G}](\mathrm{F})} \frac{\partial Q[p^{(m)}; \mathcal{S}_\mathrm{G}]}{\partial \mathbf{n}}(\mathrm{F}) \right),$$

in which the time step $\Delta\sigma$ and the constants μ_1 μ_2 and μ_3 are chosen in such a way as to satisfy the CFL condition:

$$\Delta\sigma < 1 \text{ and } \mu_i \Delta\sigma < \min(\Delta x, \Delta y), \; i = 1, 2, 3. \tag{25}$$

Observe that this CFL condition is related with the fictitious time evolution of the boundary conditions and then it is not related with the CFL condition of the real time step. In other words, we have two temporal iterations: a real time step dictated by the evolution of the Euler equations and, at each time step, we iterate a fictitious time dependent problem up to steady state to find the ghost values. The restriction (25) depends only on the type of boundary condition (Dirichlet, Neumann, etc.) and then it is not affected by a possible more restrictive CFL condition on the real time step that might be observed in other scenarios, for example the parabolic CFL condition required for viscous fluids.

The computational cost of the ghost value calculations (24) is negligible with respect to the internal nodes, since it is proportional to the number of ghost points and then the ratio between the number of ghost points and internal points vanishes when $\Delta x \to 0$.

Moreover, one can observe that the iteration on ghost values is not necessary: in fact, if we order the ghost points by their non-decreasing distance from the boundary, one can update their values using a direct method instead of iterating. In fact, the system arising from the discretization of the boundary conditions (20)–(23) results in a lower triangular system, or a block lower triangular system with a few small 2×2 blocks. The presence of these blocks is when two ghost points are such that each of them belongs to the stencil of the other, although this happens rarely.

The method can be easily adapted to the semi-implicit approach as described in the next section.

3.1 Ghost-Point Technique for Implicit Solvers

In [9] the authors propose an explicit time discretization, therefore the method suffers from the restriction on the time step imposed by the CFL condition. The aim of this work is to couple the semi-implicit discretization with the ghost-point technique proposed in [9] to solve the Euler equations on an arbitrary domain.

We denote by $\mathbf{U}_i$ and $\mathbf{U}_g$ the vectors of the numerical solution on internal and ghost points, respectively. If we develop an explicit method, we have:

$$\mathbf{U}_i^{n+1} = \mathbf{U}_i^n - \Delta t \mathbf{F}(\mathbf{U}_i^n, \mathbf{U}_g^n)_x - \Delta t \mathbf{G}(\mathbf{U}_i^n, \mathbf{U}_g^n)_y,$$

where $\mathbf{U}_g^n$ is computed solving the linear system $\mathrm{A}\mathbf{U}_g^n = \mathrm{b}(\mathbf{U}_i^n)$, arising from the discretization of boundary conditions, namely Eqs. (20)–(23), with $\mathrm{A} \in \mathbb{R}^{4N_g \times 4N_g}$.

If we develop a semi-implicit method, we have:

$$\mathbf{U}_i^{n+1} = \mathbf{U}_{\mathrm{E}_i}^n - \Delta t \mathbf{F}(\mathbf{U}_{\mathrm{E}_i}^n, \mathbf{U}_{\mathrm{E}_g}^n)_x - \Delta t \mathbf{G}(\mathbf{U}_{\mathrm{E}_i}^n, \mathbf{U}_{\mathrm{E}_g}^n)_y, \tag{26}$$

where the solution in the ghost points $\mathbf{U}_{\mathrm{E}_g}^n$ is computed by solving the linear system

$$\mathrm{A}\mathbf{U}_{\mathrm{E}_g}^n = \mathrm{b}(\mathbf{U}_{\mathrm{E}_i}^n) \tag{27}$$

for the terms that are treated explicitly, and

$$\begin{cases} \mathbf{U}_i^{n+1} = \mathbf{U}_{\mathrm{E}_i}^n - \Delta t \mathbf{F}(\mathbf{U}_{\mathrm{I}_i}^{n+1}, \mathbf{U}_{\mathrm{I}_g}^{n+1})_x - \Delta t \mathbf{G}(\mathbf{U}_{\mathrm{I}_i}^{n+1}, \mathbf{U}_{\mathrm{I}_g}^{n+1})_y \\ \mathrm{A}\mathbf{U}_{\mathrm{I}_g}^{n+1} = \mathrm{b}(\mathbf{U}_{\mathrm{E}_i}^n) \end{cases} \tag{28}$$

for the terms that are treated implicitly, for example (12).

For a fully implicit ghost-point scheme, system (28) would be substituted by

$$\begin{cases} \mathbf{U}_i^{n+1} = \mathbf{U}_{\mathrm{E}_i}^n - \Delta t \mathbf{F}(\mathbf{U}_{\mathrm{I}_i}^{n+1}, \mathbf{U}_{\mathrm{I}_g}^{n+1})_x - \Delta t \mathbf{G}(\mathbf{U}_{\mathrm{I}_i}^{n+1}, \mathbf{U}_{\mathrm{I}_g}^{n+1})_y \\ \mathrm{A}\mathbf{U}_{\mathrm{I}_g}^{n+1} = \mathrm{b}(\mathbf{U}_{\mathrm{I}_i}^{n+1}, \mathbf{U}_{\mathrm{E}_i}^n) \end{cases}, \tag{29}$$

leading to fully coupled problem (29) for internal and ghost points that must be solved at each time step. The scope of this paper is to present a boundary treatment that can be coupled with the semi-implicit method, therefore we use (28) in all numerical tests, although the approach does not rely on it and can be extended to the fully implicit Eq. (29), provided that efficient solvers are designed (since the coupling between ghost and internal nodes might become the main portion of the computational effort).

The system (28) was solved with a different technique than the one proposed in [9]. In detail, we rewrite the system of boundary conditions for each ghost point in the equivalent form:

$$Q[u; \mathcal{S}_\mathrm{G}](\mathrm{F})\mathrm{n}_x + Q[v; \mathcal{S}_\mathrm{G}](\mathrm{F})\mathrm{n}_y = 0 \tag{30}$$

$$\frac{\partial Q[p; \mathcal{S}_\mathrm{G}]}{\partial \mathbf{n}}(\mathrm{F}) = -\kappa Q[\rho; \mathcal{S}_\mathrm{G}](\mathrm{F})\,(u_\tau(\mathrm{P}))^2 \tag{31}$$

$$\frac{\partial Q[\rho; \mathcal{S}_\mathrm{G}]}{\partial \mathbf{n}}(\mathrm{F}) = \frac{\rho(\mathrm{P})}{\gamma p(\mathrm{P})} \frac{\partial Q[p; \mathcal{S}_\mathrm{G}]}{\partial \mathbf{n}}(\mathrm{F}) \tag{32}$$

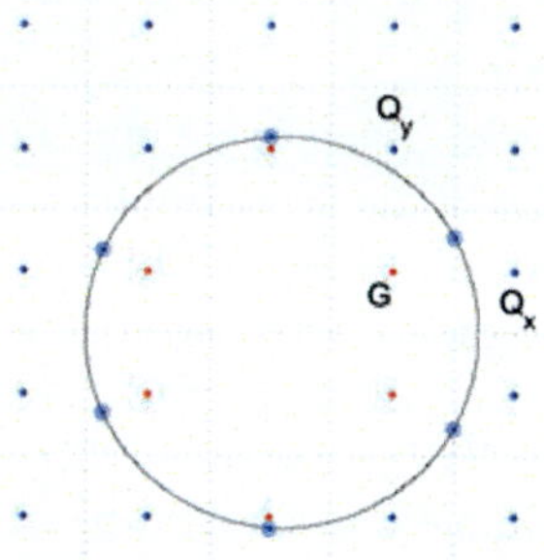

Fig. 4 Upwind neighbour points to G

$$\frac{\partial Q[u_\tau; \mathcal{S}_\mathrm{G}]}{\partial \mathbf{n}}(\mathrm{F}) = \kappa \left(-Q[u; \mathcal{S}_\mathrm{G}](\mathrm{F})\mathrm{n}_y + Q[v; \mathcal{S}_\mathrm{G}](\mathrm{F})\mathrm{n}_x\right), \tag{33}$$

where n_x and n_y are the component along x axis and y axis, respectively, of the unit normal pointing outside the domain, that we compute by the level-set function as follows:

$$\phi_x = \frac{\phi_\mathrm{G} - \phi_{\mathrm{Q}_x}}{h} \mathrm{sgn}\left(\mathrm{G}^{(x)} - \mathrm{Q}_x^{(x)}\right)$$

and

$$\phi_y = \frac{\phi_\mathrm{G} - \phi_{\mathrm{Q}_y}}{h} \mathrm{sgn}\left(\mathrm{G}^{(y)} - \mathrm{Q}_y^{(y)}\right),$$

where Q_x and Q_y are the two upwind neighbour points to G (see Fig. 4).

We have $\mathbf{n} = (n_x, n_y)$ and $\boldsymbol{\tau} = (-n_y, n_x)$, with

$$\mathrm{n}_x = \frac{\phi_x}{|\nabla\phi|} \text{ and } \mathrm{n}_y = \frac{\phi_y}{|\nabla\phi|}.$$

The non-linear terms $\left(Q[u_\tau; \mathcal{S}_\mathrm{G}](\mathrm{F})\right)^2$ and $\frac{Q[\rho; \mathcal{S}_\mathrm{G}](\mathrm{F})}{\gamma Q[p; \mathcal{S}_\mathrm{G}](\mathrm{F})}$ have been approximated with the values they assume at the closest internal point P: $(u_\tau(\mathrm{P}))^2$ and $\frac{\rho(\mathrm{P})}{\gamma p(\mathrm{P})}$, respectively.

We solve a linear system of size $4N_g \times 4N_g$ in the unknowns ρ, u, v and p in correspondence of ghost points. The value of the energy E in the ghost points is derived from the EOS.

In this chapter we use the built-in solver of MATLAB [24], although other approaches can be used, especially for larger number of grid points or for the 3D case (not addressed in this chapter). For example, a fictitious time relaxation proposed in (24) for the boundary condition can be extended to internal equation as well, following a similar approach proposed in [14] where a multigrid iterative solver is adopted to solve elliptic equations on non-rectangular domains. However, a proper analysis of the solver is out of the scope of this chapter and can be addressed in future works.

4 Numerical Simulations

In this Section we present some numerical tests to highlight the robust characteristics of the Semi-Implicit methods we have developed for the 2D compressible Euler equations on an arbitrary domain. In particular the equations are solved on a rectangular domain with an arbitrary obstacle. We have chosen two different shapes: a square obstacle (Fig. 5) and a circular obstacle (Fig. 1 on the left).

We refer to the numerical test presented in [9], in which the authors considered a simple wave that propagates around a fixed disk of center $C(x_C, y_C)$, with $x_C = 0.6$ and $y_C = 0.5$, and initial conditions:

$$(\rho(x,y,0), u(x,y,0), v(x,y,0), p(x,y,0)) = \begin{cases} (\tilde{\rho}(x,y), \tilde{u}(x,y), 0, \tilde{p}(x,y)) & \text{if } |x-0.35| < 0.25 \\ (\rho_0, 0, 0, p_0) & \text{otherwise} \end{cases},$$

where

$$\tilde{u}(x,y) = 0.5\mathrm{e}^{-\frac{(x-0.35)^2}{0.005}},$$

$$\tilde{\rho}(x,y) = \rho_0 \left(1 + \frac{\gamma-1}{2}\frac{\tilde{u}(x,y)}{c_0}\right)^{\frac{2}{\gamma-1}},$$

$$\tilde{p}(x,y) = p_0 \left(\frac{\tilde{\rho}(x,y)}{\rho_0}\right)^{\gamma},$$

with $\rho_0 = p_0 = 1$ and $c_0 = \sqrt{\frac{\gamma p_0}{\rho_0}} = \sqrt{1.4}$, from which $\gamma = 1.4$.

On the border of the obstacle (square or circular) the boundary conditions are given by the Eqs. (30), (31), (32) and (33). On the outer border of the domain (the rectangle $[0, 1]^2$) we have imposed void Neumann boundary conditions on the energy, pressure, density and v and void Dirichlet boundary conditions on u.

We adopt second-order reconstructions both in time and in space, with $\theta = 1$, and a uniform Cartesian grid with $\Delta x = \Delta y = 1/100$.

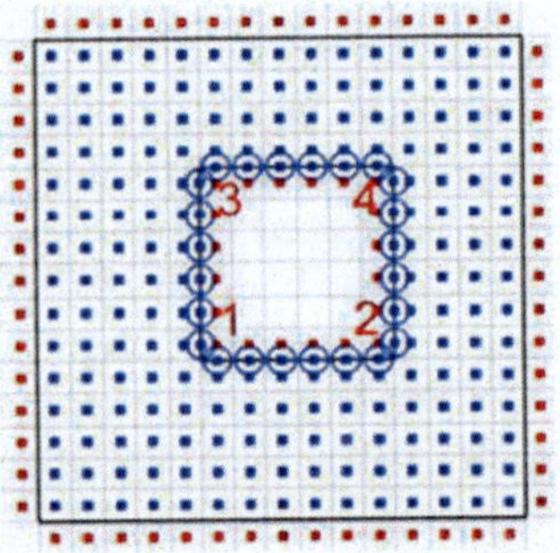

Fig. 5 Domain with a square obstacle

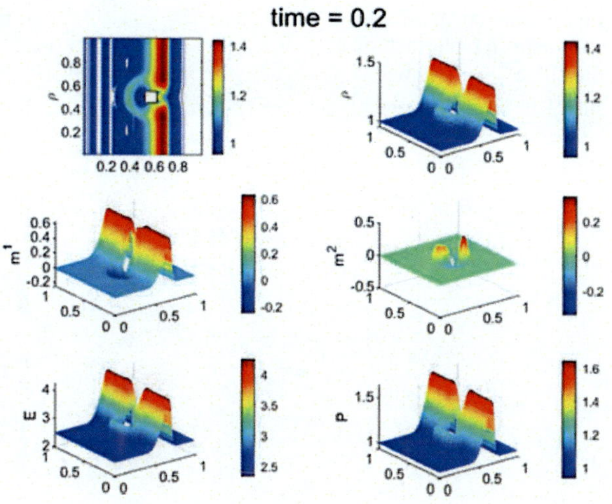

Fig. 6 Numerical solution with semi-implicit method at final time $T = 0.2$ with a square obstacle, $\varepsilon = 1$ and CFL $= 2$

4.1 *Square Obstacle*

We first solve the problem on a domain with a square obstacle (Fig. 5). In this case there is a simplification on the boundary conditions (30), (31), (32) and (33), which coincide with those imposed on the outer border, but, ρ, u, v, P and E in the ghost points at the corners of the obstacle (ghost points labelled as 1,2,3 and 4 in Fig. 5), take on two different values: one along x axis and one along y axis.

4.1.1 Test 1

We start by showing that the semi-implicit methods are stable for CFL values larger than 1. For this reason, we performed a simulation with Mach number $\varepsilon = 1$ and CFL = 2 (see Fig. 6). The simulation shows that, for the same Mach number, an increase in the CFL corresponds to an increase in the time step, which is approximately $\Delta t = 0.0063$. In fact, for CFL = 0.5 we have verified that $\Delta t = 0.0015$ approximately.

4.1.2 Test 2

We want to show that the method works well even for Mach numbers larger than 1. For this reason we have carried out a simulation for a final time $T = 0.35$ by choosing $\varepsilon = 2$ and CFL $= 2$ (see Fig. 7). A comparison with Test 1 shows that increasing the Mach number also increases the time step, which in this case is about $\Delta t = 0.01$.

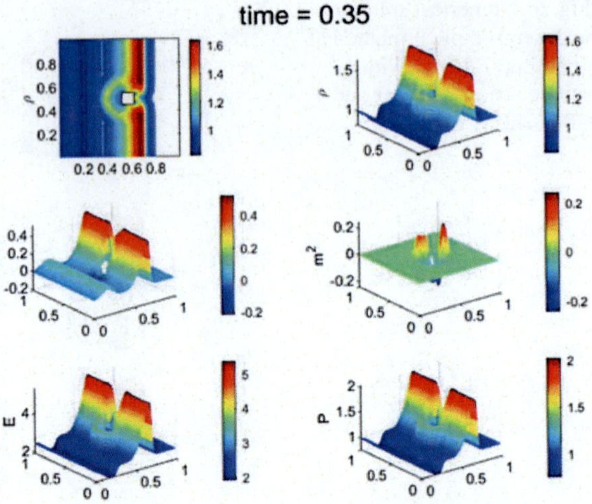

Fig. 7 Numerical solution with semi-implicit method at final time $T = 0.35$ with a square obstacle, $\varepsilon = 2$ and $\mathrm{CFL} = 2$

4.1.3 Test 3

A further increase in the $\mathrm{CFL} = 3$ leads to stability problems due to the material CFL, which almost immediately assumes a value greater than 1. However, by lowering the Mach number $\varepsilon = 1/10$, the method turns out to be more stable, although some numerical oscillations are observed in the vicinity of the obstacle, as can be observed in the density and momentum plots of Fig. 8. The oscillations are present also in the case where Δt is computed with the less restrictive condition: $\Delta t = \mathrm{CFL}_{\mathrm{imp}} \frac{\Delta x}{\max |\mathbf{u}| + \bar{c}}$, where $\bar{c} = \sqrt{\gamma p/\rho} \min\{1, 1/\varepsilon\}$ (as in [7]). In this case the material CFL is greater than 1.

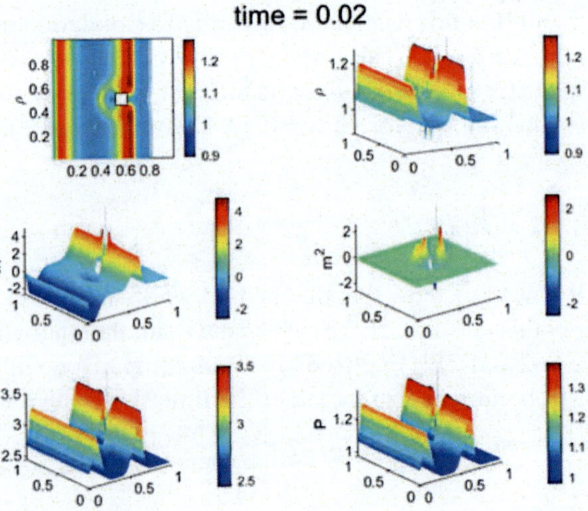

Fig. 8 Numerical solution with semi-implicit method at final time $T = 0.02$ with a square obstacle, $\varepsilon = 1/10$ and $\mathrm{CFL} = 3$

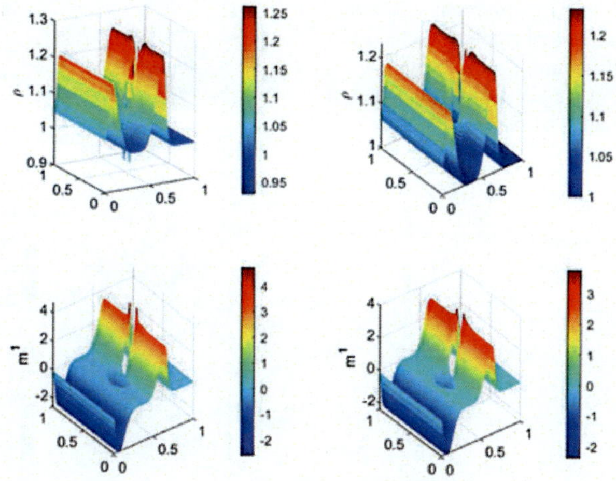

Fig. 9 Numerical solution with semi-implicit method at final time $T = 0.02$ with a square obstacle, $\varepsilon = 1/10$ and CFL $= 2$. Left: Classic semi-implicit method, Right: semi-implicit method obtained by substituting (11) with (10) for $\varepsilon < 1$

We observe that, for $\varepsilon < 1$, replacing the condition (11) with the classic condition (10) will reduce the oscillations.

With the same values ($\varepsilon = 1/10$ and CFL $= 3$), this replacement does not lead to numerical improvements, in fact, the material CFL assumes values larger than 1 and the problem is not stable. However, with $\varepsilon = 1/10$ and CFL $= 2$ it is possible to observe the efficiency of the substitution (Fig. 9). We will see greater improvements in the case of the circular domain.

4.2 *Circular Obstacle*

Finally, we have solved the problem on a domain with a circular obstacle.

4.2.1 Test 1

We show that the method works well for CFL and ε values larger than 1. Figure 10 shows results for CFL $= 1.5$ and $\varepsilon = 2$. The time step is about $\Delta t = 0.008$.

4.2.2 Test 2

Finally, we show numerical results for $\varepsilon < 1$. Also in this case, we can observe that by replacing condition (11) with condition (10) the method is more stable and the numerical oscillations at the boundary of the circular obstacle disappear (see Fig. 11).

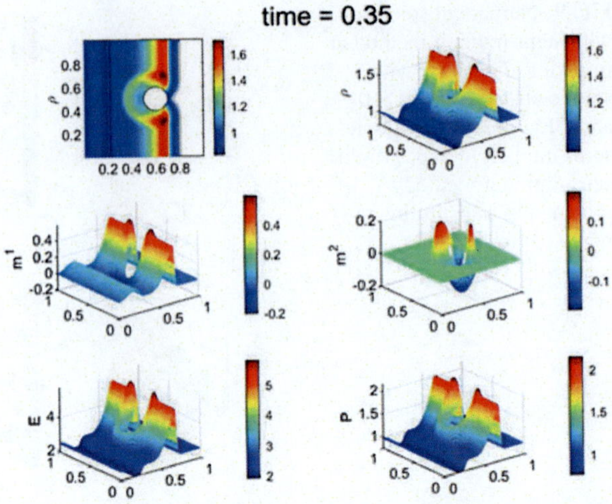

Fig. 10 Numerical solution with semi-implicit method at final time $T = 0.35$ with a circular obstacle, $\varepsilon = 2$ and CFL $= 1.5$

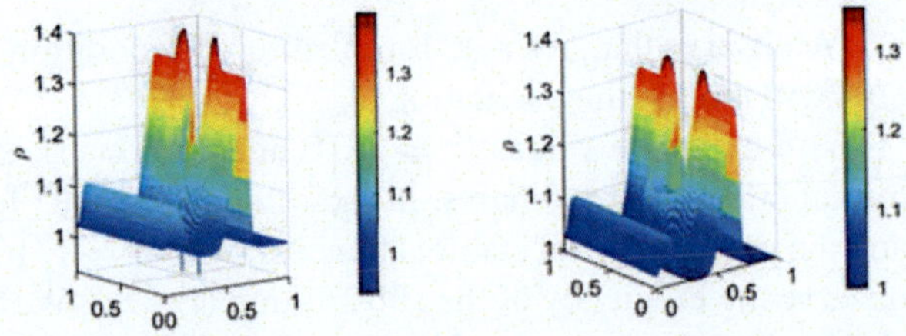

Fig. 11 Numerical solution with semi-implicit method at final time $T = 0.1$ with a circular obstacle, $\varepsilon = 1/2$ and CFL $= 1.5$. Left: Classic semi-implicit method, Right: semi-implicit method obtained by substituting (11) with (10) for $\varepsilon < 1$

4.3 Shock Tube Problems

We now apply the semi-implicit method to Riemann problems for two-dimensional Euler equations for $\varepsilon > 1$, referring to the circular explosion problem and a 2D Riemann problem reported in [17].

In the absence of obstacles in the computational domain, we have solved the equations *(a)-(d)* with respect to energy, as in [3].

4.3.1 Test 1: Circular Explosion Problem

We consider a problem with a circular symmetry, whose initial condition, in a square computational domain $[-1, 1] \times [-1, 1]$, are constant in two regions separated by a circle centered at the origin and radius 0.5:

$$(\rho, u, v, p)(x, y, t = 0) = \begin{cases} (1, 0, 0, 1) & \text{if } x^2 + y^2 \leq 0.5^2 \\ (0.125, 0, 0, 0.1) & \text{if } x^2 + y^2 > 0.5^2 \end{cases}$$

We have imposed void Neumann boundary conditions on the density, energy and pressure, and Dirichlet boundary conditions on the momentums. We show in

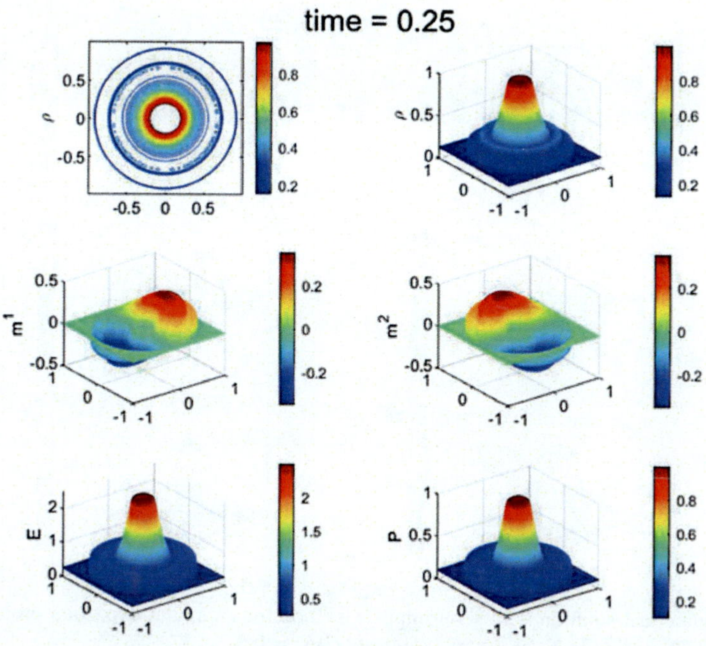

Fig. 12 Numerical solution with semi-implicit method for a circular explosion problem at final time $T = 0.25$, $\varepsilon = 1$, $\Delta x = \Delta y = 1/500$ and CFL $= 0.5$

Figs. 12 and 13 the results at final time $T = 0.25$ for CFL $= 0.5$, $\varepsilon = 1$ and $\varepsilon = 3/2$, respectively. In [17] the authors chose $\Delta x = \Delta y = 1/1000$, we instead show the results with $\Delta x = \Delta y = 1/500$.

We also show the plot of density, pressure and velocity as a function of $r = \sqrt{x^2 + y^2}$ (see Figs. 14 and 15).

4.3.2 Test 2: Two-Dimensional Riemann Problem

We consider a second test in which the initial conditions in the computational domain $[-0.5, 0.5] \times [-0.5, 0.5]$ are given by:

$$(\rho, u, v, p) = \begin{cases} (1.1, 0, 0, 1.1) & \text{if } x > 0 \text{ and } y > 0 \\ (0.5065, 0.8939, 0, 0.35) & \text{if } x \leq 0 \text{ and } y > 0 \\ (1.1, 0.8939, 0.8939, 1.1) & \text{if } x \leq 0 \text{ and } y \leq 0 \\ (0.5065, 0, 0.8939, 0.35) & \text{if } x > 0 \text{ and } y \leq 0 \end{cases}.$$

The results at final time $T = 0.2$ are shown in the Fig. 16 for $\varepsilon = 1$ (left panel) and $\varepsilon = 3/2$ (right panel), respectively. We have chosen $\Delta x = \Delta y = 1/500$.

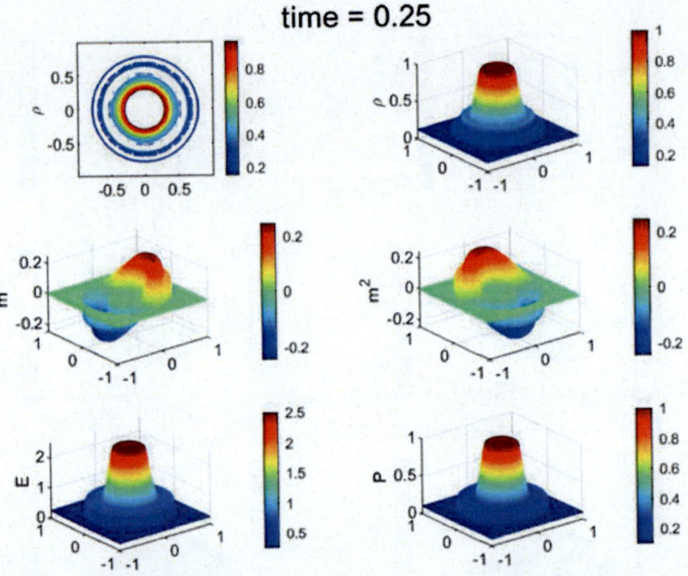

Fig. 13 Numerical solution with semi-implicit method for a circular explosion problem at final time $T = 0.25$, $\varepsilon = 3/2$, $\Delta x = \Delta y = 1/500$ and CFL $= 0.5$

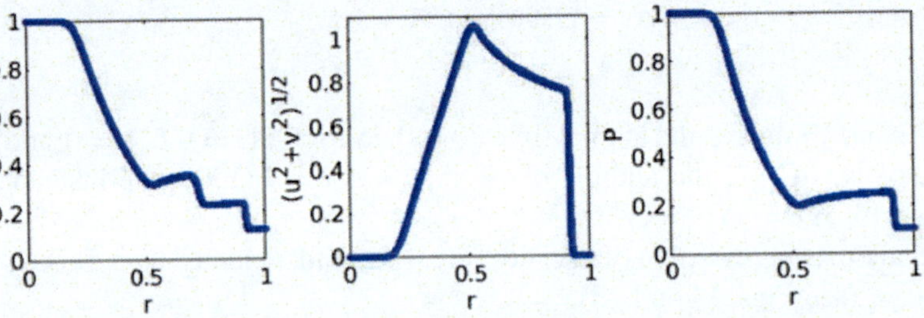

Fig. 14 Plot of density, pressure and velocity as a function of $r = \sqrt{x^2 + y^2}$ for a circular explosion problem at final time $T = 0.25$, $\varepsilon = 1$, $\Delta x = \Delta y = 1/500$ and CFL $= 0.5$

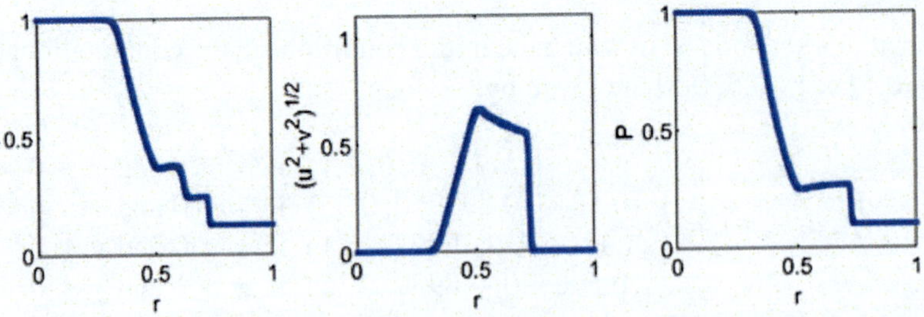

Fig. 15 Plot of density, pressure and velocity as a function of $r = \sqrt{x^2 + y^2}$ for a circular explosion problem at final time $T = 0.25$, $\varepsilon = 3/2$, $\Delta x = \Delta y = 1/500$ and CFL $= 0.5$

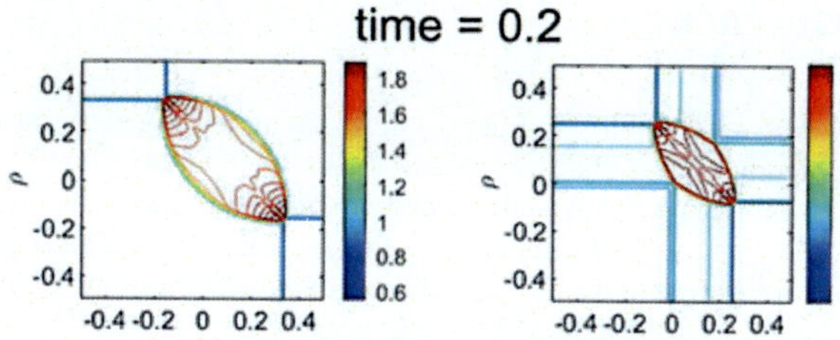

Fig. 16 Density contour plot with semi-implicit method for the Riemann problem at final time $T = 0.2$, $\varepsilon = 1$ (left panel) and $\varepsilon = 3/2$ (right panel), $\Delta x = \Delta y = 1/500$ and CFL $= 0.5$

5 Conclusions

In this chapter we have presented a finite-difference ghost-point method to solve hyperbolic equations on arbitrary domains. The equations are discretized by a semi-implicit method in order to alleviate the CFL condition, with the advantage of eliminating the restriction of explicit methods on the spatial step for the acoustic waves. They also have the advantage of not introducing excessive numerical dissipation as the fully implicit methods do.

The curved boundary is immersed in a uniform Cartesian grid and treated with a ghost-point method, obtained from the generalization of the finite-difference method for the elliptic equations on arbitrary domains proposed by one of the authors in [14] and applied to several contexts, such as incompressible fluid dynamics [10] and volcanology problems [11, 12]. The method is based on a polynomial interpolation technique to enforce boundary conditions on the orthogonal projections of the ghost points to the boundary of the domain. A 3×3 stencil with a biquadratic interpolation ensure second order accuracy of the solution and its gradient (as observed in [9]), although the arbitrary stencil width and degree of the polynomial will allow a straightforward extension to higher order of accuracy.

We have applied the technique to the Euler equations of gas dynamics past circular and square objects. Several numerical tests show the improvements on the CFL condition of the semi-implicit method with respect to the explicit approach. Numerical oscillations are observed close to the boundary for higher CFL and low Mach numbers. They are reduced when the classical (10) is implemented instead of (11).

We finally observe that it is possible to further improve the stability of semi-implicit methods by adopting a fully implicit ghost-point technique (29), by taking a suitable implicit representation of the kinetic energy and enthalpy as in [8] in order to preserve constant pressure and a constant velocity fields across a density discontinuity, or by performing an automatic step control technique [2, 20], since the error committed in the discretization method mainly depends on the time-step size, which is varied along the solution in order to minimize the computational effort.

Acknowledgements All authors acknowledge support from GNCS–INDAM (National Group for Scientific Computing, Italy). The work of A.C. has been partially supported by the CINECA ISCRA project (class C): BCMG - HP10C7YOPZ.

References

1. Abbate, E., Iollo, A., Puppo, G.: An asymptotic-preserving all-speed scheme for fluid dynamics and nonlinear elasticity. SIAM J. Sci. Comput. **41**, A2850–A2879 (2019)
2. Atkinson, K.E.: An introduction to numerical analysis, 2nd edn. Wiley, New York (1989)
3. Avgerinos, S., Bernard, F., Iollo, A., Russo, G.: Linearly implicit all Mach Number shock capturing schemes for the Euler equations. J. Comput. Phys. **393**, 278–312 (2019). https://doi.org/10.1016/j.jcp.2019.04.020
4. Baeza, A., Mulet, P., Zorío, D.: High order boundary extrapolation technique for finite difference methods on complex domains with Cartesian Meshes. J. Sci. Comput. **66**, 761–791 (2016). https://doi.org/10.1007/s10915-015-0043-2
5. Billet, G., Abgrall, R.: An adaptive shock-capturing algorithm for solving unsteady reactive flows. Comput. Fluids **32**(10), 1473–1495 (2003)
6. Boscarino, S., Filbet, F., Russo, G.: High order semi-implicit schemes for time dependent partial differential equations. J. Sci. Comput. **68**, 975–1001 (2016). https://doi.org/10.1007/s10915-016-0168-y
7. Boscarino, S., Russo, G., Scandurra, L.: All mach number second order semi-implicit scheme for the euler equations of gas dynamics. J. Sci. Comput. **77**, 850–884 (2018). https://doi.org/10.1007/s10915-018-0731-9
8. Boscheri, W., Pareschi, L.: High order pressure-based semi-implicit IMEX schemes for the 3D Navier-Stokes equations at all Mach numbers. J. Comput. Phys. **434**, 110206 (2021). https://doi.org/10.1016/j.jcp.2021.110206
9. Chertock, A., Coco, A., Kurganov, A., Russo, G.: A second-order finite-difference method for compressible fluids in domains with moving boundaries. J Comput Phys **23**(1), 230–263 (2018). https://doi.org/10.4208/cicp.OA-2016-0133
10. Coco, A.: A multigrid ghost-point level-set method for incompressible Navier-Stokes equations on moving domains with curved boundaries. J. Comput. Phys. **418**, 109623 (2020)
11. Coco, A., Currenti, G., Del Negro, C., Russo, G.: A second order finite-difference ghost-point method for elasticity problems on unbounded domains with applications to volcanology. Commun. Comput. Phys. **16**(4), 983–1009 (2014)
12. Coco, A., Currenti, G., Gottsmann, J., Russo, G., Del Negro, C.: A hydro-geophysical simulator for fluid and mechanical processes in volcanic areas. J. Math. Ind. **6**(1), 1–20 (2016)
13. Coco, A., Russo, G.: Boundary treatment in ghost point finite difference methods for compressible gas dynamics in domain with moving boundaries. In: Hyperbolic Problems: Theory, Numerics, Applications, pp. 455 (2012)
14. Coco, A., Russo, G. Finite-difference ghost-point multigrid methods on Cartesian grids for elliptic problems in arbitrary domains. *Journal of Computational Physics* 241:464-501 (2013). https://doi.org/10.1016/j.jcp.2012.11.047
15. Cordier, F., Degond, P., Kumbaro, A.: An Asymptotic-Preserving all-speed scheme for the Euler and Navier-Stokes equations. *Journal of Computational Physics* 231(17):5685-5704 (2012). https://doi.org/10.1016/j.jcp.2012.04.025
16. Dellacherie, S.: Analysis of Godunov type schemes applied to the compressible Euler system at low Mach number. J. Computa Phys. **229**(4), 978–1016 (2010). https://doi.org/10.1016/j.jcp.2009.09.044
17. Dumbser, M., Casulli, V.: A conservative, weakly nonlinear semi-implicit finite volume scheme for the compressible Navier-Stokes equations with general equation of state. Appl. Math. Comput. **272**, 479–497 (2016). https://doi.org/10.1016/j.amc.2015.08.042
18. Fernández-Fidalgo, J., Clain, S., Ramírez, L., Colominas, I., Nogueira, X.: Very high-order method on immersed curved domains for finite difference schemes with regular Cartesian grids. Comput. Methods Appl. Mech. Eng. **360**, 112782 (2020). https://doi.org/10.1016/j.cma.2019.112782
19. Godlewski, E., Raviart, P.-A.: Numerical approximation of hyperbolic systems of conservation laws. Springer (2014). https://doi.org/10.1007/978-1-0716-1344-3

20. Ilie, S., Söderlind, G., Corless, R.M.: Adaptivity and computational complexity in the numerical solution of ODEs. J. Complex **24**(3), 341–361 (2008). https://doi.org/10.1016/j.jco.2007.11.004
21. LeVeque, R.J.: Numerical Methods for Conservation Laws. Lectures in Mathematics, Birkhauser, Basel (1992)
22. LeVeque, R.J.: Finite volume methods for hyperbolic problems, vol. 31. Cambridge University Press (2002)
23. Levy, D., Puppo, G., Russo, G.: A fourth-order central WENO scheme for multidimensional hyperbolic systems of conservation laws. SIAM J. Sci. Comput. **24**(2), 480–506 (2002). https://doi.org/10.1137/S1064827501385852
24. MATLAB. version 9.7.0 (R2019b). Natick, Massachusetts: The MathWorks Inc. (2019)
25. Pareschi, L., Russo, G.: Implicit-explicit Runge-Kutta schemes and applications to hyperbolic systems with relaxation. J. Sci. Comput. **25**, 129–155 (2005). https://doi.org/10.1007/s10915-004-4636-4
26. Shu, C.: Essentially non-oscillatory and weighted essentially non-oscillatory schemes for hyperbolic conservation laws. In: Advanced Numerical Approximation of Nonlinear Hyperbolic Equations, pp. 325–432 (1998)
27. Shu, C.W.: High Order weighted essentially nonoscillatory schemes for convection dominated problems. SIAM Rev. **51**(1), 82–126 (2009). https://doi.org/10.1137/070679065
28. Shu, C.W., Osher, S.: Efficient implementation of essentially non-oscillatory shock-capturing schemes, II. J. Comput. Phys. **83**(1), 32–78 (1989). https://doi.org/10.1016/0021-9991(89)90222-2

High-Order Arbitrary-Lagrangian-Eulerian Schemes on *Crazy* Moving Voronoi Meshes

Elena Gaburro and Simone Chiocchetti

Abstract Hyperbolic partial differential equations (PDEs) cover a wide range of interesting phenomena, from *human* and *hearth*-sciences up to *astrophysics*: this unavoidably requires the treatment of many space and time scales in order to describe at the same time *observer-size* macrostructures, *multi-scale* turbulent features, and also *zero-scale* shocks. Moreover, numerical methods for solving hyperbolic PDEs must reliably handle different families of waves: smooth *rarefactions*, and discontinuities of *shock* and *contact* type. In order to achieve these goals, an effective approach consists in the combination of space-time-based *high-order* schemes, very accurate on smooth features even on coarse grids, with *Lagrangian* methods, which, by moving the mesh with the fluid flow, yield highly resolved and minimally dissipative results on both shocks and contacts. However, ensuring the high quality of *moving meshes* is a huge challenge that needs the development of innovative and unconventional techniques. The scheme proposed here falls into the family of Arbitrary-Lagrangian-Eulerian (ALE) methods, with the unique additional freedom of evolving the *shape* of the mesh elements through connectivity changes. We aim here at showing, by simple and very salient examples, the capabilities of high-order ALE schemes, and of our novel technique, based on the high-order space-time treatment of *topology changes*.

Keywords Hyperbolic equations · Direct Arbitrary-Lagrangian-Eulerian schemes · High order space-time methods · ADER discontinuous Galerkin schemes · Polygonal meshes · Voronoi tessellations · Topology changes

E. Gaburro (✉)
Inria, University of Bordeaux, CNRS, Bordeaux INP, IMB, UMR 5251, 200 Avenue de la Vieille Tour, 33405 Talence cedex, France
e-mail: elena.gaburro@inria.fr

S. Chiocchetti
Department of Civil, Environmental and Mechanical Engineering, University of Trento, Via Mesiano, 77, 38123 Trento, Italy
e-mail: simone.chiocchetti@unitn.it

G. Albi et al. (eds.), *Advances in Numerical Methods for Hyperbolic Balance Laws and Related Problems*, SEMA SIMAI Springer Series 32,
https://doi.org/10.1007/978-3-031-29875-2_5

1 Introduction

In order to reduce as much as possible the numerical errors due to nonlinear convective terms, it is possible to exploit the power of Lagrangian methods: with this kind of algorithms, the new position and configuration of each element of the mesh is recomputed at each timestep according to the local fluid velocity, so that we can closely follow the fluid flow in a Lagrangian fashion. In this way the nonlinear convective terms disappear and Lagrangian schemes exhibit *negligible numerical dissipation* at *contact* waves and material *interfaces*; moreover they results to be *Galilean and rotational invariant*, and they provide, without any additional effort, an automatic *mesh refinement* feature even when the cell count is maintained constant, simply by transporting the mesh elements wherever needed.

The use of Lagrangian methods dates back to the works of [65, 67] and then many further improvements have been introduced in literature; we cite here only some few relevant historical examples and review papers [10, 11, 16, 17, 42, 46, 48, 49, 51, 59].

However, ensuring the high quality of a *moving mesh* over long simulation times is difficult, therefore a certain degree of flexibility should be allowed in order to avoid mesh distortion, for example a slightly relaxed choice of the actual mesh velocity w.r.t the real fluid velocity, as well as the freedom of not only *moving* the control volumes, but really *evolving* their shapes and allowing topology and neighborhood changes. This led to the introduction of Arbitrary-Lagrangian-Eulerian (ALE) schemes of *direct* [6, 8, 19] and *indirect* [5, 44, 45] type. In particular, as stated and shown in [61], connectivity changes between different time level constitute a valid alternative to remeshing [5, 44, 45] for preserving or restoring mesh quality in a Lagrangian setting.

With this in mind, we present here a novel family of very *high-order* direct Arbitrary-Lagrangian-Eulerian (ALE) Discontinuous Galerkin (DG) schemes for the solution of general nonlinear hyperbolic PDE systems on moving Voronoi meshes that are *regenerated* at each timestep and which explicitly allow *topology changes* in time, in order to benefit simultaneously from high-order methods, high quality grids and substantially reduced numerical dissipation; this method has been introduced for the first time by the two authors in [28].

The *key ingredient* of our approach is the integration of a *space-time conservation formulation* of the governing PDE system over closed, non-overlapping *space-time* control volumes [8] that are constructed from the moving, *regenerated*, Voronoi-type polygonal meshes which are centroid-based dual grids of the Delaunay triangulation of a set of *generator* points: this leads to also consider what we refer to in this work as *crazy degenerate* control volumes, or *space-time sliver* elements, that only arise when adopting a space-time framework, and would not exist from a purely spatial point of view! Such degenerate elements provide a clear formal way of handling mesh connectivity changes while preserving the high-order of accuracy of the numerical method.

1.1 Goals

The goal of this book chapter is to briefly review this novel and promising approach based on high order direct ALE schemes with topology changes, and to provide numerical evidence regarding the utility of Lagrangian methods in general, and in particular of the new technique object of this work, by means of simple and illustrative benchmarks.

1.2 Structure

The rest of this chapter is organized as follows. We first introduce the class of equations of interest in Sect. 2. Next, we briefly summarize the main characteristics of the employed direct Arbitrary-Lagrangian-Eulerian scheme, focusing in particular on the space-time approach and its extension to *crazy* sliver elements, whose formation, caused by topology changes, will be also addressed in Sect. 3.2. Then, the core of this work consists in providing numerical evidence for *(i)* the key role of topology changes and sliver elements in a high-order moving mesh code, and *(ii)* the clear advantages of Lagrangian schemes on widely adopted benchmark problems. Finally, we give some conclusive remarks and an outlook towards future work in Sect. 5.

2 Hyperbolic Partial Differential Equations

In order to model a wide class of physical phenomena, we consider a very general formulation of the governing equations, namely all those which can be described by

$$\partial_t \mathbf{Q} + \nabla \cdot \mathbf{F}(\mathbf{Q}) + \mathbf{B}(\mathbf{Q}) \cdot \nabla \mathbf{Q} = \mathbf{S}(\mathbf{Q}), \tag{1}$$

where $\mathbf{Q}$ is the vector of the conserved variables, $\mathbf{F}$ the non linear flux, $\mathbf{B} \cdot \nabla \mathbf{Q}$ the nonconservative products, and $\mathbf{S}$ a nonlinear algebraic source term. Many physical models can be cast in this form, from the simple shallow water system, some multiphase flow models, the magnetohydrodynamics equations, up to the Einstein field equations of general relativity (with appropriate reformulation) or the GPR unified model of continuum mechanics, see for example [7, 13, 14, 14, 25, 26, 29, 30, 38, 54, 62]; in this work, we will present illustrative results concerning the Euler equations of gasdynamics.

3 Numerical Method

In this Section we presents a concise description of our direct Arbitrary-Lagrangian-Eulerian (ALE) Discontinuous Galerkin (DG) scheme on moving Voronoi-type meshes with topology changes; for any additional details we refer to our recent paper [28].

At the beginning of the simulation, we discretize our moving domain by a centroid-based Voronoi-type tessellation built from a set of generators (the orange points in Fig. 1), and we represent our data, the conserved variables $\mathbf{Q}$, via discontinuous high-order polynomials in each mesh polygon (we indicate the degree of the polynomial representation by P_N). Then, we let the generators move with a velocity chosen to be as close as possible to the local fluid velocity, computed mainly from a high-order approximation of their pure Lagrangian trajectories, with small corrections obtained from a flow-adaptive mesh optimization technique. The positions of the generators are being continuously updated, and thus their Delaunay triangulation may change at any timestep and the same will hold for the dual polygonal tessellation. Then, a space-time connection between two polygonal tessellations corresponding to two successive time levels has to be established in order to evolve the solution in time locally and integrate the governing PDE.

3.1 Direct Arbitrary-Lagrangian-Eulerian Schemes

The key idea of *direct* ALE methods (in contrast to *indirect* ones) consists in connecting two tessellations by means of so-called *space-time control volumes* C_i^n, and recover the unknown solution at the new timestep $\mathbf{u}_h^{n+1}$ *directly* inside the new polygon P_i^{n+1}, from the data available at the previous timestep $\mathbf{u}_h^n$ in P_i^n. This is achieved through the *integration*, over such control volumes, of the fluxes, the non-conservative products and the source terms, by means of a high-order fully discrete predictor-corrector ADER method [21, 31]. In this way, the need for any further remapping/remeshing steps is totally eliminated. By adopting the tilde symbol for

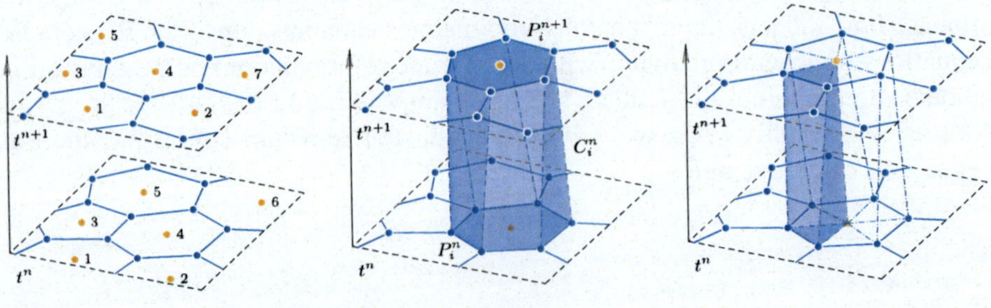

Fig. 1 Space time connectivity *without* topology changes, main space-time control volume (middle) and a standard sub-space-time control volume (right)

referring to space-time quantities, our direct ALE scheme [27, 28] reads

$$\begin{aligned}\int_{P_i^{n+1}} \tilde{\phi}_k \mathbf{u}_h^{n+1} = \int_{P_i^n} \tilde{\phi}_k \mathbf{u}_h^n - \sum_j \int_{\partial C_{ij}^n} \tilde{\phi}_k \mathcal{F}(\mathbf{q}_h^{n,-}, \mathbf{q}_h^{n,+}) \cdot \tilde{\mathbf{n}} + \int_{C_i^n} \tilde{\nabla} \tilde{\phi}_k \cdot \tilde{\mathbf{F}}(\mathbf{q}_h^n) \\ + \int_{C_i^n} \tilde{\phi}_k \left(\mathbf{S}(\mathbf{q}_h^n) - \tilde{\mathbf{B}}(\mathbf{q}_h^n) \cdot \nabla \mathbf{q}_h^n \right),\end{aligned} \tag{2}$$

where $\tilde{\phi}_k$ is a set of moving space-time basis functions, while $\mathbf{q}_h^{n,+}$ and $\mathbf{q}_h^{n,-}$ are high-order space-time extrapolated data computed through the ADER predictor. Finally, $\mathcal{F}(\mathbf{q}_h^{n,-}, \mathbf{q}_h^{n,+})$ is an ALE numerical flux function which takes into account fluxes across space-time cell boundaries ∂C_{ij}^n as well as jump terms related to nonconservative products. In particular, we adopt here a two-point path-conservative numerical flux function of Rusanov-type [53, 58]

$$\begin{aligned}\mathcal{F}(\mathbf{q}_h^{n,-}, \mathbf{q}_h^{n,+}) \cdot \tilde{\mathbf{n}} =& \frac{1}{2} \left(\tilde{\mathbf{F}}(\mathbf{q}_h^{n,+}) + \tilde{\mathbf{F}}(\mathbf{q}_h^{n,-}) \right) \cdot \tilde{\mathbf{n}}_{ij} - \frac{1}{2} s_{\max} \left(\mathbf{q}_h^{n,+} - \mathbf{q}_h^{n,-} \right) \\ &+ \frac{1}{2} \left(\int_0^1 \tilde{\mathbf{B}} \left(\Psi(\mathbf{q}_h^{n,-}, \mathbf{q}_h^{n,+}, s) \right) \cdot \mathbf{n} \, d\mathbf{x} \right) \cdot \left(\mathbf{q}_h^{n,+} - \mathbf{q}_h^{n,-} \right),\end{aligned} \tag{3}$$

where $s_{\max}$ is the maximum eigenvalue of the *ALE Jacobian matrices* evaluated on the left and right of the space-time interface and the path $\Psi = \Psi(\mathbf{q}_h^-, \mathbf{q}_h^+, s)$ is a straight-line segment path connecting $\mathbf{q}_h^{n,-}$ and $\mathbf{q}_h^{n,+}$.

We emphasize that the ALE Jacobian matrix is obtained by subtracting the local normal mesh velocity from the diagonal entries of the system matrix of the quasi-linear form of the governing equations [63] (the Jacobian of the interface-normal flux for conservative systems), thus, when the mesh velocity is sufficiently close to the local fluid velocity, the wavespeed estimates obtained from the eigenvalues are significantly reduced, leading to a lower associated numerical dissipation than what would be mandated in the Eulerian context. This, especially but not exclusively, in conjunction with complete approximate Riemann solvers [20], explains the capability of tracking material interfaces and capturing contact discontinuities which are characteristic of Lagrangian-type schemes.

Next, in order to compute the integrals with high order of accuracy, complete knowledge of the *space-time connectivity* between two consecutive timesteps is required, as opposed to only the *spatial* information at the two time levels, which would be enough for a low order scheme [61] or for indirect schemes [5, 44, 45].

When no topology changes occur, the space-time geometrical information is easily constructed by connecting via straight line segments the corresponding vertexes of each polygon, obtaining an oblique prism than can be further subdivided into a set of triangular oblique sub-prisms on which quadrature points are readily available (see Figs. 1 and 3).

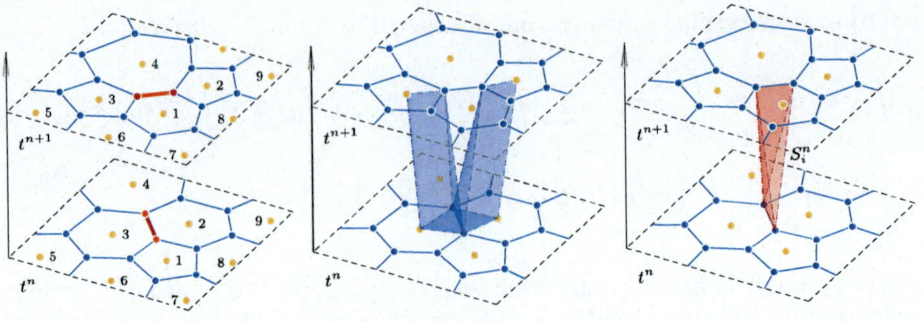

Fig. 2 Space time connectivity *with* topology changes, degenerate sub-space-time control volumes (middle) and *crazy* sliver element (right)

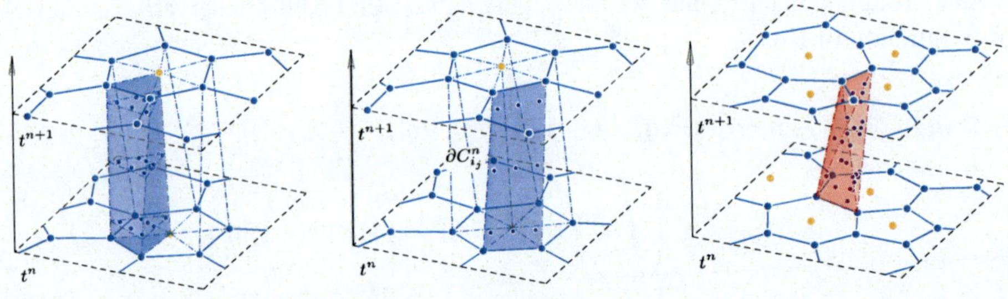

Fig. 3 Space-time quadrature points for third order methods on standard elements (left), lateral faces (middle) and *crazy* sliver elements (right)

3.2 Topology Changes and Crazy Sliver Elements

On the contrary, when a topology change occurs, as in Fig. 2, i.e. the number of edges, the shape, and the neighbors of a polygon evolve within two consecutive timesteps, the space-time connection between the mesh elements gives raise to degenerate elements of two types: *(i) degenerate sub*-space-time control volumes, where either the top or bottom faces are degenerate triangles that are collapsed to a segment; *(ii)* and also *crazy sliver* space-time elements S_i^n. The first type of degenerate elements does not pose any problems, and was already treated in [32]. Instead, space-time sliver elements are a completely new type of control volume. In particular, they do not exist neither at time t^n, nor at time t^{n+1}, since they coincide with an edge of the tessellation at the old *and* at the new time levels, and for this reason have zero area in space at the two bounding time levels. However, they have a *non-negligible volume* in space-time. The difficulties related to this kind of elements are due to the fact that for them an initial condition is not clearly defined at time t^n, and that contributions across these elements should not be lost at time t^{n+1}, in order to ensure conservation. All the details on how to successfully extend our direct ALE scheme also to *crazy* elements can be found in our recent paper [28].

We would like to emphasize that topology changes are fundamental for long time simulations in the ALE framework, in order to avoid explicit data remap steps, and our *crazy* sliver elements represent a *novel and formally grounded* way to allow for a relatively simple space-time connection around a change of connectivity. The numerical results shown in Sect. 4.1 provide a clear proof of the necessity of topology changes already on the simple situation of the long time evolution of a stationary isentropic smooth vortex.

3.3 ADER-ALE Algorithm: The Predictor Step

The *predictor* step represents an essential ingredient for obtaining high-order in time in a fully-discrete one-step procedure: it yields a *local* solution of the governing Eq. (1) *in the small* $\mathbf{q}_h^n$, inside each space-time element, including the *crazy* elements. The solution is *local* in the sense that it is obtained by only considering the initial data in each polygon, the governing equations and the geometry of C_i^n, without taking into account interactions between C_i^n and its neighbors. Such local solution is computed for each standard space-time control volume C_i^n and for each *crazy* control volume S_i^n, in the form of a high order polynomial *in space and in time*, which serves as a predictor solution, to be used for evaluating all the integrals in the *corrector* step (2), i.e. the final update of the solution from t^n to t^{n+1}.

3.4 A Posteriori Sub-cell FV Limiter

High-order schemes that can be seen as linear in the sense of Godunov [34], may develop spurious oscillations in presence of discontinuities. In order to prevent this phenomenon, in the case of a DG discretization we adopt an *a posteriori* limiting procedure based on the MOOD paradigm [15, 31, 47]: we first apply our unlimited ALE-DG scheme everywhere, and then (*a posteriori*), at the end of each timestep, we check the reliability of the obtained solution in each cell against physical and numerical admissibility criteria, such as floating point exceptions, violation of positivity or violation of a relaxed discrete maximum principle (and see [35, 39] for further criteria). Next, we mark as *troubled* those cells where the DG solution cannot be accepted. For the troubled cells we now repeat the time evolution by employing, instead of the DG scheme, a more robust finite volume (FV) method. Moreover, in order to maintain the accurate resolution of our original high-order DG scheme, which would be lost when switching to a FV scheme, the FV scheme is applied on a *finer sub-cell grid* that accounts for recovering the optimal accuracy of the numerical method performing a reconstruction step.

4 Numerical Examples

In order to provide simple and clear numerical evidence of the effectiveness of the proposed ALE scheme with topology changes we consider here the well known Euler equations, that can be cast in the form (1) by choosing

$$\mathbf{Q} = \begin{pmatrix} \rho \\ \rho u \\ \rho v \\ \rho E \end{pmatrix}, \quad \mathbf{F} = \begin{pmatrix} \rho u & \rho v \\ \rho u^2 + p & \rho u v \\ \rho u v & \rho v^2 + p \\ u(\rho E + p) & v(\rho E + p) \end{pmatrix}, \qquad \mathbf{B} = \mathbf{0}, \qquad \mathbf{S} = \mathbf{0}. \tag{4}$$

The vector of conserved variables $\mathbf{Q}$ is composed of the fluid density ρ, the momentum density vector $\rho\mathbf{v} = (\rho u, \rho v)$ and the total energy density ρE; next, the fluid pressure p is computed using the equation of state for an ideal gas

$$p = (\gamma - 1)\left(\rho E - \frac{1}{2}\rho \mathbf{v}^2\right), \tag{5}$$

where γ (in this work taken to be $\gamma = 7/5$) is the ratio of specific heats. For this choice of equation of state, the adiabatic speed of sound takes the form $c = \sqrt{\gamma p/\rho}$.

In what follows we will present numerical results regarding the following notable features of Lagrangian schemes and of our direct Arbitrary-Lagrangian-Eulerian method with variable topology:

i. Flows characterized by strong differential rotations, for example vortices, can be studied over very long periods only by conceding to the element motion the additional freedom of introducing topology changes, see Sect. 4.1;
ii. The use of sliver elements allows to clearly define the space-time evolution of the solutions in-between discrete time levels and achieves high-order of accuracy also in presence of many topology changes, see Sect. 4.1.1;
iii. Lagrangian schemes sharply capture *shock* waves thanks to the automatic refinement obtained at the shock locations without needing to increment the number of mesh elements but simply because the element density increases wherever needed, see Sect. 4.2;
iv. Lagrangian schemes minimize dissipation of *contact* discontinuities, by applying reduced numerical dissipation when using approximate Riemann solvers. In a pure Lagrangian context, schemes capable of capturing stationary discontinuities exactly will do the same also for moving interfaces (since the mesh motion is specified to follow such features). Moreover, even when such hard constraints are relaxed in Arbitrary-Lagrangian-Eulerian methods and even using simpler solvers like the Rusanov flux, the bounding wavespeed estimates and the associated numerical dissipation can be much lower than what would be mandated in the Eulerian context, see Sect. 4.3;

v. Lagrangian schemes discretely preserve the Galilean and rotational invariance properties of the governing equations, so that they can better capture any events (such as the explosion-type problems reported in this work) that may occur in superposition to a high-speed background flow, see Sect. 4.3.

4.1 Long Time Evolution of a Shu-type Vortical Equilibrium

As a first test we consider a smooth isentropic vortex flow defined as similarly to [36]. The initial computational domain is the square $\Omega = [0; 10] \times [0; 10]$ and boundary conditions are of wall (slip) type everywhere. The initial condition is given by some perturbations δ that are superimposed onto a homogeneous background state $\mathbf{Q}_0 = (\rho, u, v, p) = (1, 0, 0, 1)$, assuming that the entropy perturbation is zero, i.e. $\delta S = 0$. The perturbations for density and pressure are

$$\delta\rho = (1 + \delta T)^{\frac{1}{\gamma-1}} - 1, \quad \delta p = (1 + \delta T)^{\frac{\gamma}{\gamma-1}} - 1, \tag{6}$$

with the temperature fluctuation $\delta T = -\frac{(\gamma-1)\epsilon^2}{8\gamma\pi^2} e^{1-r^2}$ and the vortex strength $\epsilon = 5$. The velocity field is specified by

$$\begin{pmatrix} \delta u \\ \delta v \end{pmatrix} = \frac{\epsilon}{2\pi} e^{\frac{1-r^2}{2}} \begin{pmatrix} -(y-5) \\ (x-5) \end{pmatrix}. \tag{7}$$

This is a stationary equilibrium of the system so the exact solution coincides with the initial condition at any time.

Preserving this kind of vortical solution over long simulation times with minimal dissipation is a nontrivial task in a moving-mesh context. To achieve this result, we propose the use of a very high-order scheme (here an ADER-DG method of order 4) in a Lagrangian framework. We remark that the combination cannot be used with fixed topology, or advanced remapping techniques, because the quality of the moving mesh subject to this constraint quickly deteriorates, as is clearly apparent in Fig. 4, where the simulation has to be stopped after about half a vortex rotation period. This highlights the well-known fact that, for long time evolution, the mesh connectivity must be somehow updated. In this work this is naturally achieved by means of space-time topology changes.

Further, Fig. 5 demonstrates that the treatment of topology changes via high-order integration over *crazy* sliver elements is actually quite effective. Indeed one can note that the solution is visually the same at the beginning of the simulation and 500 seconds after, even on a rather coarse mesh of only 957 polygonal elements. Moreover, we take advantage of this test case to also emphasize the high precision of the mesh movement. The Voronoi-type polygonal cells, as well as the generator points, in fact can be observed to orbit along perfectly circular trajectories, as evidenced in Figs. 5 and 6.

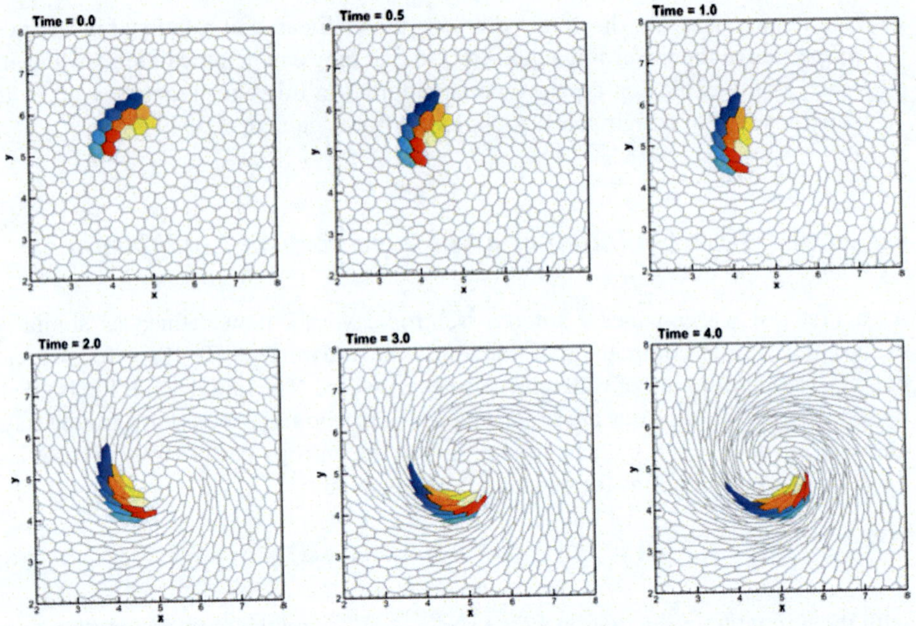

Fig. 4 Mesh evolution corresponding to the solution of the stationary rotating vortex of Sect. 4.1 solved on a moving grid with *fixed* topology. The mesh quality rapidly deteriorates: elements are stretched, the timestep size is reduced, and even mesh-tangling occurs, which means that the simulation may stop entirely

4.1.1 Order of Convergence

Finally, this stationary test case allows to show numerically the order of convergence of the proposed ALE-DG scheme with topology changes, reported in Table 1 up to order 4. Furthermore, we present a quantitative comparison with the scheme applied in a purely Eulerian setting (i.e. on a fixed mesh) and with the classical direct ALE approach with fixed topology. For the purpose of this test, we consider the domain $\Omega = [-10; 30] \times [-10; 30]$, covered with a Voronoi-type tessellation obtained as the centroid-based dual of a Delaunay mesh generated by Ruppert's algorithm [57]. We report our results at time $t = 4$ (the time at which the ALE simulations with fixed topology terminate due to mesh tangling) and $t = 10$ (a long simulation time at which differences in mesh configuration become very significant).

It should be stressed that, due to the absence of discontinuous features or strong background flows, this test problem is not intended to highlight the capabilities of moving mesh algorithms, but rather to show the high order convergence of the method on smooth flows, while highlighting the necessity for a changing mesh topology.

On the results, we note that the ALE method applied to a fixed mesh topology, in addition to early termination around time $t = 4$, as shown in Fig. 4, also suffers an increase in numerical errors, to the point that the correct order of convergence cannot be obtained when the mesh is severely tangled. Instead, the ALE algorithm presented

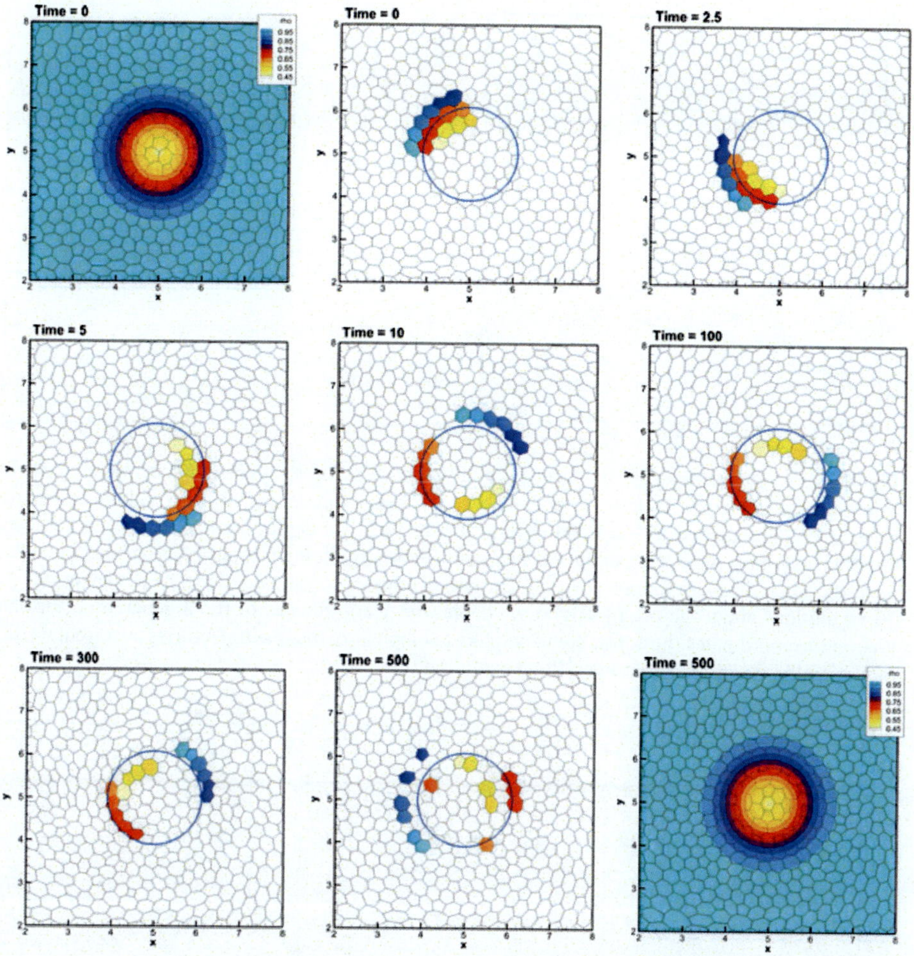

Fig. 5 Stationary rotating vortex solved with our fourth order ALE-DG scheme. Density contours at $t = 0$ and $t = 500$ and position of a bunch of highlighted elements at different times. Note that the solution is well preserved for more than *eighty* complete rotation periods of the yellow elements and generator trajectories are perfectly circular

in this work, not only deals with topology changes without accuracy losses, but in fact the mesh motion allows to gradually optimize the shape of the elements with respect to the flow field. This gradual optimization procedure, translates into lower errors at large times with respect to the Eulerian scheme, for which the mesh is fixed to its initial generic configuration. We refer also to Fig. 7 for a visual illustration of the different mesh motion approaches.

Finally, we would like to emphasize that in Table 1 we show the numerical errors obtained at large computational times ($t = 4$ and $t = 10$), when computations have been carried out for thousands of timesteps and thousands of *crazy* sliver elements have appeared (the total number is indicated in the Table), hence showing that the

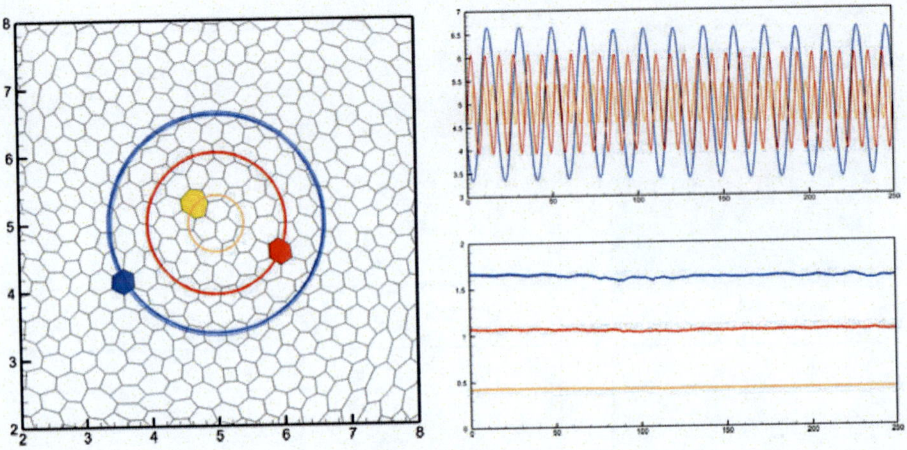

Fig. 6 Stationary rotating vortex solved with our fourth order P_3 ALE-DG scheme on a moving Voronoi-type mesh of 957 elements with dynamical change of connectivity and with the generators trajectories computed with fourth order of accuracy. Left: we depict the trajectories (in Cartesian coordinates) of the generators of 3 mesh elements (those highlighted respectively in blue, violet and red) from time $t = 0$ up to time $t = 250$. During this time interval the red mesh element completes 30 revolutions about the origin. Right: we depict the y coordinates of the 3 generators (top) and their radial coordinates (bottom). We would like to emphasize that the trajectories are circular (their radius is almost constant) for a very long evolution time

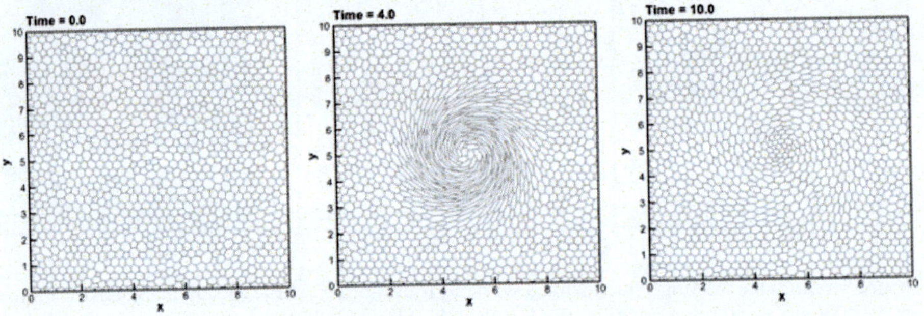

Fig. 7 Stationary vortex test case. We show here an example of a mesh employed for the convergence test case of Sect. 4.1.1 and of its evolution due to different Lagrangian schemes. In particular, on the left we show the initial mesh, in the middle the mesh obtained with a standard direct ALE scheme with fixed topology at time $t = 4$, i.e. just before the simulation terminates due to mesh tangling, and on the right the mesh obtained at time $t = 10$ with our ALE algorithm dealing with topology changes, which has gradually adapted to the fluid flow, optimizing the element shapes and allowing an increased precision for the DG scheme

numerical method is genuinely high order accurate also when sliver elements are present.

Table 1 Stationary vortex test case with final time $t = 4$ and $t = 10$. We report here the order of convergence, on the variable ρ in the L_2 norm, for our DG scheme up to order 4 in the Eulerian case (left), for a standard Lagrangian method with fixed topology (middle) and for our ALE scheme with topology changes (right). In the last case we also report the total number of *crazy* sliver elements that have been originated during the simulations: the high order of convergence is maintained also when many sliver elements appear in the mesh. At large times ($t = 10$), the ALE scheme with topology changes, produces numerical errors that are comparable or smaller than those obtained with the corresponding Eulerian method on a fixed mesh

		Eulerian			ALE fixed		ALE+sliver				
		t = 4	t = 10		t = 4		t = 4		t = 10		
	h	$\epsilon(\rho)$	$\epsilon(\rho)$	$\mathcal{O}(L_2)$	$\epsilon(\rho)$	$\mathcal{O}(L_2)$	Sliv	$\epsilon(\rho)$	Sliv	$\epsilon(\rho)$	$\mathcal{O}(L_2)$
$P_1 \rightarrow \mathcal{O}2$	5.9E-1	4.7E-2	8.3E-2	–	1.4E-1	–	111	4.8E-2	293	9.0E-2	–
	4.4E-1	2.8E-2	4.6E-2	2.0	1.1E-1	1.0	192	2.4E-2	514	4.3E-2	2.6
	2.9E-1	1.0E-2	1.8E-2	2.4	4.6E-2	2.1	420	9.3E-3	1119	1.5E-2	2.6
	2.2E-1	4.9E-3	8.0E-3	2.8	2.4E-2	2.3	789	4.6E-3	2111	6.5E-3	2.9
$P_2 \rightarrow \mathcal{O}3$	5.9E-1	6.7E-3	1.1E-2	–	2.7E-2	–	97	7.3E-3	277	1.0E-2	–
	4.4E-1	2.8E-3	3.9E-3	3.6	1.3E-2	2.5	181	2.9E-3	498	3.2E-3	4.0
	2.9E-1	9.6E-4	1.1E-3	3.3	6.8E-3	1.6	401	9.7E-4	1066	9.0E-4	3.2
	2.2E-1	4.0E-4	4.1E-4	3.4	3.3E-3	2.5	745	4.3E-4	1981	4.1E-4	2.8
$P_3 \rightarrow \mathcal{O}4$	1.7E-0	5.6E-2	8.0E-2	–	5.8E-2	–	2	5.3E-2	4	6.8E-2	–
	1.1E-0	1.2E-2	2.2E-2	3.2	2.8E-2	1.8	10	1.3E-2	41	1.9E-2	3.2
	8.7E-1	4.7E-3	6.6E-3	4.4	2.0E-2	1.3	36	5.8E-3	110	6.1E-3	4.2
	5.9E-1	1.1E-3	1.3E-3	4.2	5.4E-3	3.3	93	1.1E-3	257	1.3E-3	3.9

4.2 Sedov Explosion Problem

This test problem is a classic benchmark in the literature [43] and describes the evolution of a strong blast wave that is generated at the origin $\mathbf{O} = (x, y) = (0, 0)$ of the computational domain $\Omega(0) = [0; 1.2] \times [0; 1.2]$. The difficulty of this benchmark is mainly due to the near zero pressure outer state that may induce positivity-preservation problems. An exact solution based on self-similarity arguments is available from [60].

The initial condition consists in a uniform density $\rho_0 = 1$ and a near zero pressure p_0 imposed everywhere except in the cell V_{or} containing the origin $\mathbf{O}$ where it is given by

$$p_{or} = (\gamma - 1)\rho_0 \frac{E_{tot}}{|V_{or}|}, \quad \text{with } E_{tot} = 0.979264, \tag{8}$$

with E_{tot} being the total energy concentrated in the cell containing the coordinate $\mathbf{x} = \mathbf{0}$. We set $p_0 = 10^{-9}$ and solve this numerical test with a fourth order P_3 DG

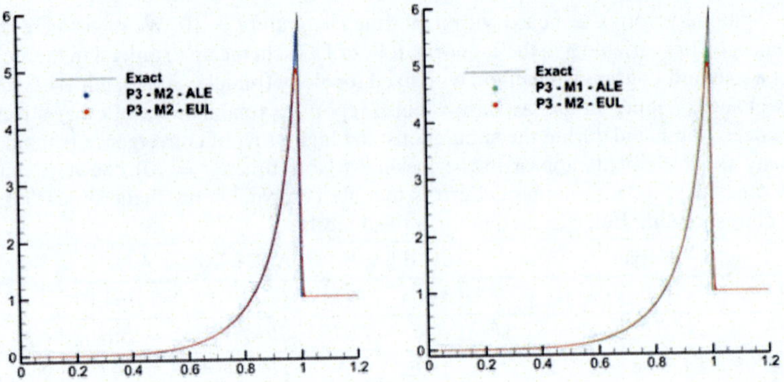

Fig. 8 Sedov explosion problem. Comparison between the exact solution (black), the solution obtained with a fourth order Eulerian P_3 DG method on the fine mesh $M2$ (red) and with our P_3 ALE-DG scheme both on $M2$ (blue) and $M1$ (green). Our ALE scheme is more accurate than the Eulerian one even using coarser meshes

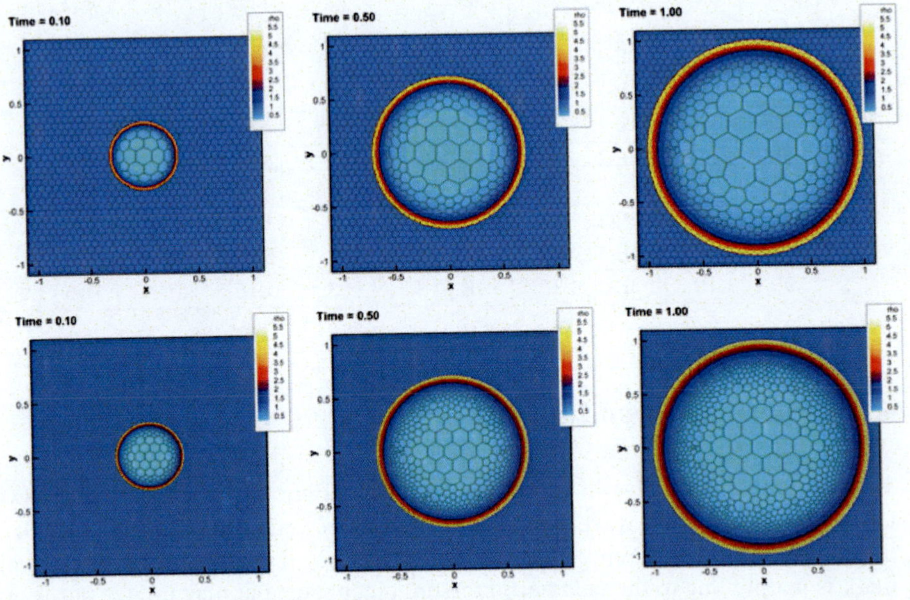

Fig. 9 Sedov explosion problem. In this figure we show the density evolution and the corresponding mesh movement at different output times computed with our P_3 ALE-DG scheme on the mesh $M1$ (top) and $M2$ (bottom)

scheme; we employ a coarse mesh $M1$ made of 1345 polygonal cells and a finer mesh $M2$ of 6017 polygonal elements.

The density profiles are shown in Fig. 8 for various output times $t = 0.1, 0.5, 1.0$. The obtained results are in perfect agreement with the reference solution and the symmetry is very good despite using an unstructured grid, as opposed to a regular one built in polar coordinates. Also, one can note that the regularization procedure

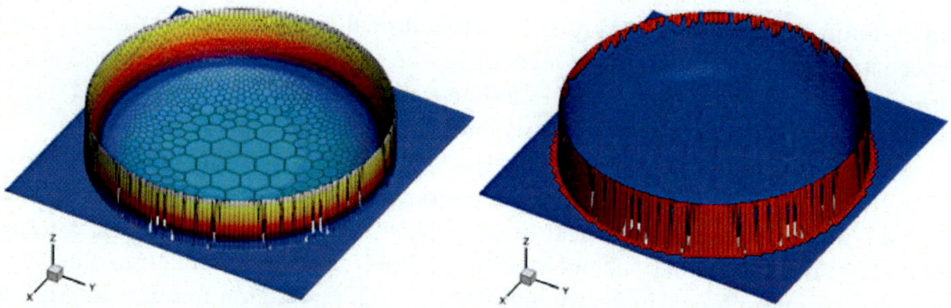

Fig. 10 Sedov explosion problem. $3D$ density profile computed with our P_3 ALE-DG scheme on $M2$. In particular, on the right, we highlight in red the so-called *troubled* cells marked by our detector on which the *a posteriori* FV limiter has been employed. We make use of this image to further emphasize the robustness of our ALE schemes with topology changes also in the presence of strong shock waves and near-zero pressure outer states

applied to the mesh elements does not compromise the natural expansion of the central cells expected in such an explosion problem, as can be noticed in Fig. 9. Moreover, one can refer to Fig. 8 for a comparison between our numerical solution (scatter plot) and the exact solution: the position of the shock wave and the density peak are perfectly captured.

In particular, we have chosen this test case in order to emphasize that Lagrangian schemes show a superior resolution w.r.t. Eulerian ones even when both are compared at very high-order of accuracy, and furthermore that our direct ALE scheme results more accurate then the Eulerian method, even on a mesh ($M1$) that is coarser by a factor of two with respect to the finer mesh $M2$.

Finally, we refer to Fig. 10 for the behavior of our *a posteriori* sub-cell finite volume limiter, which activates only where the shock wave is located and is able to avoid any spurious oscillations or positivity problems, as can be noticed from the precise $3D$ density profile shown in Fig. 10.

4.3 Traveling Sod-type Explosion Problem

The explosion problem can be seen as a multidimensional extension of the classical Sod test case. Here, we consider as computational domain a square of dimension $[-1.1; 1.1] \times [-1.1; 1.1]$ covered with a mesh made of 4105 Voronoi-type elements, and the initial condition is composed of two different states, separated by a discontinuity at radius $r_d = 0.5$

$$\begin{cases} \rho_L = 1, \quad \mathbf{u}_L, \quad p_L = 1, & \|\mathbf{x}\| \le r_d \\ \rho_R = 0.125, \quad \mathbf{u}_R, \quad p_R = 0.1, & \|\mathbf{x}\| > r_d. \end{cases} \tag{9}$$

In addition, we aim at capturing the evolution of this explosion over a very high speed moving background (much higher than the speed of sounds): we impose $\mathbf{u}_L = \mathbf{u}_R = 40$, so that at the final simulation time $t_f = 0.25$ the square $[-1; 1] \times [-1; 1]$ will have been displaced by 5 times its initial size.

We would like to underline that this test problem involves three different waves, therefore it allows each ingredient of our Lagrangian scheme to be properly checked. Indeed, we have *(i)* one cylindrical shock wave that is running towards the external boundary: our high-order scheme does not exhibit spurious oscillations thanks to the *a posteriori* sub-cell finite volume limiter; *(i)* a rarefaction fan traveling in the opposite direction, which is well captured thanks to the high-order of accuracy of the DG scheme; and *(iii)* an outward-moving contact wave, which is well resolved thanks to the Lagrangian nature of our scheme, in which, since the mesh moves together with the fluid flow, we can introduce a minimal dissipation when computing approximate Riemann fluxes.

In addition, the high speed moving background allows to show the *translational invariance* property of the Lagrangian schemes that indeed perfectly captures the three waves even when the explosion solution is moving at high speed, while the

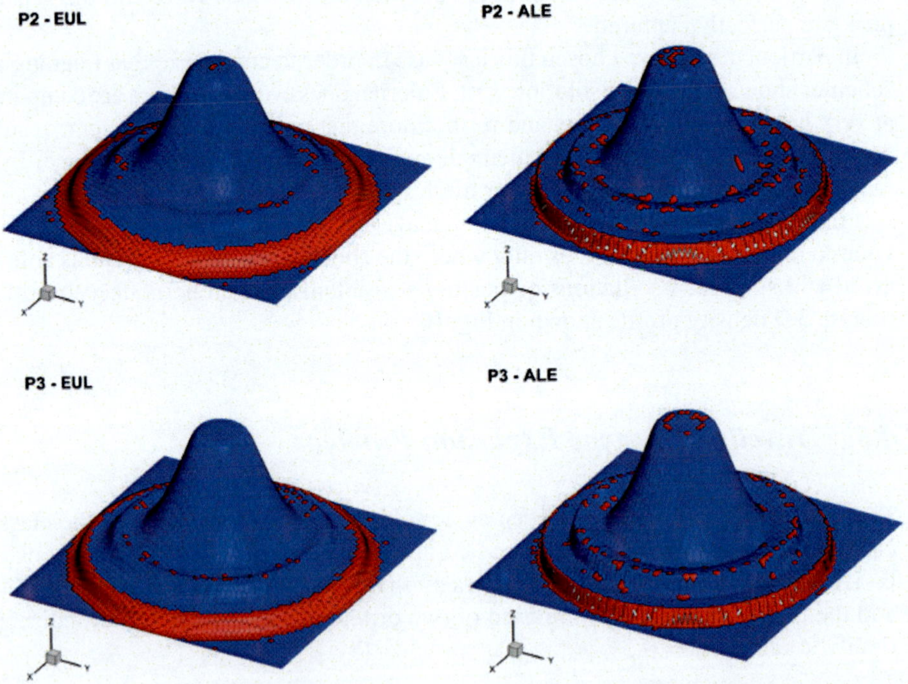

Fig. 11 Traveling Sod explosion problem. $3D$ density profile (z-axis) and limiter activation (red cells), over a domain located in $[8.9; 11.1] \times [-1.1; 1.1]$ at the final time $t_f = 0.25$, obtained with P_2 and P_3 ADER-DG schemes run on a fixed Eulerian mesh (left) and our direct ALE framework with topology changes (right). The difference on the numerical dissipation between the Eulerian and the Lagrangian schemes is quite evident. We clarify that these results are obtained solving the classical Sod explosion problem over a high speed moving background

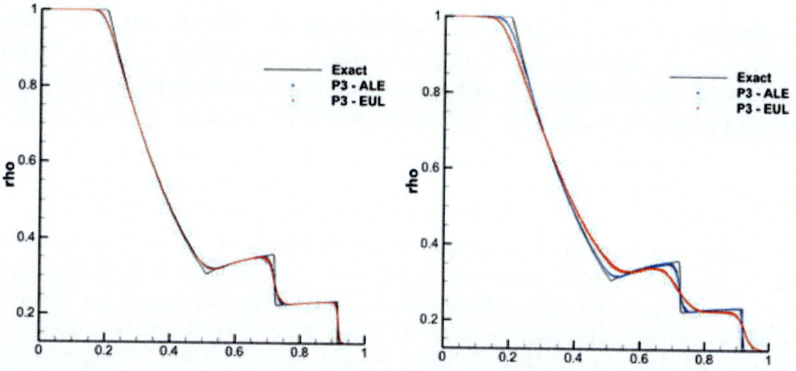

Fig. 12 Sod explosion problem: fixed background (left) and high speed traveling background (right). Comparison between the P_3 DG schemes on fixed Eulerian meshes (red) and in the moving-mesh ALE framework (blue). This numerical results clearly explain that the Lagrangian schemes allow to obtain minimally dissipative results not only in a vanishing background flow, but even in a high speed one, and therefore that Lagrangian methods discretely preserve the Galilean invariance of the equations. On the contrary the influence of strong background flows on the solution obtained with Eulerian schemes is immediately apparent

Eulerian scheme is heavily affected by the increased numerical dissipation. Numerical evidence of the above statements can be found in Fig. 11; moreover, in Fig. 12 we show that for this mild explosion it is really the background motion that requires the use of Lagrangian schemes, which, while still useful, would be instead not fundamental on a fixed background.

Finally, we want to remark that, despite the very high dissipation associated with the high base convective speed, the overall symmetry of the solution, even in the Eulerian case, is not entirely compromised, thanks to the use of polygonal elements (see [9] for further discussion on the benefits of adopting polygonal meshes).

5 Conclusion and Outlook

The accuracy of our results clearly show that the new combination of very high-order schemes with regenerated meshes, that allow topology changes, may open new perspectives in the fundamental research field of Lagrangian methods.

We would like to remark that the chosen simple test cases can be seen as prototypes of classical difficulties in astrophysical applications. Indeed, we have proposed here a method able to deal with long time simulation of vortical phenomena, as those necessary for the study of gas clouds evolving around black holes and neutron stars, and events, like explosions or interactions with near zero pressure states, occurring in superposition with high speed background flows, as for supersonic or relativistic jets originating from proto-planetary nebulae, binary stars or nuclei of active galaxies.

Future developments of this work will mainly concern the improvements of its robustness and effectiveness through mesh optimization and smoothing techniques [2, 18, 50, 56, 66] and structure preserving algorithms [1, 12, 22, 23, 33, 37, 41] so that future applications will effectively target in particular supersonic flows in aerodynamics [3, 64] and astrophysics [30, 52, 55], as well as fluid-structure interaction problems [4, 24, 40].

Acknowledgements E. Gaburro is member of the CARDAMOM team at the Inria center of the University of Bordeaux in France and S. Chiocchetti is member of the INdAM GNCS group in Italy. E. Gaburro gratefully acknowledges the support received from the European Union's Horizon 2020 Research and Innovation Programme under the Marie Skłodowska-Curie Individual Fellowship *SuPerMan*, grant agreement No. 101025563. S. Chiocchetti acknowledges the support obtained by the Deutsche Forschungsgemeinschaft (DFG) via the project DROPIT, grant no. GRK 2160/2, and from the European Union's Horizon Europe Research and Innovation Programme under the Marie Skłodowska-Curie Postdoctoral Fellowship MoMeNTUM, grant agreement No. 101109532.

References

1. Abgrall, R., Öffner, P., Ranocha, H.: Reinterpretation and extension of entropy correction terms for residual distribution and discontinuous galerkin schemes: Application to structure preserving discretization. J. Comput. Phys. **453**, 110955 (2022)
2. Anderson, R.W., Dobrev, V.A., Kolev, T.V., Rieben, R.N., Tomov, V.Z.: High-order multi-material ale hydrodynamics. SIAM J. Sci. Comput. **40**(1), B32–B58 (2018)
3. Antoniadis, A., Tsoutsanis, P., Drikakis, D.: High-order schemes on mixed-element unstructured grids for aerodynamic flows. In: 42nd AIAA Fluid Dynamics Conference and Exhibit, p. 2833 (2012)
4. Basting, S., Quaini, A., Canic, S., Glowinski, R.: Extended ale method for fluid-structure interaction problems with large structural displacements. J. Comput. Phys. **331**, 312–336 (2017)
5. Bo, W., Shashkov, M.J.: Adaptive reconnection-based arbitrary Lagrangian Eulerian method. J. Comput. Phys. **299**, 902–939 (2015)
6. Boscheri, W., Dumbser, M., Zanotti, O.: High order cell-centered Lagrangian-type finite volume schemes with time-accurate local time stepping on unstructured triangular meshes. J. Comput. Phys. **291**, 120–150 (2015)
7. Boscheri, W., Chiocchetti, S., Peshkov, I.: A cell-centered implicit-explicit lagrangian scheme for a unified model of nonlinear continuum mechanics on unstructured meshes. J. Comput. Phys. **451**, 110852 (2022)
8. Boscheri, W., Dumbser, M.: Arbitrary-lagrangian-eulerian one-step weno finite volume schemes on unstructured triangular meshes. Commun. Comput. Phys. **14**(5), 1174–1206 (2013)
9. Boscheri, W., Dumbser, M., Gaburro, E.: Continuous finite element subgrid basis functions for discontinuous galerkin schemes on unstructured polygonal voronoi meshes. Commun. Comput. Phys. **32**(1), 259–298 (2022)
10. Caramana, E.J., Shashkov, M.J.: Elimination of artificial grid distorsion and hourglass-type motions by means of Lagrangian subzonal masses and pressures. J. Comput. Phys. **142**, 521–561 (1998)
11. Carré, G., Del Pino, S., Després, B., Labourasse, E.: A cell-centered Lagrangian hydrodynamics scheme on general unstructured meshes in arbitrary dimension. J. Comput. Phys. **228**, 5160–5183 (2009)
12. Castro, M., Gallardo, J.M., López-GarcÍa, J.A., Parés, C.: Well-balanced high order extensions of godunov's method for semilinear balance laws. SIAM J. Numer. Anal. **46**(2), 1012–1039 (2008)

13. Chiocchetti, S., Müller, C.: A solver for stiff finite-rate relaxation in baer–nunziato two-phase flow models. In: Droplet Interactions and Spray Processes, pp. 31–44. Springer (2020)
14. Chiocchetti, S., Peshkov, I., Gavrilyuk, S., Dumbser, M.: High order ader schemes and glm curl cleaning for a first order hyperbolic formulation of compressible flow with surface tension. J. Comput. Phys. **426**, 109898 (2021)
15. Clain, S., Diot, S., Loubère, R.: A high-order finite volume method for systems of conservation laws-multi-dimensional optimal order detection (MOOD). J. Comput. Phys. **230**(10), 4028–4050 (2011)
16. Després, B.: Numerical methods for Eulerian and Lagrangian conservation laws. Birkhäuser (2017)
17. Dobrev, V.A., Ellis, T.E., Kolev, T.V., Rieben, R.N.: High order curvilinear finite elements for axisymmetric Lagrangian hydrodynamics. Comput. Fluids**83**, 58–69 (2013)
18. Dobrev, V., Knupp, P., Kolev, T., Mittal, K., Rieben, R., Tomov, V.: Simulation-driven optimization of high-order meshes in ale hydrodynamics. Comput. Fluids **208**, 104602 (2020)
19. Dumbser, M.: Arbitrary-Lagrangian-Eulerian ADER-WENO finite volume schemes with time-accurate local time stepping for hyperbolic conservation laws. Comput. Methods Appl. Mech. Eng. **280**, 57–83 (2014)
20. Dumbser, M., Balsara, D.S.: A new efficient formulation of the HLLEM Riemann solver for general conservative and non-conservative hyperbolic systems. J. Comput. Phys. **304**(C), 275–319 (2016)
21. Dumbser, M., Balsara, D.S., Toro, E.F., Munz, C.-D.: A unified framework for the construction of one-step finite volume and discontinuous galerkin schemes on unstructured meshes. J. Comput. Phys. **227**(18), 8209–8253 (2008)
22. Dumbser, M., Chiocchetti, S., Peshkov, I.: On numerical methods for hyperbolic pde with curl involutions. In: Continuum Mechanics, Applied Mathematics and Scientific Computing: Godunov's Legacy, pp. 125–134. Springer (2020)
23. Dumbser, M., Fambri, F., Gaburro, E., Reinarz, A.: On GLM curl cleaning for a first order reduction of the CCZ4 formulation of the Einstein field equations. J. Comput. Phys. 109088 (2019)
24. Dürrwächter, J., Kurz, M., Kopper, P., Kempf, D., Munz, C.-D., Beck, A.: An efficient sliding mesh interface method for high-order discontinuous Galerkin schemes. Comput. Fluids **217**, 104825 (2021)
25. Fambri, F., Dumbser, M., Zanotti, O.: Space-time adaptive ADER-DG schemes for dissipative flows: Compressible Navier-Stokes and resistive MHD equations. Comput. Phys. Commun. **220**, 297–318 (2017)
26. Gabriel, A.-A., Li, D., Chiocchetti, S., Tavelli, M., Peshkov, I., Romenski, E., Dumbser, M.: A unified first-order hyperbolic model for nonlinear dynamic rupture processes in diffuse fracture zones. Philosop. Trans. R. Soc. A **379**(2196), 20200130 (2021)
27. Gaburro, E.: A unified framework for the solution of hyperbolic pde systems using high order direct arbitrary-Lagrangian-Eulerian schemes on moving unstructured meshes with topology change. Arch. Comput. Methods Eng. **28**(3), 1249–1321 (2021)
28. Gaburro, E., Boscheri, W., Chiocchetti, S., Klingenberg, C., Springel, V., Dumbser, M.: High order direct arbitrary-Lagrangian-Eulerian schemes on moving voronoi meshes with topology changes. J. Comput. Phys. **407**, 109167 (2020)
29. Gaburro, E., Castro, M.J., Dumbser, M.: A well balanced diffuse interface method for complex nonhydrostatic free surface flows. Comput. Fluids **175**, 180–198 (2018)
30. Gaburro, E., Castro, M.J., Dumbser, M.: A well balanced finite volume scheme for general relativity. SIAM J. Sci. Comput. **43**(6), B1226–B1251 (2021)
31. Gaburro, E., Dumbser, M.: A posteriori subcell finite volume limiter for general PNPM schemes: applications from gasdynamics to relativistic magnetohydrodynamics. J. Sci. Comput. **86**(3), 1–41 (2021)
32. Gaburro, E., Dumbser, M., Castro, M.J.: Direct Arbitrary-Lagrangian-Eulerian finite volume schemes on moving nonconforming unstructured meshes. Comput. Fluids **159**, 254–275 (2017)

33. Gaburro, E., Öffner, P., Ricchiuto, M., Torlo, D.: High order entropy preserving ADER-DG schemes. Appl. Math. Comput. **440**, 127644 (2023)
34. Godunov, S.K.: Finite difference methods for the computation of discontinuous solutions of the equations of fluid dynamics. Math. USSR: Sbornik **47**, 271–306 (1959)
35. Guermond, J.-L., Nazarov, M., Popov, B., Tomas, I.: Second-order invariant domain preserving approximation of the euler equations using convex limiting. SIAM J. Sci. Comput. **40**(5), A3211–A3239 (2018)
36. Hu, C., Shu, C.W.: A high-order weno finite difference scheme for the equations of ideal magnetohydrodynamics. J. Comput. Phys. **150**, 561–594 (1999)
37. Käppeli, R., Mishra, S.: Well-balanced schemes for the Euler equations with gravitation. J. Comput. Phys. **259**, 199–219 (2014)
38. Kemm, F., Gaburro, E., Thein, F., Dumbser, M.: A simple diffuse interface approach for compressible flows around moving solids of arbitrary shape based on a reduced Baer-Nunziato model. Comput. Fluids **204**, 104536 (2020)
39. Kenamond, M., Kuzmin, D., Shashkov, M.: A positivity-preserving and conservative intersection-distribution-based remapping algorithm for staggered ale hydrodynamics on arbitrary meshes. J. Comput. Phys. **435**, 110254 (2021)
40. Kikinzon, E., Shashkov, M., Garimella, R.: Establishing mesh topology in multi-material cells: Enabling technology for robust and accurate multi-material simulations. Comput. Fluids **172**, 251–263 (2018)
41. Klingenberg, C., Puppo, G., Semplice, M.: Arbitrary order finite volume well-balanced schemes for the euler equations with gravity. SIAM J. Sci. Comput. **41**(2), A695–A721 (2019)
42. Liu, W., Cheng, J., Shu, C.W.: High order conservative Lagrangian schemes with Lax-Wendroff type time discretization for the compressible Euler equations. J. Comput. Phys. **228**, 8872–8891 (2009)
43. Loubère, R., Maire, P.H., Váchal, P.: 3D staggered Lagrangian hydrodynamics scheme with cell-centered Riemann solver-based artificial viscosity. Int. J. Numer. Methods Fluids **72**, 22–42 (2013)
44. Loubère, R., Maire, P.H., Shashkov, M.J.: ReALE: a reconnection Arbitrary-Lagrangian-Eulerian method in cylindrical geometry. Comput. Fluids **46**, 59–69 (2011)
45. Loubère, R., Maire, P.H., Shashkov, M.J., Breil, J., Galera, S.: ReALE: a reconnection-based arbitrary-Lagrangian-Eulerian method. J. Comput. Phys. **229**, 4724–4761 (2010)
46. Loubère, R., Shashkov, M.J.: A subcell remapping method on staggered polygonal grids for arbitrary-lagrangian-eulerian methods. J. Comput. Phys. **23**, 155–160 (2004)
47. Loubere, R., Dumbser, M., Diot, S.: A new family of high order unstructured mood and ader finite volume schemes for multidimensional systems of hyperbolic conservation laws. Commun. Comput. Phys. **16**(3), 718–763 (2014)
48. Maire, P.H.: A unified sub-cell force-based discretization for cell-centered Lagrangian hydrodynamics on polygonal grids. Int. J. Numer. Methods Fluids **65**, 1281–1294 (2011)
49. Morgan, N.R., Archer, B.J.: On the origins of lagrangian hydrodynamic methods. Nucl. Technol. **207**(sup1), S147–S175 (2021)
50. Morgan, N.R., Liu, X., Burton, D.E.: Reducing spurious mesh motion in lagrangian finite volume and discontinuous Galerkin hydrodynamic methods. J. Comput. Phys. **372**, 35–61 (2018)
51. Munz, C.D.: On Godunov-type schemes for Lagrangian gas dynamics. SIAM J. Numer. Anal. **31**, 17–42 (1994)
52. Olivares, H., Peshkov, I.M., Most, E.R., Guercilena, F.M., Papenfort, L.J.: New first-order formulation of the Einstein equations exploiting analogies with electrodynamics. Phys. Rev. D **105**(12), 124038 (2022)
53. Parés, C.: Numerical methods for nonconservative hyperbolic systems: a theoretical framework. SIAM J. Numer. Anal. **44**(1), 300–321 (2006)
54. Peshkov, I., Dumbser, M., Boscheri, W., Romenski, E., Chiocchetti, S., Ioriatti, M.: Simulation of non-newtonian viscoplastic flows with a unified first order hyperbolic model and a structure-preserving semi-implicit scheme. Comput. Fluids **224**, 104963 (2021)

55. Peshkov, I., Romenski, E., Dumbser, M.: Continuum mechanics with torsion. Continuum Mech. Thermodyn **31**(5), 1517–1541 (2019)
56. Re, B., Dobrzynski, C., Guardone, A.: An interpolation-free ale scheme for unsteady inviscid flows computations with large boundary displacements over three-dimensional adaptive grids. J. Comput. Phys. **340**, 26–54 (2017)
57. Ruppert, J.: A new and simple algorithm for quality 2-dimensional mesh generation. Proceedings of the 4th ACM-SIAM Symposium on Discrete Algorithms, pp. 83–92 (1993)
58. Rusanov, V.V.: Calculation of Interaction of Non-Steady Shock Waves with Obstacles. J. Comput. Math. Phys. USSR **1**, 267–279 (1961)
59. Scovazzi, G.: Lagrangian shock hydrodynamics on tetrahedral meshes: a stable and accurate variational multiscale approach. J. Comput. Phys. **231**, 8029–8069 (2012)
60. Sedov, L.I.: Similarity and Dimensional Methods in Mechanics. Academic Press, New York (1959)
61. Springel, V.: E pur si muove: Galilean-invariant cosmological hydrodynamical simulations on a moving mesh. Monthly Notices R. Astronom. Soc. **401**(2), 791–851 (2010)
62. Tavelli, M., Chiocchetti, S., Romenski, E., Gabriel, A.-A., Dumbser, M.: Space-time adaptive ader discontinuous Galerkin schemes for nonlinear hyperelasticity with material failure. J. Comput. Phys. **422**, 109758 (2020)
63. Toro, E.F.: Riemann Solvers and Numerical Methods for Fluid Dynamics: A Practical Introduction. Springer Science & Business Media (2013)
64. Tsoutsanis, P., Kokkinakis, I.W., Könözsy, L., Drikakis, D., Williams, R.J.R.,Youngs, D.L.: Comparison of structured-and unstructured-grid, compressible and incompressible methods using the vortex pairing problem. Comput. Methods Appl. Mech. Eng. **293**, 207–231 (2015)
65. von Neumann, J., Richtmyer, R.D.: A method for the calculation of hydrodynamics shocks. J. Appl. Phys. **21**, 232–237 (1950)
66. Wang, L., Persson, P.-O.: A high-order discontinuous Galerkin method with unstructured space-time meshes for two-dimensional compressible flows on domains with large deformations. Comput. Fluids **118**, 53–68 (2015)
67. Wilkins, M.L.: Calculation of elastic-plastic flow. Methods Comput. Phys. 3 (1964)

Overview on Uncertainty Quantification in Traffic Models via Intrusive Method

Elisa Iacomini

Abstract We consider traffic flow models at different scales of observation. Starting from the well known hierarchy between microscopic, kinetic and macroscopic scales, we will investigate the propagation of uncertainties through the models using the stochastic Galerkin approach. Connections between the scales will be presented in the stochastic scenario and numerical simulations will be performed.

Keywords Traffic flow · Stochastic Galerkin · Hierarchical models · BGK models · Aw-Rascle-Zhang model · Follow-the-leader model

1 Introduction

In the last decades, several traffic models have been developed and investigated at different spacial and temporal scales.

Starting from the natural idea of tracking every single vehicle, several follow-the leader models grew up for computing positions, velocities and accelerations of each car by means of systems of ordinary differential equations (ODEs) [3, 22]. Zooming out, the approaches vary from kinetic [16, 27, 33], which provides a statistical description of traffic taking into account cars-to-cars interactions and mass distribution of traffic, to macroscopic fluid-dynamics [2, 20, 25], focusing on average quantities by means of partial differential equations (PDEs), in particular conservation laws.

Hierarchies and links between the those scales have been widely studied, in particular mathematical connections between different types of models, especially microscopic and kinetic ones which converge to macroscopic models in certain suitable limit, [1, 8, 14, 16, 34].

E. Iacomini (✉)
Institut für Geometrie und Praktische Mathematik RWTH Aachen, Templergraben 55,52056 Aachen, Germany
e-mail: iacomini@igpm.rwth-aachen.de

G. Albi et al. (eds.), *Advances in Numerical Methods for Hyperbolic Balance Laws and Related Problems*, SEMA SIMAI Springer Series 32,
https://doi.org/10.1007/978-3-031-29875-2_6

The choice of the scale of observation mainly depends on the aim of the modeling, i.e. forecast traffic evolution on highways or traffic jams at a junction, on the number of involved vehicles and so on. Unfortunately, due to the particular structure of the corresponding models, they are usually studied in rather separate works.

Therefore, one of the main focus of this work is to provide a uniform setting to study uncertainty quantification in traffic at different scales.

Indeed, in recent works it has been pointed out how traffic is exposed to the presence of various sources and types of uncertainty, also at different scales of observation [35, 36]. For example, it is well known that real data may be affected by noise and errors in the measurements, or that the reaction time of drivers and cars is not a deterministic event. Therefore, investigating how to include and to model the uncertainty at different scales in a consistent way, and how it affects the existing hierarchy are the main goals of this work.

We will introduce the uncertainty in the initial data at a microscopic, mesoscopic and macroscopic scale respectively and we will analyze the obtained stochastic models. In order to study the propagation of input uncertainty through the models, several approaches have been proposed in the literature and can be classified in non-intrusive, e.g. based on sampling (Monte-Carlo) or based on collocation [21], and intrusive methods [12, 13], where the stochastic quantities are described by a series of orthogonal functions, known as generalized polynomial chaos (gPC) expansion [6, 24, 37, 38]. The idea is then to substitute the series in the governing equations and project, using a Galerkin projection, in order to obtain a system of deterministic coefficients. From the coefficients, stochastic moments can be recovered at each point in space and time without performing the simulation several times, as in the Monte-Carlo simulations. The detailed knowledge of the stochastic moments, allowed by the intrusive methods both in space and time, led us to follow this approach, namely the stochastic Galerkin method.

Many challenges arise here, since some desired properties of the original system are not necessarily transferred to the intrusive formulation [7]. In particular, we refer to the hyperbolicity in the macroscopic models, [11, 29]. In fact, the deterministic Jacobian of the projected system differs from the random Jacobian of the original system. For more details we refer to [10]. Therefore, applying the intrusive stochastic Galerkin method to hyperbolic equations is an active field of research [18, 19, 23].

The paper is organized as follows: uncertainty and the stochastic Galerkin method are introduced in Sect. 2. In Sect. 3 stochastic microscopic models at first and second order are presented. Section 4 deals with the mesoscopic scale, i.e. stochastic kinetic models are considered, in particular the BGK model. In Sect. 5 the macroscopic models are described and recovered from the previous ones and the main properties are illustrated. Numerical tests will be provided. We conclude the work with some comments and an overview on future directions.

2 Stochastic Galerkin Approach

In order to describe the presence of uncertainty, we introduce a (possibly multi-dimensional) random variable ω. Let ω be defined on the probability space $(\Omega_\omega, \mathcal{F}(\Omega), \mathbb{P})$. Then we denote by $\xi = \xi(\omega) : \Omega_\omega \to \Omega \subset \mathbb{R}^d$ a (possibly d-dimensional) real-valued random variable. Assume further that ξ is absolutely continuous with respect to the Lebesgue measure on $\mathbb{R}^d$ and denote by $p_\Xi(\xi) : \Omega \to \mathbb{R}_+$ the probability density function of ξ. For simplicity we assume that the uncertainty enters only in the initial data, so that it affects the initial configuration.

To study the arising stochastic models, we briefly recall the intrusive approach we will employ, namely the stochastic Galerkin method introduced by [12].

Here, a random field $u(t, \xi)$, namely the stochastic input, can be expressed by a spectral expansion [24] under the assumption of being sufficiently regular and in particular $L^2_p(\Omega)$

$$u_i(t, \xi) = \sum_{k=0}^{\infty} \hat{u}_k(t)\phi_k(\xi) \tag{1}$$

where $\phi_k \in L^2(\Omega, p_\Xi)$ are basis functions, typically chosen orthonormal with respect to the weighted scalar product, and $\{\hat{u}_k(t)\}_{k=0}^{\infty}$ is a set of coefficients:

$$\hat{u}_k(t) = \int_\Omega u(t, \xi)\phi_k(\xi)p_\Xi(\xi)d\xi. \tag{2}$$

The previous expansion is truncated at K to obtain an approximation with $K + 1$ moments. The projection of $u(t, \cdot)$ to the span of the $K + 1$ base functions is denoted by

$$G_K(u(t, \cdot))(\xi) := \sum_{k=0}^{K} \hat{u}_k(t)\phi_k(\xi) \quad a.e.\ \xi \in \Omega. \tag{3}$$

The expansion (1) is called generalized polynomial chaos expansion (gPC) [24]. Under mild condition on the probability measure the truncated expansion converges in the sense $||G_K(u(t, \cdot))(\xi) - u(t, \cdot)|| \to \infty$ for $K \to \infty$ as shown in [6]. A weak approximation of stochastic systems is obtained substituting the truncated expansion (3) into the system itself and projecting the resulting expression onto the subspace of $L^2(\Omega, \mathbb{P})$ spanned by the basis $\{\phi_k(\xi)\}_{k=0}^{K}$.

Moreover let us recall, as in [11, 24] the Galerkin product for any finite $K > 0$ and any $u, z \in L^2(\Omega, p_\Xi)$, $\hat{u} = (\hat{u}_i)_{i=0}^{K}$, $\hat{z} := (\hat{z}_i)_{i=0}^{K}$ and for all $i, j, \ell = 0, \ldots, K$:

$$\mathcal{G}_K[u,z](t,\xi) := \sum_{k=0}^{K} (\hat{u} * \hat{z})_k(t)\phi_k(\xi),$$

$$(\hat{u} * \hat{z})_k(t) := \sum_{i,j=0}^{K} \hat{u}_i(t)\hat{z}_j(t)\mathcal{M}_\ell,$$

$$(\mathcal{M}_\ell)_{i,j} := \int_\Omega \phi_i(\xi)\phi_j(\xi)\phi_\ell(\xi)p_\Xi(\xi)d\xi.$$

Note that $\mathcal{M}_\ell$ is a symmetric matrix of dimension $(K+1) \times (K+1)$ for any fixed $\ell \in \{0, \ldots, K\}$. Moreover, we have $\hat{u} * \hat{z} = \mathcal{P}(\hat{u})\hat{z}$ for $\mathcal{P} \in \mathbb{R}_+^{K+1\times K+1}$ and $\hat{u} \in \mathbb{R}_+^{K+1}$ defined by

$$\mathcal{P}(\hat{u}) := \sum_{\ell=0}^{K} \hat{u}_\ell \mathcal{M}_\ell. \tag{4}$$

The Galerkin product is symmetric, but not associative [7, 21, 31]. However, the Galerkin product defined by $\mathcal{G}_K$ is not the only possible projection of the product of random variables u, z on the subspace $span\{\phi_0, \ldots, \phi_K\}$, nevertheless here we stick to this choice.

A challenge occurs since only the gPC modes corresponding to the initial data are known. To determine them for $t > 0$ we derive a differential equation called stochastic Galerkin formulation, that describes their propagation in time and space. Recovering the stochastic Galerkin formulation at different scales of observation is the aim of the following sections.

3 Microscopic Scale

Studying individual vehicles and their interactions is the purpose of microscopic models. Starting from cars' positions and velocities, the trajectory of vehicles is reconstructed by means of dynamical systems, in the form of ODEs systems. We assume that N vehicles are moving along a single-lane infinite road where overtaking is not possible. This means that cars are ordered and have to follow the first one, namely the *leader*. Cars that are not the leader are termed *followers*. We indicate the position $x_i(t)$ and the velocity $v_i(t)$ of each vehicle $i = 1, \ldots, N$ at different time t.

The most simple dynamics can be described by a first order system of ODEs where the velocity is given as a known function $s(\cdot)$, which depends on the distance between the position of two consecutive vehicles, i.e. $s\left(\frac{L}{x_{i+1}-x_i}\right)$ where $\Delta x_i = x_{i+1} - x_i$ and L is the length of the cars. For simplicity we assume all the vehicles have the same length.

However, one of the main difficulties for drivers might be to estimate their distance from the vehicle in front, which would then influence the velocity itself. Thus, we introduce the random variable ξ, as stated in Sect. 2, in the initial data as $\Delta x_i(\xi) =$

$x_{i+1} - x_i + \xi$. The uncertainty then affects also the positions, since the velocity is a function of Δx_i. Therefore we are interested in the evolution of $x_i(t, \xi) : \mathbb{R}^+ \times \Omega$, for $i = 1, \ldots, N$:

$$\begin{cases} \dot{x}_i(t,\xi) = v_i(t,\xi) & i = 1, \ldots N \\ v_i(t,\xi) = s\left(\frac{L}{x_{i+1} - x_i + \xi}\right) & i = 1, \ldots, N-1 \\ v_N = \bar{s}. \end{cases} \tag{5}$$

Note that the leader, the N-th vehicle, has its own dynamics, i.e. an assigned speed value $\bar{s}$.

Following the method presented in Sect. 2, the random field $x_i(t, \cdot)$ can be expressed by a spectral expansion $x_i(t,\xi) = \sum_{k=0}^{\infty} \hat{x}_{i_k}(t)\phi_k(\xi)$, where $\hat{x}_{i_k}$ is the k-coefficient of the vehicle i. The stochastic Galerkin formulation of (5) reads as:

$$\begin{cases} \dot{\hat{x}}_{i_k} = \hat{v}_{i_k} & i = 1, \ldots N \\ \hat{v}_{i_k} = \widehat{s_{i_k}}\left(\frac{L}{\Delta x_i}\right) & i = 1, \ldots, N-1 \\ \hat{v}_N = \bar{s}\; e_1 \end{cases} \tag{6}$$

where $\widehat{s_{i_k}} = \int_\Omega s\left(\frac{L}{x_{i+1} - x_i + \xi}\right)\Phi_k(\xi)p(\xi)d\xi$, $e_1 = (1, 0, \ldots, 0)^T$, since $\bar{s}$ is a deterministic value. Note that the following approximation holds when s is linear: $\widehat{s_{i_k}}\left(\frac{L}{\Delta x_i}\right) \approx s\left(\frac{L}{\Delta \hat{x}_{i_k}}\right)$, where $\Delta \hat{x}_{i_k} = \hat{x}_{i+1_k} - \hat{x}_{i_k}$.

The system (6) is now deterministic with no explicit dependence on the random variable ξ. We will show in Sect. 5.1 how we can recover the macroscopic stochastic model starting from (6).

Furthermore, the acceleration term might be also taken into account, which leads to consider second order models:

$$\begin{cases} \dot{x}_i(t,\xi) = v_i(t,\xi) & i = 1, \ldots N \\ \dot{v}_i(t,\xi) = a(x_{i+1}(t,\xi), x_i(t,\xi), v_{i+1}(t,\xi), v_i(t,\xi)) & i = 1, \ldots, N-1 \\ \dot{v}_N = \bar{a} \end{cases} \tag{7}$$

where $\bar{a}$ describes the dynamics of the leader, independent from the other vehicles. The second equation describes the acceleration which depends on positions and velocities of two consecutive vehicles. Here we consider the formula described in [2, 26], namely $a = C\frac{v_{i+1}(t,\xi) - v_i(t,\xi)}{\Delta x_i^2(t,\xi)} + \frac{A}{t_r}(s(\frac{L}{\Delta x_i(t,\xi)}) - v_i(t,\xi))$. In particular, this choice of the acceleration term takes into account the difference with the velocity of the car in front, weighted with the distance between the vehicles, and the attitude to travel with an optimal velocity given by s.

The stochastic Galerkin approach can be applied to (7) in the same way. Explicitly, the expanded dynamics of system (7) with that particular choice of acceleration function reads as:

$$\begin{cases} \sum_{k=0}^{\infty} \dot{\hat{x}}_{i_k}(t)\phi_k(\xi) = & \sum_{k=0}^{\infty} \hat{v}_{i_k}(t)\phi_k(\xi) \\ \sum_{k=0}^{\infty} \dot{\hat{v}}_{i_k}(t)\phi_k(\xi) = & C\left(\sum_{k=0}^{\infty} \Delta\dot{\hat{x}}_{i_k}\phi_k(\xi)\right) * \left(\sum_{k=0}^{\infty} \Delta\hat{x}_{i_k}\phi_k(\xi)\right)^{-2} + \\ & \frac{A}{t_r}\left(\widehat{s_{i_k}} - \sum_{k=0}^{\infty} \hat{v}_{i_k}(t)\phi_k(\xi)\right) \\ \dot{v}_N = & \bar{a} \end{cases} \tag{8}$$

where $\Delta\dot{\hat{x}}_{i_k} = \hat{v}_{i+1_k} - \hat{v}_{i_k}$. Then, employing the notation introduced above, the projection of (8) corresponds to

$$\begin{cases} \dot{\hat{x}}_{i_k}(t) = \hat{v}_{i_k}(t) & i = 1, \dots N \\ \dot{\hat{v}}_{i_k}(t) = C\left(\mathcal{P}^{-2}(\Delta\hat{x}_{i_k})\Delta\dot{\hat{x}}_{i_k}\right) + \frac{A}{t_r}\left(\widehat{s_{i_k}} - \sum_{k=0}^{\infty} \hat{v}_{i_k}(t)\right) & i = 1, \dots N-1 \\ \dot{v}_N = \bar{a}. \end{cases} \tag{9}$$

4 Mesoscopic Scale

Kinetic traffic flow models provide a statistical description of traffic, taking into account both car-to-car interactions and mass distribution of traffic. Such a mixture is particularly suitable for taking uncertain parameters into account. Therefore several works have been done to study the uncertainty at the mesoscopic scale, starting from [36] to more recent contributions [33–35, 39].

Here we consider a kinetic traffic models class of BGK (Bhatnagar, Gross and Krook [4]) type, which got recently new insights thanks to the works done in [15, 16]. In the former, the derivation of a well-posed model linked to the hierarchy of the scales has been provided in the deterministic framework. The latter instead focuses on the stochastic scenario.

To begin with, we introduce the desired speed $w_i = v_i + h(\rho_i)$, where $h = h(\rho_i) : \mathbb{R}^+ \to \mathbb{R}^+$ is an increasing, differentiable function of the density called hesitation function [9] which satisfies $h(\rho) \geq 0, h'(\rho) \geq 0$. The variable we will consider in the following is the mass distribution function of the flow, namely $g(t, x, w) : \mathbb{R}^+ \times \mathbb{R} \times W \to \mathbb{R}^+$, such that $g dx dw$ gives the number of vehicles at time $t \in \mathbb{R}^+$ with position in $[x, x + dx] \subset \mathbb{R}$ and desired speed in $[w, w + dw] \subset W$. Note that macroscopic variables like density and flux function can be recovered from g, namely $\rho(t, x) = \int_W g(t, x, w) dw$ and $q = \int_W w g(t, x, w) dw$.

Due to the difficulties in estimating the real distribution of vehicles, we consider a stochastic initial kinetic distribution function $g_0(x, w, \xi)$, where the uncertainty described by the random variable ξ enters in the initial data.

We are interested in the evolution of the random field $g(t, x, w, \xi) : \mathbb{R}^+ \times \mathbb{R} \times W \times \Omega \to \mathbb{R}^+$ governed by the BGK-kinetic equation:

$$\partial_t g(t, x, w, \xi) + \partial_x\Big[(w - h(\rho(t, x, \xi)))g(t, x, w, \xi)\Big] = \frac{1}{\varepsilon}\Big(M_g(w; \rho(t, x, \xi)) - g(t, x, w, \xi)\Big), \tag{10}$$

$$g(0, x, w, \xi) = g_0(x, w, \xi), \tag{11}$$

where $M_g(w; \rho)$ is the distribution at the equilibrium.

Applying the intrusive method introduced in Sect. 2, the random field $g(t, x, w, \xi)$ can be approximated by the spectral expansion truncated at K, $g(t, x, w, \xi) = \sum_{k=0}^{K} \widehat{g}_i(t, x, w)\phi_i(\xi)$, and the stochastic Galerkin formulation for (10) reads as

$$\partial_t \widehat{g}_i(t, x, w) + \partial_x \Big(\big(wId - \mathcal{P}(h(\widehat{\rho}(t, x))) \big) \widehat{g}(t, x, w) \Big)_i = \frac{1}{\varepsilon} \left(\widehat{M}_i(w; \widehat{\rho}(t, x)) - \widehat{g}_i(t, x, w) \right), \tag{12}$$

$$\widehat{g}_i(0, x, w) = \int_{\Omega} g_0(t, x, w, \xi)\phi_i(\xi) p_{\Xi}(\xi) d\xi \quad \text{for } i = 0, \dots, K \tag{13}$$

where $(\mathcal{P}(h(\widehat{\rho}))\widehat{g})_i = \sum_{j=0}^{K} \int_{\Omega} h\left(\sum_{\ell=0}^{K} \widehat{\rho}_\ell \phi_\ell(\xi)\right) \widehat{g}_j \phi_j(\xi)\phi_i(\xi) p_{\Xi}(\xi) d\xi$ and w is a deterministic variable.

Further, we define for $i = 0, \dots, K$

$$\widehat{M}_i(w; \widehat{\rho}(t, x)) := \int_{\Omega} M_g\Big(w; \sum_{\ell=0}^{K} \widehat{\rho}_\ell(t, x)\phi_\ell(\xi)\Big)\phi_i(\xi) p_{\Xi} dw d\xi. \tag{14}$$

Thus we recovered a stochastic Galerkin system, namely (12), for the BGK model. As in [15], it might be employed to detect and forecast regions of high risk of congestions or traffic instabilities. For a detailed analysis and numerical tests we refer to [15].

5 Macroscopic Scale

Macroscopic models describe traffic flow in terms of aggregate quantities as density, $\rho = \rho(x, t)$ and mean velocity $v = v(x, t)$ of vehicles at a location $x \in \mathbb{R}$ and time $t > 0$. In contrast to the kinetic scale, each reference to the detailed level of vehicles' description is completely lost.

The natural assumption that the total mass is conserved along the road leads to impose that ρ and v satisfy: $\partial \rho(x, t) + \partial_x (\rho(x, t)v(x, t)) = 0$, $\rho(x, 0) = \rho_0(x)$.

However, another relationship has to be provided in order to close the equation, i.e. we have a single equation for two fields. Depending on the closure we distinguish between first and second order models.

In first order models the velocity is given as a function of the density. Among them, one of the most relevant model was introduced by Lighthill, Whitham and Richards (LWR) [20, 28],where a typical choice for the velocity function is $v(x, t) = V(\rho(x, t)) = 1 - \rho$.

In second order models instead an equation describing the variation of the velocity in time is added to the system. The prototype for second order macroscopic model is given by the ARZ model (Aw, Rascle [2] and Zhang [40]).

However, a challenge occurs when we take into account the uncertainty which affects vehicular traffic. In the following, we are indeed interested in studying the stochastic scenario, where the uncertainty enters only in the initial data.

In particular, we assume ρ_0 to depend on the space and on the random variable ξ introduced in Sect. 2, i.e. $\rho_0(x,\xi)$, for both first and second order macroscopic models. Under the same assumptions as in Sect. 2, the stochastic LWR model reads as follow:

$$\partial_t \rho(t,x,\xi) + \partial_x(\rho(t,x,\xi)\; V(\rho(t,x,\xi))) = 0 \tag{15}$$

$$\rho(0,x,\xi) = \rho_0(x,\xi). \tag{16}$$

Then we follow the same procedure described in Sect. 2 to get the stochastic Galerkin formulation of the system, which reads as:

$$\partial_t \widehat{\rho} + \partial_x \left(\mathcal{P}(\widehat{\rho}(t,x))\widehat{V}_{eq}(\widehat{\rho}(t,x))\right) = \vec{0}\,, \tag{17}$$

with $\vec{0} = (0,\ldots,0)^T$ vector of $K+1$ components.
Note that an arbitrary but consistent gPC expansion $\widehat{V}_{eq}$ is required. For example, the corresponding gPC expansion of $V_{eq} = 1-\rho$ is $\widehat{V}_{eq} = e_1 - \widehat{\rho}$, with unit vector $e_1 = (1,0,\ldots,0)^T$.

On the other hand, stochastic second order macroscopic models are more challenging to treat. Indeed, beside a more complicated structure, also the properties of the deterministic model have to be preserved, in particular the hyperbolicity of the system has to be ensured.
As second order model, we consider the ARZ model as stated before. In order to write it in a conservative form, an auxiliary variable is usually introduced, namely $z(t,x) = \rho(v + h(\rho))$. In the deterministic case the system reads as:

$$\partial_t \rho + \partial_x(z - \rho h(\rho)) = 0, \tag{18}$$

$$\partial_t z + \partial_x \left(\frac{z^2}{\rho} - zh(\rho)\right) = \frac{1}{\varepsilon}\left((\rho V_{eq}(\rho) + \rho h(\rho)) - z\right) \tag{19}$$

$$\rho(0,x) = \rho_0(x),\; z(0,x) = z_0(x). \tag{20}$$

As before, we introduce the uncertainty in the initial data, and more precisely in the initial density. This affects also the second equation, which means we end up with a stochastic system depending on $\rho(t,x,\xi), z(t,x,\xi)$.

Remark 1 A naive approach would be to substitute the truncated expansion for ρ and v into the stochastic system and then use the Galerkin projection onto the space spanned by the basis functions leads to a loss of hyperbolicity [10]. Indeed, in this case the jacobian of the flux has not necessarily real eigenvalues and a full set of eigenvectors.

In order to compute the gPC expansion for the term $\frac{z^2}{\rho}$, the Riemann invariant $w = v + h(\rho)$ is taken into account, such that $z = \rho w$, which leads to $\widehat{z} = \mathcal{P}(\widehat{\rho})\widehat{w}$ as in [10]. According to this, the term $\frac{z^2(t,x,\xi)}{\rho(t,x\xi)} = z(t,x,\xi)w(t,x,\xi)$ and the corresponding gPC expansion is $\widehat{z} * \widehat{w} = \mathcal{P}(\widehat{z})\mathcal{P}^{-1}(\widehat{\rho})\widehat{z}$. Thus the stochastic Galerkin formulation for the ARZ model reads as:

$$\partial_t \widehat{\rho} + \partial_x \left(\hat{z} - \mathcal{P}(\widehat{\rho})\widehat{h}(\widehat{\rho})\right) = \vec{0} \tag{21}$$

$$\partial_t \widehat{z} + \partial_x \left(\mathcal{P}(\widehat{z})\mathcal{P}^{-1}(\widehat{\rho})\widehat{z} - \mathcal{P}(\widehat{z})\widehat{h}(\widehat{\rho})\right) = \vec{0}\,. \tag{22}$$

In order to ensure the hyperbolicity of the system as proved in [10, Thm 2], the basis functions have to fulfill the following properties

(A1) The matrices $\mathcal{M}_\ell$ and $\mathcal{M}_k$ commute for all $\ell, k = 0, \ldots, K$.
(A2) The matrices $\mathcal{P}(\widehat{u})$ and $\mathcal{P}(\widehat{z})$ commute for all $\widehat{u}, \widehat{z} \in \mathbb{R}^{K+1}$.
(A3) There is an eigenvalue decomposition $\mathcal{P}(\widehat{u}) = V\mathcal{D}(\widehat{u})V^T$ with constant eigenvectors V.

It has been shown that for example the one–dimensional Wiener–Haar basis and piecewise linear multiwavelets fulfill the previous assumptions, but, Legendre and Hermite polynomials do *not* fulfill those requirements.

5.1 *From Micro to Macro*

Connections between microscopic and macroscopic traffic flow models are already well established. Indeed there are several works in the literature investigating the limit for first and second order models as [1, 8]. The natural question now is if and how the uncertainty influences the relationship between the scales.

Proposition 1 *Let ξ be a random variable as in Sect. 3, with N cars of fixed length L. Assume that $s(\frac{L}{\Delta x}) = v(\rho)$. Then the stochastic ODEs system (5) converges to the stochastic LWR model (15) for $L \to 0$ and $N \to \infty$.*

Proof First of all we recall that the uncertainty enters only in the initial data, as in Sects. 3–5. Then, we define the stochastic local density, according to the deterministic case [1]:

$$\rho_i^{(N)}(t,\xi) = \frac{L}{\Delta x_i(t,\xi)} = \frac{L}{x_{i+1}(t,\xi) - x_i(t,\xi)} \qquad i = 1, \ldots, N-1 \tag{23}$$

with an abuse of notation, where we assume the uncertainty to be represented by ξ both at the micro and at the macro level. Note that $\rho_i^{(N)}(t,\xi)$ is the same term which appears in the velocity function.

From the definition of local density

$$\frac{d}{dt}\frac{1}{\rho_i^{(N)}(t,\xi)} = \frac{1}{L}\left(\frac{d}{dt}x_{i+1}(t) - \frac{d}{dt}x_i(t) + \frac{d}{dt}\xi\right) \tag{24}$$

$$= \frac{1}{L}\left(s\left(\frac{L}{x_{i+2}(t) - x_{i+1}(t) + \xi}\right) - s\left(\frac{L}{x_{i+1}(t) - x_i(t) + \xi}\right)\right) \tag{25}$$

$$= \frac{s\left(\rho_{i+1}^{(N)}(t,\xi)\right) - s\left(\rho_i^{(N)}(t,\xi)\right)}{L}. \tag{26}$$

To compute the limit, we follow the procedure as in [5, 17, 32]. We consider the Lagrangian coordinate y, and we define

$$\rho^{(N)}(y,t,\xi) = \rho_i^{(N)}(t,\xi) \quad \text{if } y \in [iL, (i+1)L) \tag{27}$$

so we get:

$$\frac{d}{dt}\frac{1}{\rho^{(N)}(y,t,\xi)} = \frac{s\left(\rho^{(N)}(y+L,t,\xi)\right) - s\left(\rho^{(N)}(y,t,\xi)\right)}{L}. \tag{28}$$

By defining

$$L = \frac{1}{N} \tag{29}$$

and assuming that there exists a function $\rho(y,t,\xi)$ such that $\rho(y,t,\xi) = \lim_{N\to\infty}\rho^{(N)}(y,t,\xi)$, (28) can be rewritten:

$$\frac{d}{dt}\frac{1}{\rho(y,t,\xi)} = \partial_y s(\rho(y,t,\xi)). \tag{30}$$

While y represents the continuous number of cars, we are interested in the position of a car itself, namely $x(y,t)$:

$$\partial_y x(y,t,\xi) = \lim_{L\to 0}\frac{x(y+L,t,\xi) - x(y,t,\xi)}{L} = \frac{1}{\rho(y,t,\xi)} \tag{31}$$

so that

$$x(y,t,\xi) = \int_0^y \frac{1}{\rho(z,t,\xi)}dz, \tag{32}$$

assuming $x(0,t,\xi) = 0$.

Then:

$$\partial_t x(y,t,\xi) = \int_0^y \partial_t \frac{1}{\rho(z,t,\xi)} dz = \int_0^y \partial_z s(\rho(z,t,\xi)) dz = s(\rho(y,t,\xi)). \quad (33)$$

On the other hand, y can be recovered as:

$$y(x,t,\xi) = \left(\int_0^x \frac{1}{\rho(z,t,\xi)} dz \right)^{-1} \quad (34)$$

and we can compute the derivative with respect to x

$$\partial_x y(x,t,\xi) = \rho(y(x,t,\xi),t,\xi). \quad (35)$$

Moreover

$$\frac{d}{dt} y(x(y,t,\xi),t,\xi) = 0 \quad (36)$$

since $y(x(y,t+\Delta t,\xi),t+\Delta t,\xi) = y(x(y,t,\xi),t,\xi)$, for $\Delta t \geq 0$. Explicitly

$$\frac{d}{dt} y(x(y,t,\xi),t,\xi) = \partial_x y(x,t,\xi)\partial_t x(y,t,\xi) + \partial_t y(x,t,xi) = 0 \quad (37)$$

$$\partial_t y(x,t,xi) = -s(\rho(y,t,\xi))\rho(y(x,t,\xi),t,\xi). \quad (38)$$

We define $\rho(x,t,\xi) = \rho(y(x,t,\xi),t,\xi)$ and we compute the time derivative exploiting (35)–(38)

$$\begin{aligned}
\partial_t \rho(x,t,\xi) &= \frac{d}{dt}\rho(y(x,t,\xi),t,\xi) = && (39)\\
&= \partial_y \rho(y(x,t,\xi),t,\xi)\partial_t y(x,t,\xi) + \partial_t \rho(y(x,t,\xi),t,\xi) && (40)\\
&= -s(\rho(x,t,\xi))\partial_x \rho(x,t,\xi) + \rho(x,t,\xi)\partial_x s(\rho(x,t,\xi)) && (41)\\
&= -\partial_x \left(s(\rho(x,t,\xi))\rho(x,t,\xi)\right). && (42)
\end{aligned}$$

Thus we recover the stochastic LWR model (15). □

For second order models, a similar procedure can be applied.

5.2 *From Meso to Macro*

Kinetic and macroscopic scales are closely related, indeed one can recover the macroscopic quantities from the kinetic models, as stated in Sect. 4. Here we are interested in investigating the connections between the models described in Sects. 4 and 5. We recall that we consider a particular kinetic model, the BGK model, where the interaction kernel is given by a linear term which described a relaxation towards the equilibrium, namely the right hand side term of (10).

In order to understand the intrinsic connection to the macroscopic scale, let us consider traffic flow at equilibrium conditions. This means that at the kinetic level $M_g = g$ and the desired velocity is the velocity itself, $w = v$, which implies $h = 0$. Under these assumptions, integrating (10) with respect to w we immediately get the stochastic LWR if w coincides with V_{eq}.

As far as concerns the second order macroscopic models, the link between the kinetic BGK model and the ARZ has been proved in the deterministic case in [16], while in [15] the link has been extended also to the stochastic framework. For completeness, we recall here the main result where a gPC formulation of the fluid model obtained by the stochastic BGK model (10) is derived.

Further, this model is compared with the stochastic ARZ model. The theorem shows that under assumption (43) the derived gPC model is equivalent to the stochastic model of [10]. Therein, it has also been shown that the partial differential equation is hyperbolic.

Theorem 1 [15, Thm 2.2] *Let $K > 0, \varepsilon > 0$. Assume the base functions $\{\phi_0, \dots, \phi_K\}$ fulfill (A1)–(A3) and assume that*

$$\int_W \widehat{M}_i\,(w; \widehat{\rho}(t,x))\;dw = \widehat{\rho}_i(t,x), \tag{UM1}$$

$$\int_W w\,\widehat{M}_i\,(w; \widehat{\rho}(t,x))\,dw = \Big(\mathcal{P}(V_{eq}(\widehat{\rho}(t,x)))\widehat{\rho}(t,x) + \mathcal{P}(h(\widehat{\rho}(t,x)))\widehat{\rho}(t,x)\Big)_i. \tag{UM2}$$

Let $\widehat{g}_i$ be a strong solution to (12) and (14) for $i = 0, \dots, K$. Further, assume that for $i = 0, \dots, K$ and $(t,x) \in \mathbb{R}^+ \times \mathbb{R}$

$$\int_W w^2\,\widehat{g}_i(t,x,w)dw = (\mathcal{P}(\widehat{q}(t,x))\mathcal{P}^{-1}(\widehat{\rho}(t,x))\widehat{q}(t,x))_i, \tag{43}$$

where $(\widehat{\rho}, \widehat{q})_i$ are the first and second moment of $\widehat{g}_i$ and $\mathcal{P}$ is defined by (4). Then, the functions $(\widehat{\rho}, \widehat{q})$ formally fulfill pointwise in $(t,x) \in \mathbb{R}^+ \times \mathbb{R}$ and for all $i = 0, \dots, K$ the second–order traffic flow model

$$\partial_t \widehat{\rho}_i(t,x) + \partial_x \left[\widehat{q}_i(t,x) - (\mathcal{P}(\widehat{\rho}(t,x))\widehat{\rho}(t,x))_i\right] = 0 \tag{44a}$$

$$\partial_t \widehat{q}_i(t,x) + \partial_x \Big[(\mathcal{P}(\widehat{q}(t,x))\mathcal{P}^{-1}(\widehat{\rho}(t,x))\widehat{q}(t,x))_i - (\mathcal{P}(\widehat{\rho}(t,x))\widehat{q}(t,x))_i\Big] = \tag{44b}$$

$$\frac{1}{\varepsilon}\Big(\Big(\mathcal{P}(V_{eq}(\widehat{\rho}(t,x)))\widehat{\rho}(t,x) + \mathcal{P}(h(\widehat{\rho}(t,x)))\widehat{\rho}(t,x)\Big)_i - \widehat{q}_i(t,x)\Big) \tag{44c}$$

$$\widehat{\rho}_i(0,x) = \int_W \widehat{g}_{0,i}(t,x,w)dw, \tag{44d}$$

$$\widehat{q}_i(0,x) = \int_W w\,\widehat{g}_{0,i}(t,x,w)dw. \tag{44e}$$

The system (44) *is hyperbolic for* $\widehat{\rho}_i > 0$.
Let the random fields $(\rho, q) = (\rho, q)(t, x, \xi) : \mathbb{R}^+ \times \mathbb{R} \times \Omega \to \mathbb{R}^2$ *be a pointwise a.e. solution with second moments w.r.t. to* ξ *of the stochastic Aw–Rascle–Zhang system with random initial data:*

$$\partial_t \rho + \partial_x (q - \rho h(\rho)) = 0, \tag{45a}$$

$$\partial_t q + \partial_x \Big(\frac{q^2}{\rho} - q h(\rho) \Big) = \frac{1}{\varepsilon} \left(\rho V_{eq}(\rho) + \rho h(\rho) - q \right), \tag{45b}$$

$$\rho(0, x, \xi) = \rho_0(x, \xi), \; q(0, x, \xi) = q_0(x, \xi). \tag{45c}$$

Under the previous assumptions on the base functions $\{\phi_0, \ldots, \phi_K\}$ *and provided that for all* $i = 0, \ldots, K$

$$\int_\Omega \rho_0(x, \xi)\phi_i(\xi) p_\Xi d\xi = \int_W \widehat{g}_{0,i}(t, x, w) dw, \; \int_\Omega q_0(x, \xi)\phi_i(\xi) p_\Xi d\xi = \int_W w \widehat{g}_{0,i}(t, x, w) dw, \tag{46}$$

we have

$$G_K \left(\rho(t, x, \cdot) \right) (\xi) = \sum_{i=0}^{K} \widehat{\rho}_i(t, x) \phi_i(\xi) \text{ and } G_K \left(q(t, x, \cdot) \right) (\xi) = \sum_{i=0}^{K} \widehat{q}_i(t, x) \phi_i(\xi), \tag{47}$$

where $(\widehat{\rho}, \widehat{q})$ *fulfill equation* (44).

For the proof we refer to [15].

5.3 Numerical Test

Numerically, we focus on the macroscopic scale, in order to exploit the strength of the stochastic Galerkin approach. Indeed, thanks to the stochastic Galerkin formulation, the stochastic quantities can be recovered at each time step in every point of the space grid, solving only once the coefficients system. In particular, in the following we focus on the fundamental diagram. To this aim, we run simulations for the stochastic LWR model and we reconstruct an approximation of the stochastic fundamental diagram from the Galerkin coefficients of the density, namely $\tilde{f}(t, x, \xi) = \mathcal{P}(\widehat{\rho}(t, x)) \widehat{V_{eq}}$, where $\widehat{V_{eq}}_i = (e_1)_i - \widehat{\rho}_i$ for $i = 0, \ldots, K$.

We employ the local Lax-Friederichs scheme to solve the PDE for each coefficient (17). The numerical parameters are as follows. We consider the space interval $x \in [a, b] = [0, 2]$ and define the uniform spatial grid of size $\Delta x = 1 \cdot 10^{-2}$. Moreover, let $T_f = 1$ be the final time of the simulations and Δt the time step, which is chosen in such a way that the CFL condition is fulfilled. By N_t we denote the number of the time steps needed to reach T_f. The random variable ξ is assumed to be uniform distributed on (0, 1), i.e., $p_\Xi = 1$ and $\Omega = (0, 1)$. As basis functions we consider the Haar basis. The prototype of the initial data is a Riemann problem:

$$\rho_0(x,\xi) = \begin{cases} \rho_l \equiv \xi \sim \mathcal{U}(u_1, u_2) & x < 1 \\ \rho_r & x \geq 1 \end{cases}, \tag{48}$$

with $\rho_l, \rho_r, u_1, u_2 \in [0, 1]$.

Moreover, according to the results presented in [10, 15] where the numerical convergence with respect to K is studied, we choose $K = 15$.

In order to understand how the uncertainty in the initial data affects the shape of the fundamental diagram at $T_f = 1$, we consider several initial data, fixing $\rho_l \sim \mathcal{U}(0.75, 0.95)$ and varying $\rho_r \in [0, 1]$. We mainly focus on the rarefaction case since the density is more spread, with respect to the shock case, and therefore the reconstruction of the fundamental diagram is more accurate. In Fig. 1 we plot the mean, i.e. the 0-coefficient, of the fundamental diagram. One may recognize the typical form of the Greenshields flux function. However, a cloud of points is also present approximately for $\rho \in (0.35, 0.83)$, which means that the same density value could lead to different velocities. Moreover, we note that this scatter behavior affects only region where there is the transition between free flow and congested flow. Furthermore, for $\rho > 0.85$ we still observe a scattering in the flux, right box in Fig. 1, while in the free flow there is none, left box in Fig. 1.

It is important to note that the scattered fundamental diagram we observe here, is typical of traffic dynamics [30], and it can be recovered from a first order macroscopic

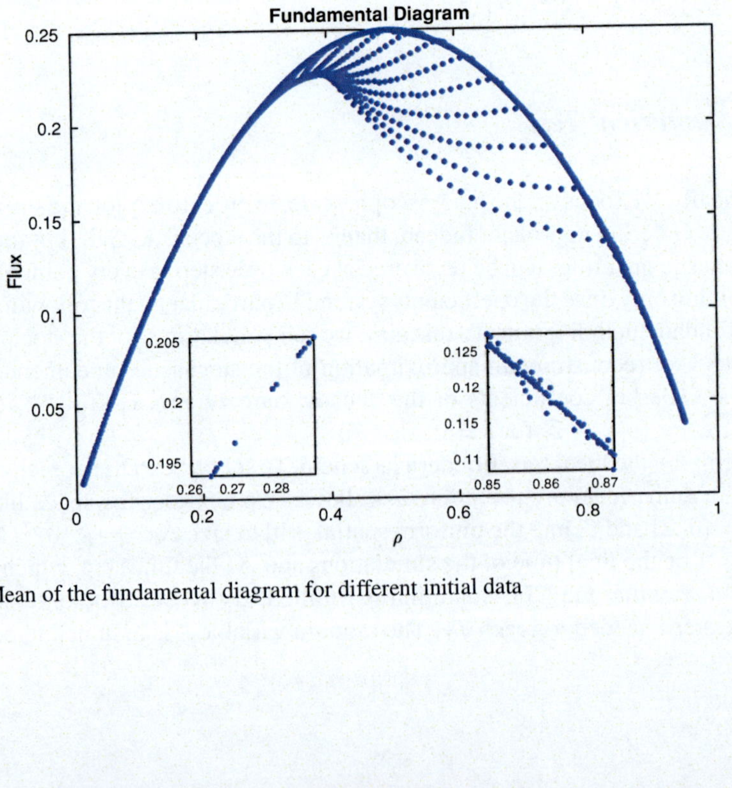

Fig. 1 Mean of the fundamental diagram for different initial data

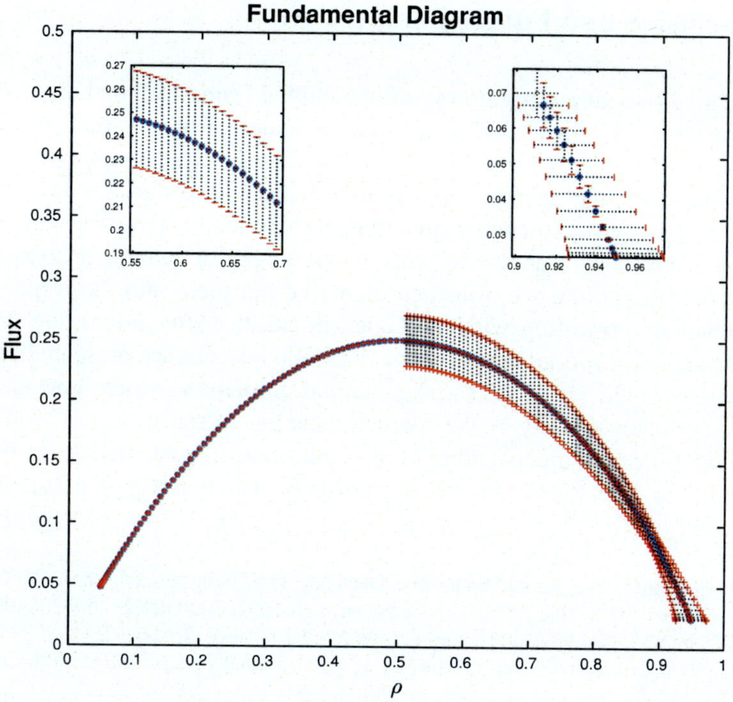

Fig. 2 Fundamental diagram with mean and variance

model with uncertainty only in the initial data. A deterministic first order model fails in reproducing such a scattered dynamics.

In order to investigate the uncertainty more in details, the variance is also taken into account. The approximation of the variance is computed by the sum of all the coefficients but the first one, squared, i.e. for the density $Var(\rho) \approx \sum_{k=1}^{K} \widehat{\rho}_k^2$. In Fig. 2, the blue dots stand for the mean of the reconstructed fundamental diagram while the red bars indicate the mean plus and minus the variance, both for density and flux. It is very interesting to note that while for $\rho \in [0.5, 0.8]$ a very small variance in the density corresponds to high variance in the flux, for a more congested traffic the scenario is the opposite, right box in Fig. 2. This can be explained as follows: in the highly congested traffic situation the velocity is close to 0 even for some variation of the density values and this causes a very small variation in the flux with respect to the density. On the other hand, switching from free flow to congested regime, very small changes in the density values lead to higher uncertainty in the flux value since the velocity has a large impact close to the maximum of the flux, left box in Fig. 2.

6 Conclusion and Future Perspectives

The presented overview aims at providing a unified framework on how to deal with uncertainty in traffic flow models at different scales of observation. In particular, starting from the microscopic models, to the kinetic equation and finally to the macroscopic ones, the uncertainty was introduced in the initial data in a consistent way. Moreover, the stochastic quantities were treated in the same way following the intrusive stochastic Galerkin approach: first we performed the gPC expansion of the stochastic quantities, we truncated them and put them into the respective evolution equations projecting with the Galerkin ansatz. After presenting the proper stochastic Galerkin formulations for the different systems, we presented some connections between the scales and some numerical simulations which show the intrinsic influence of the uncertainty on the characteristic law of traffic.

However, a deeper understanding of the link between the Galerkin coefficients at different scales, in particular between microscopic and macroscopic, is still an open question.

Acknowledgements The author thanks the Deutsche Forschungsgemeinschaft (DFG, German Research Foundation) for the financial support through 20021702/GRK2326, 333849990/IRTG-2379, HE5386/19-1,22-1,23-1 and under Germany's Excellence Strategy EXC-2023 Internet of Production 390621612. The author is member of the "National Group for Scientific Computation (GNCS-INDAM)".

References

1. Aw, A., Klar, A., Rascle, M., Materne, T.: Derivation of continuum traffic flow models from microscopic follow-the-leader models. SIAM J. Appl. Math. **63**(1), 259–278 (2002)
2. Aw, A., Rascle, M.: Resurrection of "second order" models of traffic flow?. SIAM J. Appl. Math. **60**, 916–938 (2000)
3. Bando, M., Hasebe, K., Nakayama, A., Shibata, A., Sugiyama, Y.: Dynamical model of traffic congestion and numerical simulation. Phys. Rev. E. **51**(2), 10–35 (1995)
4. Bhatnagar, P.L., Gross, E.P., Krook, M.: A model for collision processes in gases. i. small amplitude processes in charged and neutral one-component systems. Phys. Rev. **94**(3), 511 (1954)
5. Burger, M., Göttlich, S., Jung, T.: Derivation of a first order traffic flow model of Lighthill-Whitham-Richards type. IFAC-PapersOnLine. **51**(9), 49–54 (2018)
6. Cameron, R.H., Martin, W.T.: The orthogonal development of non-linear functionals in series of Fourier-Hermite functionals. Ann. Math. **48**(2), 385–392 (1947)
7. Debusschere, B.J., Najm, H.N., Pébay, P.P., Knio, O.M., Ghanem, R.G., Maître, O.P.L.: Numerical challenges in the use of polynomial chaos representations for stochastic processes. SIAM J. Sci. Comput. **26**(2), 698–719 (2004)
8. Di Francesco, M., Fagioli, S., Rosini, M.D.: Many particle approximation of the Aw-Rascle-Zhang second order model for vehicular traffic. Math. Biosci. Eng. **14**(1), 127–141 (2017)
9. Fan, S., Herty, M., Seibold, B.: Comparative model accuracy of a data-fitted generalized Aw-Rascle-Zhang model. Netw. & Heterog. Media. **9**(2), 239 (2014)
10. Gerster, S., Herty, M., Iacomini, E.: Stability analysis of a hyperbolic stochastic Galerkin formulation for the Aw-Rascle-Zhang model with relaxation. Math. Biosci. Eng. MBE. **18**(4), 4372–4389 (2021)

11. Gerster, S., Herty, M., Sikstel, A.: Hyperbolic stochastic Galerkin formulation for the p-system. J. Comput. Phys. **395**, 186–204 (2019)
12. Ghanem, R.G., Spanos, P.D.: Stochastic Finite Elements: A Spectral Approach, 1st edn. Springer, New York (1991)
13. Gottlieb, D., Hesthaven, J.S.: Spectral methods for hyperbolic problems. J. Comput. Appl. Math. **128**(1), 83–131 (2001)
14. Helbing, D., Hennecke, A., Shvetsov, V., Treiber, M.: MASTER: macroscopic traffic simulation based on a gas-kinetic, non-local traffic model. Transp. Res. Part B Methodol. **35**(2), 183–211 (2001)
15. Herty, M., Iacomini, E.: Uncertainty quantification in hierarchical vehicular flow models. Kinet. & Relat. Model. **15**(2), 239 (2022)
16. Herty, M., Puppo, G., Roncoroni, S., Visconti, G.: The BGK approximation of kinetic models for traffic. Kinet. & Relat. Model. **13**(2), 279 (2020)
17. Holden, H., Risebro, N.H.: Follow-the-Leader models can be viewed as a numerical approximation to the Lighthill-Whitham-Richards model for traffic flow. Netw. & Heterog. Media. **13**(3), 409 (2018)
18. Jin, S., Xiu, D., Zhu, X.: A well-balanced stochastic Galerkin method for scalar hyperbolic balance laws with random inputs. J. Sci. Comput. **67**, 1198–1218 (2016)
19. Kusch, J., Alldredge, G.W., Frank, M.: Maximum-principle-satisfying second-order intrusive polynomial moment scheme. SMAI J. Comput. Math. **5**, 23–51 (2019)
20. Lighthill, M.J., Whitham, G.B.: On kinematic waves II. A theory of traffic flow on long crowded roads. In: Proceedings of the Royal Society of London. Series A. Math. Phys. Sci. **229**(1178), 317–345 (1955)
21. Maître, O.P.L., Knio, O.M.: Spectral Methods for Uncertainty Quantification, 1st edn. Springer, Netherlands (2010)
22. Newell, G.F.: Nonlinear effects in the dynamics of car following. Oper. Res. **9**(2), 209–229 (1961)
23. Pettersson, P., Iaccarino, G., Nordström, J.: A stochastic Galerkin method for the Euler equations with Roe variable transformation. J. Comput. Phys. **257**, 481–500 (2014)
24. Pettersson, P., Iaccarino, G., Nordström, J.: Polynomial Chaos Methods for Hyperbolic Partial Differential Equations. Springer International Publishing, Switzerland (2015)
25. Piccoli, B., Tosin, A.: Vehicular traffic: a review of continuum mathematical models. In: Meyers, R. (ed.) Encyclopedia of Complexity and Systems Science, pp. 9727–9749. Springer, New York, NY (2009)
26. Piu, M., Puppo, G.: Stability analysis of microscopic models for traffic flow with lane changing. Networks and Heterogeneous Media (2022)
27. Puppo, G., Semplice, M., Tosin, A., Visconti, G.: Kinetic models for traffic flow resulting in a reduced space of microscopic velocities. Kinet. & Relat. Model. **10**(3), 823 (2016)
28. Richards, P.I.: Shock waves on the highway. Oper. Res. **4**(1), 42–51 (1956)
29. Schlachter, L., Schneider, F.: A hyperbolicity-preserving stochastic Galerkin approximation for uncertain hyperbolic systems of equations. J. Comput. Phys. **375**, 80–98 (2018)
30. Siebel, F., Mauser, W.: On the fundamental diagram of traffic flow. SIAM J. Appl. Math. **66**, 1150–1162 (2005)
31. Sullivan, T.J.: Introduction to Uncertainty Quantification, 1st edn. Texts in Applied Mathematics. Springer, Switzerland (2015)
32. Tordeux, A., Costeseque, G., Herty, M., Seyfried, A.: From traffic and pedestrian follow-the-leader models with reaction time to first order convection-diffusion flow models. SIAM J. Appl. Math. **78**(1), 63–79 (2018)
33. Tosin, A., Zanella, M.: Kinetic-controlled hydrodynamics for traffic models with driver-assist vehicles. Multiscale Model. & Simul. **17**(2), 716–749 (2019)
34. Tosin, A., Zanella, M.: Boltzmann-type description with cutoff of follow-the-leader traffic models. In: Trails in Kinetic Theory, pp. 227–251. Springer (2021)
35. Tosin, A., Zanella, M.: Uncertainty damping in kinetic traffic models by driver-assist controls. Math. Control. & Relat. Fields. **11**(3), 681 (2021)

36. Wegener, R., Klar, A.: A kinetic model for vehicular traffic derived from a stochastic microscopic model. Transp. Theory Stat. Phys. **25**(7), 785–798 (1996)
37. Wiener, N.: The homogeneous chaos. Am. J. Math. **60**(4), 897–936 (1938)
38. Xiu, D., Karniadakis, G.E.: The Wiener-Askey polynomial chaos for stochastic differential equations. SIAM J. Sci. Comput. **24**, 619–644 (2002)
39. Zanella, M.: Structure preserving stochastic galerkin methods for fokker-planck equations with background interactions. Math. Comput. Simul. **168**, 28–47 (2020)
40. Zhang, H.M.: A non-equilibrium traffic model devoid of gas-like behavior. Transp. Res. Part B Methodol. **36**(3), 275–290 (2002)

A Study of Multiscale Kinetic Models with Uncertainties

Liu Liu

Abstract There have been many challenges in studying multiscale kinetic models with uncertainties. Quantifying uncertainties in the models such as arisen from collision kernels, initial or boundary data, forcing terms, and developing efficient computational methods have become an important task in industrial applications. In this article, we will report some of the recent progress on multiscale kinetic models with random inputs. We will discuss from the aspects of both mathematical theory, by using the hypocoercivity of kinetic operators, and numerical computation. Two categories of intrusive and non-intrusive methods will be addressed, in particular the stochastic Galerkin method for problems with relatively low-dimensional random inputs and stochastic collocation method in a multi-fidelity setting for more challenging, higher-dimensional problems. Numerical experiments for several kinetic models such as the Boltzmann, bipolar Boltzmann-Poisson, linear transport and epidemic kinetic equations will be shown as examples.

Keywords Multiscale kinetic models · Uncertainty quantification · Sensitivity analysis · Stochastic Galerkin method · Multi-fidelity method

1 Introduction

Kinetic equations are widely used in many important areas such as rarefied gas, plasma physics, astrophysics, and have also reached out to new realms such as semiconductor device modeling [61], environmental, social and biological sciences [66]. They describe the non-equilibrium dynamics of a gas or system composed of a large number of particles. The Boltzmann equation, as a typical example, is used to model different phenomena ranging from rarefied gas flows found in hypersonic aerodynamics, gases in vacuum technologies, or fluids inside microelectromechanical devices [9], to the description of social and biological phenomena [67, 73]. There has been

L. Liu (✉)
The Chinese University of Hong Kong, Hong Kong, Hong Kong
e-mail: lliu@math.cuhk.edu.hk

G. Albi et al. (eds.), *Advances in Numerical Methods for Hyperbolic Balance Laws and Related Problems*, SEMA SIMAI Springer Series 32,
https://doi.org/10.1007/978-3-031-29875-2_7

great development of efficient and accurate numerical methods for solving kinetic equations and in particular the Boltzmann equation contributed by researchers working in different fields [4, 7, 24, 40, 68]. We refer to [13, 15, 71, 72, 81] for recent monographs, collections and surveys.

One main computational challenge for kinetic problems is that these models often encounter multiple temporal and spatial scales, characterized by the Knudsen number ε that may vary dramatically in orders of magnitude in the computational domain, covering from fluid, transition to rarefied regimes. One effective computational paradigm to tackle this challenge is *Asymptotic-Preserving (AP)* schemes [41, 43], which has been very popular in kinetic and hyperbolic communities in the last two decades.

In simple words, an AP scheme preserves the discrete analogy of the asymptotic passage from a microscopic model to its macroscopic model, without numerically resolving the small Knudsen number thus can reduce the computational cost significantly. There have been many activities in the last two decades on the development of AP schemes for deterministic kinetic equations, see review articles [12, 38].

On the other hand, Uncertainty Quantification (UQ) has drawn many attentions over the past decade. Quantifying uncertainties in the inputs of models is important to assess, validate and improve the underlying models, allowing us to obtain more reliable predictions for the outputs and better risk assessment. Kinetic equations, usually derived from N-body Newton's equations via the mean field limit [5, 9], typically contain an integral operator modeling interactions between particles. The collision kernel or scattering cross-section in this integral describes the transition rate during particle collisions. Calculating the collision kernel from first principles is extremely complicated and it is even harder for complex particle systems. Thus, heuristic approximations are frequently used. Other sources of uncertainties may come from inaccurate measurements for the initial or boundary data, forcing or source terms etc.

In spite of the vast amount of existing research on the Boltzmann and related equations–both at the theoretical level and in its numerical approximation, however, the study of kinetic models with random uncertainties has only started in recent years [39, 42, 74, 78, 79, 82], see the recent collection [45] and the survey [70]. We also refer to [60, 69, 77, 79] for related researches in computational fluid dynamics, hyperbolic conservation laws and mention the work [62] on biomedical applications of kinetic tumour growth modeling with clinical uncertainties, where the distribution of uncertainties are indeed unknown and should be estimated from real-life data in a model-dependent fashion.

This book chapter will present some recent results on UQ for multiscale kinetic equations, from the perspectives of both mathematical theory and numerical methods. In Sect. 2, we will conduct the sensitivity analysis and use hypocoercivity theory to study the regularity and long-time behavior of the solution to a class of collisional kinetic models with different scalings and uncertainties; in particular the Boltzmann equation with random collision kernel and/or initial data will be discussed. Regarding numerical methods for UQ problems, in Sect. 3, we introduce a typical intrusive method, namely the stochastic Galerkin (SG) method and stochastic AP schemes in this framework, including the error estimates obtained from hypocoercivity theory for the SG system. An example of the bipolar Boltzmann-Poisson equation with random parameters will be shown. On the contrary, when the dimension of random parameters is high, the SG method may not be efficient given the complexity of our kinetic models. In Sect. 4, we review non-intrusive methods and focus on the stochastic collocation (SC) method in a multi-fidelity framework. In particular, we will give examples of adapting the bi-fidelity SC method to solve the Boltzmann and linear transport equation under diffusive scaling, as well as the epidemic kinetic transport system with random parameters.

2 Mathematical Theory for Uncertain Kinetic Equations

Consider the Boltzmann equation with random uncertainties and different scalings [8],

$$\begin{cases} \partial_t f + \dfrac{1}{\varepsilon^\alpha} v \cdot \nabla_x f = \dfrac{1}{\varepsilon^{1+\alpha}} Q(f, f), \\ f(0, x, v, z) = f_I(x, v, z), \qquad x \in \mathbb{T}^{d_x},\ v \in \mathbb{R}^{d_v},\ z \in I_z, \end{cases} \tag{1}$$

where $f = f(t, x, v, z)$ is the particle density distribution that depends on time t, particle position $x \in \mathbb{T}^{d_x}$ of a periodic box, velocity $v \in \mathbb{R}^{d_v}$ and a random variable z that characterizes the random inputs and lies in the domain $I_z \subset \mathbb{R}^{d_z}$, with the distribution $\pi(z)$ that is assumed known. Here ε is the dimensionless Knudsen number, the constant $\alpha = 1$ refers to the incompressible Navier-Stokes scaling; $\alpha = 0$ corresponds to the Euler (or acoustic) scaling. The operator Q models the binary collisional interactions between particles and is given by

$$Q(f, g) = \int_{\mathbb{R}^{d_v} \times \mathbb{S}^{d_v - 1}} B(|v - v^*|, \cos\theta, z)(f' g'^* - f g^*)\, dv^* d\sigma, \tag{2}$$

where $f' = f(v')$, $g^* = g(v^*)$, $g'^* = g(v'^*)$ (v' and v'^* are the post-collisional velocities), and $\mathbb{S}^{d_v-1}$ is a $(d_v - 1)$ dimensional unit sphere. The collision kernel $B = B(|v - v^*|, \cos\theta, z)$ is a non-negative function depending on the modulus of relative velocity $|v - v^*|$ and cosine of the deviation angle θ; it is assumed to be uncertain and depends on the random parameter z.

Hypocoercivity theory is a useful and important tool to study the stability and long-time behavior of the solution for kinetic equations, see for example [6, 20, 34, 64, 83, 85] for deterministic equations. Unlike the parabolic equations, in which the elliptic operator is coercive, a kinetic operator is degenerately dissipative (hence the name hypocoercivity), making it difficult to obtain the exponential decay to the global equilibrium in the usual energy or Sobolev norm. According to [85], the study for the dissipative evolution equation such as (1) involves (i) a degenerate dissipative operator, and (ii) a conservative streaming operator $\mathcal{J} = v \cdot \nabla_x$ such that the combination of both operators implies convergence to a uniquely determined equilibrium state. The key for the analysis is to construct a new Lyapunov functional, which is equivalent to the Sobolev norm and can be shown to decay to zero exponentially in time.

By extending the hypocoercivity theory developed for general collisional nonlinear kinetic equations for deterministic problems [6, 64] to the random uncertain setting, in [57] the authors established the regularity and long-time behavior of the solution with random initial data, and random collision kernel in particular for the Boltzmann equation, under suitable assumptions. The framework provided in [6, 57] is quite general, ranging from the Boltzmann and Landau equations, to the semi-classical relaxation type quantum Boltzmann equation and linear transport equations. We also refer to [44, 53, 55] for the study of linear kinetic equations with uncertainties, [56] for the uncertain bipolar semiconductor Boltzmann-Poisson system and [11] on the analysis for the mixture Boltzmann equations.

2.1 Theoretical Framework: The Perturbative Setting

For kinetic problems with small scalings, in the deterministic setting, it was initiated by Bardos, Golse and Levermore [1] to derive the fluid limits which include incompressible Navier-Stokes, compressible Euler equations and acoustic system from the DiPerna-Lions renormalized solutions [19]. See for example [30, 52]. Here we will study the solution in the perturbative setting, which guarantees that the solution will be classical, thus allows one to conduct estimates in the Sobolev space [33].

Now consider the linearization of the solution to kinetic equation (1) around the equilibrium:

$$f = \mathcal{M} + \varepsilon\sqrt{\mathcal{M}}h, \tag{3}$$

where $\mathcal{M}$ is the global equilibrium given by $\mathcal{M}(v) = \frac{1}{(2\pi)^{\frac{d_v}{2}}} e^{-\frac{|v|^2}{2}}$. Insert ansatz (3) into Eq. (1), then the random fluctuation h satisfies

$$\begin{cases} \partial_t h + \dfrac{1}{\varepsilon^{\alpha}} v \cdot \nabla_x h = \dfrac{1}{\varepsilon^{1+\alpha}} \mathcal{L}(h) + \dfrac{1}{\varepsilon^{\alpha}} \mathcal{F}(h, h), \\ h(t = 0) = h_I, \end{cases} \tag{4}$$

where the linearized and the nonlinear collision operators are defined by, respectively,

$$\mathcal{L}(h) = \mathcal{M}^{-1/2} \left[Q(\sqrt{\mathcal{M}}h, \mathcal{M}) + Q(\mathcal{M}, \sqrt{\mathcal{M}}h) \right],$$

$$\mathcal{F}(h, h) = \mathcal{M}^{-1/2} \left[Q(\sqrt{\mathcal{M}}h, \sqrt{\mathcal{M}}h) + Q(\sqrt{\mathcal{M}}h, \sqrt{\mathcal{M}}h) \right].$$

The linearized operator $\mathcal{L}$ is acting on $L_v^2 = \{f \mid \int_{\mathbb{R}^d} f^2\, dv < \infty\}$, with the kernel denoted by $N(\mathcal{L}) = \text{Span}\{\varphi_1, \ldots, \varphi_n\}$, and $\{\varphi_i\}_{1\leq i\leq n}$ is an orthonormal family of polynomials in v corresponding to the manifold of local equilibria for the linearized kinetic models. The orthogonal projection on $N(\mathcal{L})$ in L_v^2 is defined by

$$\Pi_{\mathcal{L}}(h) = \sum_{i=1}^{n} \left(\int_{\mathbb{R}^d} h\varphi_i\, dv \right) \varphi_i, \tag{5}$$

where $\Pi_{\mathcal{L}}$ is the projection on the 'fluid part' and $\mathbb{I} - \Pi_{\mathcal{L}}$ is the projection on the kinetic part, with $\mathbb{I}$ the identity operator.

2.2 *Convergence to the Global Equilibrium*

Due to the small scaling, if one directly applies the estimates in [64], typically the v-derivatives contribute to the energy norm by a factor of $1/\epsilon$, which prevents us to obtain a uniform exponential decay for the v-derivatives. As initiated by Guo [33], one needs to study the v derivatives of the microscopic part of h solved by equation (4). For the deterministic problem, the author in [6] constructs a new energy norm to capture the structure of $\mathcal{L}$ on its orthogonal part, which, when combined with the previous strategy, leads to a uniform exponential decay for solutions close to the global equilibrium. The result is uniform in ϵ, thus gives a strong convergence in time to the incompressible Navier-Stokes equations as ϵ goes to zero, under some assumptions on the initial conditions. In [57], we show that the general framework put forth in [6] for deterministic problems can be adopted to conduct sensitivity analysis for uncertain collisional kinetic problems with small scalings, which gives rise to an exponential convergence of the random solution toward the (deterministic) global equilibrium, under suitable conditions on the collision kernel. We will briefly review the results in the following.

2.2.1 Notations and Assumptions

First, we introduce some notations of multi-indices and Sobolev norms. For two multi-indices j and l in $\mathbb{N}^d$, define $\partial_l^j = \partial/\partial v_j\, \partial/\partial x_l$. For $i \in \{1, \ldots, d\}$, denote by $c_i(j)$ the value of the i-th coordinate of j and by $|j|$ the l^1 norm of the multi-index, that is, $|j| = \sum_{i=1}^{d} c_i(j)$. Define the multi-index δ_{i_0} by: $c_i(\delta_{i_0}) = 1$ if $i = i_0$ and 0 otherwise. We use the notation $\partial_z^\alpha h = \partial^\alpha h$ for z-derivatives, and $||\cdot||_\Lambda := ||\,||\cdot||_{\Lambda_v}\,||_{L_x^2}$. The Sobolev norms on $H_{x,v}^s$ and H_Λ^s are defined by

$$||h||_{H_{x,v}^s}^2 = \sum_{|j|+|l|\leq s} ||\partial_l^j h||_{L_{x,v}^2}^2, \qquad ||h||_{H_\Lambda^s}^2 = \sum_{|j|+|l|\leq s} ||\partial_l^j h||_\Lambda^2.$$

Define the sum of Sobolev norms of the z derivatives by

$$||h||^2_{H^{s,r}_{x,v}} = \sum_{|m|\le r} ||\partial^m h||^2_{H^s_{x,v}}, \qquad ||h||^2_{H^{s,r}_\Lambda} = \sum_{|m|\le r} ||\partial^m h||^2_{H^s_\Lambda},$$

$$||h||^2_{H^{s,r}_x L^2_v} = \sum_{|m|\le r} ||\partial^m h||^2_{H^s_x L^2_v}.$$

Note that these norms are all functions of z. Define the norms in the (x, v, z) space

$$||h(x, v, \cdot)||^2_{H^s_{x,v} H^r_z} = \int_{I_z} ||h||_{H^{s,r}_{x,v}} \pi(z)\, dz,$$

in addition to the sup norm in z variable, $||h||_{H^s_{x,v} L^\infty_z} = \sup_{z\in I_z} ||h||_{H^s_{x,v}}$.

Among several assumptions for the operators in (4), we list the linearized and nonlinear collision operator $\mathcal{F}$, see details in [57, Sect. 2.2].

Assumption on the Linearized Term $\mathcal{L}$:
One of the key properties for the linearized operator $\mathcal{L}$ is its *local coercivity* property: For each $z \in I_z$, there exists $\lambda > 0$ such that $\forall h \in L^2_v$,

$$\langle \mathcal{L}(h), h \rangle_{L^2_v} \le -\lambda\, ||h^\perp||^2_{\Lambda_v},$$

where $h^\perp = h - \Pi_{\mathcal{L}}(h)$ is the remainder of the projection of h onto Null($\mathcal{L}$) (the null space of $\mathcal{L}$), namely $\Pi_{\mathcal{L}}(h)$, and Λ_v-norm is collision operator specific. See details in [6, 64] on other assumptions for operators $\mathcal{L}$ and $\mathcal{F}$.

Assumption on the Nonlinear Term $\mathcal{F}$:
$\mathcal{F} : L^2_v \times L^2_v \to L^2_v$ is a bilinear symmetric operator such that for all multi-indexes j and l such that $|j| + |l| \le s$, $s \ge 0$, $m \ge 0$,

$$\left| \langle \partial^m \partial^j_l \mathcal{F}(h, h),\ f \rangle_{L^2_{x,v}} \right| \le \begin{cases} \mathcal{G}^{s,m}_{x,v,z}(h, h)\, ||f||_\Lambda, & \text{if } j \ne 0, \\ \mathcal{G}^{s,m}_{x,z}(h, h)\, ||f||_\Lambda, & \text{if } j = 0. \end{cases}$$

Sum up $m = 0, \ldots, r$, then $\exists\, s_0 \in \mathbb{N}$, $\forall s \ge s_0$, there exists a z-independent $C_{\mathcal{F}} > 0$ such that for all z,

$$\sum_{|m|\le r} (\mathcal{G}^{s,m}_{x,v,z}(h, h))^2 \le C_{\mathcal{F}}\, ||h||^2_{H^{s,r}_{x,v}}\, ||h||^2_{H^{s,r}_\Lambda},$$

$$\sum_{|m|\le r} (\mathcal{G}^{s,m}_{x,z}(h, h))^2 \le C_{\mathcal{F}}\, ||h||^2_{H^{s,r}_x L^2_v}\, ||h||^2_{H^{s,r}_\Lambda}.$$

In a shared spirit of the work by Guo [33] who studies the fluid part and the microscopic part of the solution independently, for the deterministic problems the author in [6] constructs a new functional that is equivalent to the standard Sobolev norm and obtains an exponential decay in $H^s_{x,v}$. Following a similar fashion, in our stochastic problems, define $||\cdot||_{\mathcal{H}^s_{\epsilon_\perp}}$ by

$$||\cdot||^2_{\mathcal{H}^s_{\epsilon_\perp}} = \sum_{|j|+|l|\le s,\,|j|\ge 1} b^{(s)}_{j,l}\,||\partial^j_l(\mathbb{I}-\Pi_{\mathcal{L}})\cdot||^2_{L^2_{x,v}} + \sum_{|l|\le s}\alpha^{(s)}_l\,||\partial^0_l\cdot||^2_{L^2_{x,v}}$$
$$+ \sum_{|l|\le s,\,i,c_i(l)>0} \epsilon\, a^{(s)}_{i,l}\,\langle\partial^{\delta_i}_{l-\delta_i}\cdot,\,\partial^0_l\cdot\rangle_{L^2_{x,v}}, \tag{6}$$

and the corresponding Sobolev norms

$$||h||^2_{\mathcal{H}^{s,r}_{\epsilon_\perp}} = \sum_{|m|\le r}||\partial^m h||^2_{\mathcal{H}^s_{\epsilon_\perp}}, \qquad ||h||_{\mathcal{H}^{s,r}_{\epsilon_\perp}L^\infty_z} = \sup_{z\in I_z}||h||_{\mathcal{H}^{s,r}_{\epsilon_\perp}}.$$

The following Theorem on the regularity of the perturbed solution h is from [57], and it indicates that the solution f of (3) exponentially decays to the global Maxwellian.

Theorem 1 *For all $s \ge s_0$, $\exists\,(b^{(s)}_{j,l})$, $(\alpha^{(s)}_l)$, $(a^{(s)}_{i,l}) > 0$ and $0 \le \epsilon_d \le 1$, such that for all $0 \le \epsilon \le \epsilon_d$,*
(1) $||\cdot||_{\mathcal{H}^s_{\epsilon_\perp}} \sim ||\cdot||_{H^s_{x,v}}$;
(2) Assume $||h_{in}||_{H^s_{x,v}L^\infty_z} \le C_I$, then the solution h of (4) is in $H^s_{x,v}$ for all z, we have

$$||h||_{H^{s,r}_{x,v}L^\infty_z} \le C_I\,e^{-\tau_s t}, \qquad ||h||_{H^s_{x,v}H^r_z} \le C_I\,e^{-\tau_s t}. \quad \text{for } \alpha = 1;$$

$$||h||_{H^{s,r}_{x,v}L^\infty_z} \le C_I\,e^{-\varepsilon\tau_s t}, \qquad ||h||_{H^s_{x,v}H^r_z} \le C_I\,e^{-\varepsilon\tau_s t}. \quad \text{for } \alpha = 0,$$

where C_I, τ_s are positive constants independent of ε.

2.2.2 The Boltzmann Equation with Random Collision Kernels

As an example of the general theory, we discuss the Boltzmann equation with random collision kernel and/or random initial data, given by (1). The collision operator Q conserves mass, momentum and energy. Its solution formally satisfies the celebrated Boltzmann's H theorem [8]:

$$-\frac{d}{dt}\int_{\mathbb{R}^d} f\log f\,dv = -\int_{\mathbb{R}^d} Q(f,f)\log(f)\,dv \ge 0. \tag{7}$$

The equilibrium distribution is given by the Maxwellian distribution

$$\mathcal{M}(\rho_\infty, u_\infty, T_\infty) = \frac{\rho_\infty}{(2\pi T_\infty)^{N/2}}\exp\left(-\frac{|u_\infty - v|^2}{2T_\infty}\right), \tag{8}$$

where ρ_∞, u_∞, T_∞ are the density, mean velocity and temperature of the gas

$$\rho_\infty = \int_{\Omega\times\mathbb{R}^d} f(v)\,dxdv, \qquad u_\infty = \frac{1}{\rho_\infty}\int_{\Omega\times\mathbb{R}^d} vf(v)\,dxdv,$$

$$T_\infty = \frac{1}{N\rho_\infty}\int_{\Omega\times\mathbb{R}^d} |u_\infty - v|^2 f(v)\,dxdv,$$

which are all determined by the initial datum due to the conservation properties. We will consider hard potentials with B satisfying Grad's angular cutoff, that is,

$$B(|v-v_*|, \cos\theta, z) = \phi(|v-v_*|)\,b(\cos\theta, z), \quad \phi(\xi) = C_\phi\,\xi^\gamma, \text{ with } \gamma \in [0,1],$$

$$\forall \eta \in [-1,1], \quad |b(\eta, z)| \le C_b, \quad |\partial_\eta b(\eta, z)| \le C_b, \quad |\partial_z^k b(\eta, z)| \le C_b^*, \ \forall 0 \le k \le r. \tag{9}$$

where b is non-negative and not identically equal to 0.

For the Boltzmann equation, Theorem 1 shows that the uncertainties from the initial datum will eventually diminish and the solution will exponentially decay in time to the deterministic global equilibrium, with a decay rate of $O(e^{-t})$ under the incompressible Navier-Stokes scaling and $O(e^{-\varepsilon t})$ under the acoustic scaling. We remark that the hypocoercivity analysis performed in [10, 57] are carried out for solutions near the global equilibrium, *i.e.* in a *perturbative* setting, so that the solution can be defined under suitable Sobolev norms, as in the case for deterministic equations [21, 83, 85]. For general initial data, due to the possibility of singularity formation of the solution, Sobolev estimates are impossible, even in the deterministic case, for the study of long-time behavior. This remains an open problem.

3 Stochastic Galerkin Method: An Intrusive Scheme

Among popular intrusive numerical methods in UQ, one of the standard and efficient approaches is the generalized polynomial chaos approach in the stochastic Galerkin (referred as gPC-SG) framework [26, 32, 36, 37, 48, 88]. Compared with the classical Monte-Carlo method, the gPC-SG approach enjoys a spectral accuracy in the random space–if the solution is smooth enough–while the Monte-Carlo method converges with only half-th order accuracy. This makes it very efficient if the dimension of the random space is not too high, compared with the classical Monte-Carlo method.

In the SG method, one approximates h by

$$h(t,x,v,z) \approx \sum_{|\mathbf{k}|=1}^{K} h_{\mathbf{k}}(t,x,v)\psi_{\mathbf{k}}(z) := h^K(t,x,v,z), \tag{10}$$

where $\mathbf{k} = (k_1, \ldots, k_n)$ is a multi-index with $|\mathbf{k}| = k_1 + \cdots + k_n$. The orthonormal gPC basis functions $\{\psi_{\mathbf{k}}(z)\}$ satisfy

$$\int_{I_z} \psi_{\mathbf{k}}(z)\psi_{\mathbf{j}}(z)\pi(z)dz = \delta_{\mathbf{kj}}, \qquad 1 \le |\mathbf{k}|,\ |\mathbf{j}| \le K,$$

where $\pi(z)$ is the probability distribution function of random variable z, which is assumed known in our problem. By inserting the ansatz (10) into (4) and conducting a standard Galerkin projection, one obtains the SG system for h_k for each $1 \le k \le K$:

$$\begin{cases} \partial_t h_k + \dfrac{1}{\varepsilon^\alpha} v \cdot \nabla_x h_k = \dfrac{1}{\varepsilon^{1+\alpha}} \mathcal{L}_k(h^K) + \dfrac{1}{\varepsilon^\alpha} \mathcal{F}_k(h^K, h^K), \\ h_k(0, x, v) = h_k^0(x, v). \end{cases} \tag{11}$$

Here on left hand side all the modes are decoupled, while on the right hand side $\mathcal{L}_k$ and $\mathcal{F}_k$ couple all modes h_k.

Regarding the trends to deterministic or uncertain equilibrium of kinetic models, we mention different possible performances of SG methods: A standard implementation of SG methods may lead to a loss of accuracy and reduced efficiency when the long time computation of the system is considered, a phenomenon that has been studied in [25], where the accuracy of standard SG schemes is destroyed even in simple relaxation problems with *deterministic* large time behavior. On the other hand, in [27] the authors carefully constructed a new micro-macro based SG method for nonlinear Fokker-Planck equations with random inputs, see [27, Sect. 4.1], where the *uncertain* local equilibrium is preserved in the large time behavior, with a spectral accuracy for the gPC approximation. In this article, we will focus on the case of deterministic Maxwellian equilibrium and study the long-time behavior of numerical solution to the SG system.

3.1 Error Analysis of the gPC-SG System

Due to its Galerkin formulation, mathematical analysis of the SG methods can be conducted conveniently. Indeed many of the analytical methods well-established in kinetic theory can be easily adopted or extended to study the SG system of the random kinetic equations. For example, the study of regularity, and hypocoercivity analysis can be used to analyze the SG system. For the SG approximation of uncertain Boltzmann equation with multiple scalings, in [57] we established the spectral convergence and long-time exponential error decay, under the condition that the random perturbation of the collision kernel is in the order of the Knudsen number ε. Later in [10], by carefully handling the linearized collision operator for the SG system, we were able to establish a spectral gap estimate and remove the $O(\varepsilon)$ restriction in [57] on the random perturbation for the collision kernel. In [11] we obtained similar results under different norm spaces for the more challenging multi-species Boltzmann equation. There a new decomposition of the system was introduced such that the hypodissipative and regularising properties of the new kinetic operators for high-order z derivatives of the distribution function can be proved and used in a similar way as in the deterministic problem. Thus [10, 57] as a whole have provided powerful theoretical tools to study and conduct the local sensitivity analysis for uncertain collisional kinetic equations and their numerical approximations.

Again, we take the example of the Boltzmann equation with uncertain collision kernel and/or initial data. In [57], the following energy estimate, regularity of the solution solved by the gPC-SG system and error estimates on the SG method are shown:

Theorem 2 *Assume the collision kernel B satisfies (9) and is linear in z, with the form off*

$$b(\cos\theta, z) = b_0(\cos\theta) + b_1(\cos\theta)z\,, \tag{12}$$

with $|\partial_z b| \le O(\epsilon)$. We also assume the technical condition $||\psi_k||_{L^\infty} \le Ck^p$ for $\forall k \ge 1$, with a parameter $p > 0$. Let $q > p+2$, define the energy E^K by

$$E^K(t) = E^K_{s,q}(t) = \sum_{k=1}^{K} ||k^q h_k||^2_{H^s_{x,v}}, \tag{13}$$

with the initial data satisfying $E^K(0) \le \eta$. Then for all $s \ge s_0$, $0 \le \epsilon_d \le 1$, such that for $0 \le \epsilon \le \epsilon_d$, if h^K is a gPC solution of (11) in $H^s_{x,v}$, we have:

$$E^K(t) \le \eta\, e^{-\tau t}, \quad \text{for } \alpha = 1;$$

$$E^K(t) \le \eta\, e^{-\epsilon\tau t}, \quad \text{for } \alpha = 0,$$

where η, τ are all positive constants that only depend on s and q, independent of K and z.

From there, we conclude that, for h^K solved by the SG system (11), $||h^K||_{H^s_{x,v}L^\infty_z}$ also decays exponentially in time, with the same rate as $E^K(t)$, namely

$$||h^K||_{H^s_{x,v}L^\infty_z} \le \eta\, e^{-\tau t}, \quad \text{for } \alpha = 1;$$

$$||h^K||_{H^s_{x,v}L^\infty_z} \le \eta\, e^{-\epsilon\tau t}, \quad \text{for } \alpha = 0.$$

Finally, using all the above results, we obtain the error estimates on the SG method for the uncertain Boltzmann equation.

Theorem 3 *Suppose the assumptions on the collision kernel and basis functions are satisfied (see details in [57]), then*

$$||h - h^K||_{H_z^s} \leq C\frac{e^{-\lambda t}}{K^r}, \quad for\, \alpha = 1;$$

$$||h - h^K||_{H_z^s} \leq C\frac{e^{-\varepsilon\lambda t}}{K^r}, \quad for\, \alpha = 0,$$

with the constants $C, \lambda > 0$ independent of K and ϵ.

3.2 Stochastic AP Schemes

Furthermore, for numerical approximation of multiscale kinetic equations, the SG methods allow one to extend the deterministic Asymptotic-Preserving framework to the stochastic case. Stochastic asymptotic-preserving (s-AP) schemes, first introduced in the SG setting [49], require that as the Knudsen number $\varepsilon \to 0$, the SG system for the microscopic model with randomness automatically becomes a SG approximation for the limiting uncertain macroscopic equation. The SG method yields systems of deterministic equations that resemble the original kinetic equation, although in vector forms. Thus one can easily use the deterministic AP scheme to numerical solve the random counterpart, and allowing minimum "intrusion" to the legacy deterministic codes [42].

We will use the bipolar semiconductor Boltzmann-Poisson system as an example to demonstrate the efficiency of the s-AP scheme. We refer to [56] for details. In semiconductor devices, electrical currents originate from the transport of electrons and holes. $f_n(x, v, t)$, $f_p(x, v, t)$ represent the existence probability of an electron and a hole, respectively, at position $x \in \mathbb{R}^d$, with the velocity $v \in \mathbb{R}^d$, where d is the dimension, at time $t \geq 0$. The Boltzmann equations that give the evolution of the distribution functions are modelled by [50, 80]

$$\epsilon \partial_t f_n + (v \cdot \nabla_x f_n - E \cdot \nabla_v f_n) = \frac{1}{\epsilon} Q_n(f_n) + \epsilon I_n(f_n, f_p), \tag{14}$$

$$\epsilon \partial_t f_p + (\beta v \cdot \nabla_x f_p + E \cdot \nabla_v f_p) = \frac{1}{\epsilon} Q_p(f_p) + \epsilon I_p(f_n, f_p), \tag{15}$$

$$\gamma\, \Delta_x \Phi = n - p - C(x), \qquad E = -\nabla_x \Phi. \tag{16}$$

where $\beta = m_e^*/m_h^*$ is the ratio of the effective masses of electrons and holes, $\Phi = \Phi(t, x)$ represents the electric potential, $E = E(t, x)$ is the self-consistent electric field given by the Poisson equation (16), and γ is some scaled Debye length, $C(x)$ is the doping profile. The densities of the electron and the hole are

$$n = \int_{\mathbb{R}^d} f_n\, dv, \qquad p = \int_{\mathbb{R}^d} f_p\, dv.$$

The linear collision operators are given by

$$Q_i(f_i) = \int_{\mathbb{R}^d} \sigma_i(x, v, w)(M_i(v) f_i(w) - M_i(w) f_i(v)) dw, \qquad i = n, p,$$

with M_n and M_p the normalized Maxwellian distribution of the electrons and holes. The recombination-generation operators are described by [50, 80]

$$I_n(f_n, f_p) = \int_{\mathbb{R}^d} \sigma_I(x, v, w) \left[M_n(v) - M_p(w) f_n(v) f_p(w) \right] dw, \tag{17}$$

$$I_p(f_n, f_p) = \int_{\mathbb{R}^d} \sigma_I(x, w, v) \left[M_n(w) - M_p(v) f_n(w) f_p(v) \right] dw, \tag{18}$$

where σ_I is the generation-recombination kernel and rotationally invariant.

We show one numerical example done in [56]. Assume the uncertain doping profile

$$c(x, z) = \left[1 - (1 - m) \left(\tanh\left(\frac{x - x_1}{s} \right) - \tanh\left(\frac{x - x_2}{s} \right) \right) \right] (1 + 0.5z),$$

with z following a uniform distribution on $[-1, 1]$. Let the random collision kernels be

$$\sigma_1 = \sigma_2 = 2 + z, \qquad \sigma_I(v, w) = \frac{1}{\sqrt{\pi}} e^{-(v-w)^2}.$$

We assume equilibrium boundary conditions in x,

$$f_i(x_L, v, t) = M_i(v),\ v > 0; \qquad f_i(x_R, v, t) = M_i(v),\ v < 0,$$

and the initial distributions $f_i(x, v, t = 0) = M_i(v)$, for $i = n, p$. The collision and generation-recombination kernels are given by

$$\sigma_1(v, w) = \sigma_2(v, w) = 2, \qquad \sigma_I(v, w) = \frac{1}{\sqrt{\pi}} e^{-(v-w)^2},$$

and $\beta = 0.9$, $\gamma = 0.002$, $\Phi(x_L) = 0$, $\Phi(x_R) = 5$. Here

$$c(x) = 1 - (1 - m) \left[\tanh\left(\frac{x - 0.3}{0.02} \right) - \tanh\left(\frac{x - 0.7}{s} \right) \right],$$

with $m = (1 - 0.001)/2$.

We adopt the s-AP scheme developed for the bipolar semiconductor Boltzmann equations in [56] and show the numerical results in Figs. 1 and 2. A satisfactory agreement between gPC-SG solutions and the reference solutions, computed by using

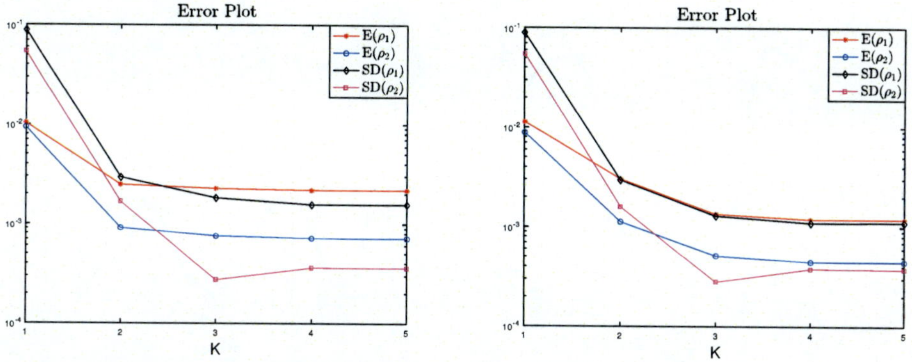

Fig. 1 Error plots for mean and standard deviation of $\rho_1, \rho_2, \varepsilon = 10^{-3}$ and $\varepsilon = 10^{-4}$ respectively. Final time is $T = 0.005$

high-order stochastic collocation method, is clearly seen. From Fig. 1, we observe fast exponential convergence of the error of the gPC-SG method, with respect to an increasing K.

4 Multi-fidelity Method: A Non-intrusive Scheme

When dealing with kinetic equations, especially for models containing high-dimensional random parameters, non-intrusive sampling methods, such as Monte Carlo (MC) sampling and stochastic collocation method [16–18, 28, 39, 63, 86], have several advantages over an intrusive gPC approach, since they allow us to adopt existing, robust deterministic numerical solvers that satisfy certain relevant physical properties [13, 15]. Fast algorithms and parallelization techniques can also be used, which are essential to reduce the computational complexity [14].

One of the challenges central to uncertainty quantification for kinetic equations is the simulation cost. Most of the existing algorithms are mainly developed based on the direct resolution of the main reference model, the so-called high-fidelity model. For many complex systems, in particular, the kinetic equations with multidimensional random inputs, an accurate high-fidelity deterministic simulation can be so time-consuming and memory demanding that only a few high-fidelity simulations can be afforded. Many stochastic algorithms require repetitive implementations of the deterministic solver, the overall accurate stochastic simulation can be difficult and even computationally infeasible. However, there usually exist some approximate, less complex low-fidelity models which compared to the high-fidelity models, usually contain simplified physics and/or are simulated on a coarser physical mesh, and consequently, own a cheaper computational cost. Although their accuracy may not be high, the low-fidelity models are designed in such a way that they can resolve or capture certain important features of the underlying problem and produce reliable

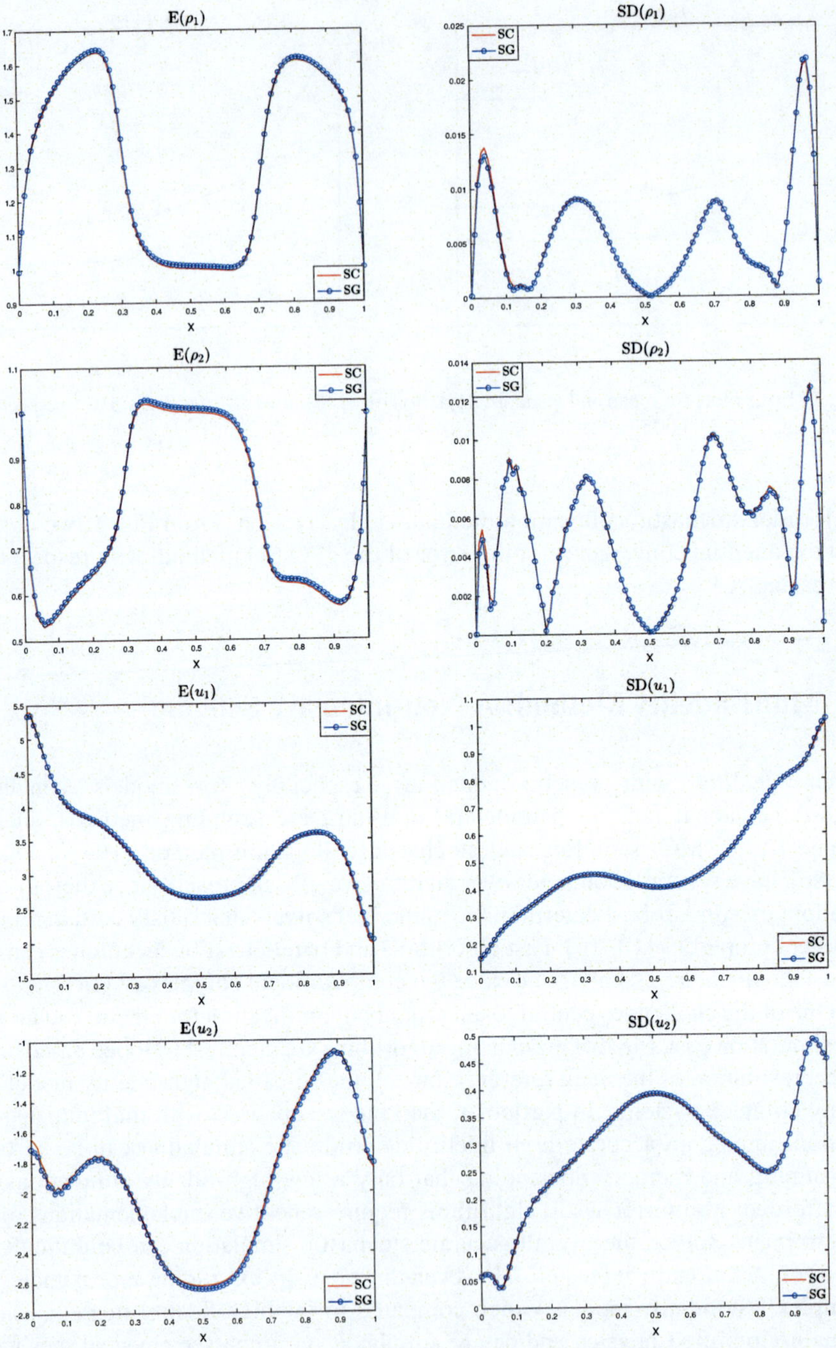

Fig. 2 Random doping and initial data. $\Delta x = 0.01$, $\Delta t = 2 \times 10^{-6}$, $N_v = 16$. Red solid line: reference solutions by the SC method with 16 quadrature points. Blue line with circles: gPC-SG method with $K = 4$

and qualitative predictions. Recently, there has been a surging interest in developing efficient uncertainty quantification algorithms by leveraging the strengths of multiple models where costs and fidelity, to be intended as the capacity of correctly describing the problem under consideration, vary. This approach is known as the multi-fidelity method in the literature [22, 75, 76]. In the following, we will mainly focus on the bi-fidelity method in a stochastic collocation framework [90].

4.1 A Bi-fidelity Stochastic Collocation (BFSC) Algorithm

First, we introduce notations in the following. We let $u(x, t; z)$ be the solution of a complex system subject to uncertainty, where $x \in \mathcal{D}$ and t are the spatial and temporal variables, and $z \in \Omega \subset \mathbb{R}^{d_z}$ is a d_z-dimensional random variable. Here Ω is the support of z, where the probability distribution $p(z)$ is defined. For the sake of simplicity, we denote $u(x, t; z)$ by $u(z)$. Assume the high-fidelity solutions $u^H(z)$ and low-fidelity solutions $u^L(z)$ are available, solved by corresponding numerical schemes of both models. Set N the number of affordable low-fidelity simulation runs, which is often very large due to the reduced complexity of the model. On the other hand, M denotes the number of high-fidelity simulation runs that can be afforded and is typically very small, i.e. the setting is such that $N \gg M$. Let finally $\gamma_k = \{z_1, \ldots, z_k\}$, $k \geq 1$ be a set of sample points in Ω.

We denote by $u^L(\gamma_k) = [u^L(z_1), \ldots, u^L(z_k)]$ the low-fidelity snapshot matrix corresponding to the solution of the low fidelity model for the sample point z_k. To this matrix we can associate a corresponding low-fidelity approximation space spanned by the set of sample points γ_k,

$$U^L(\gamma_k) = \text{span}\{u^L(\gamma_k)\} = \text{span}\{u^L(z_1), \ldots, u^L(z_k)\}.$$

Similarly, the high-fidelity snapshot matrix, obtained from the sample set γ_k, and the corresponding high-fidelity approximation space spanned by the solutions computed at nodes z_k, are defined as follows:

$$u^H(\gamma_k) = [u^H(z_1), \ldots, u^H(z_k)], \qquad U^H(\gamma_k) = \text{span}\{u^H(\gamma_k)\}.$$

The main idea of the BFSC method is to construct an inexpensive surrogate $u^B(z; \gamma_M)$ of the high-fidelity solution in the following non-intrusive manner

$$u^H(z) \approx u^B(z; \gamma_M) = \sum_{k=1}^{M} c_k(z) u^H(z_{i_k}), \tag{19}$$

where M is expected to be the number of high-fidelity samples and where correspondingly $z_{i_k} \in \gamma_M$ is a subset of size M of the sample space of size M_E. In other words, we approximate the solution of the high fidelity model in the space spanned by $u^H(z_{i_k}),\ k = 1, \ldots, M$. When constructing such algorithm one seeks for M to be as small as possible, since large M means more high-fidelity simulations and consequently prohibitive computational efforts. Thus, the central idea of the BFSC algorithm is to use cheap low-fidelity models to learn the coefficients $c_i(z)$ in (19), then apply the same approximation rule to a limited number and carefully selected, of high-fidelity samples to construct the bi-fidelity approximations of high-fidelity samples.

The BFSC algorithm for approximating the high-fidelity solution consists of offline and online stages. In the offline stage, we employ the cheap low-fidelity model to explore the parameter space to find the most important parameter points, i.e. a small number of samples permitting to give a suitable approximate solution. During the online stage, we learn the approximation rule from the low-fidelity model for any given z, and apply it to construct the bi-fidelity approximation. We outline the key ideas in Algorithm 1.

Algorithm 1 BFSC method

1. **Offline stage**

   a. Select a sample set $\Gamma_N = \{z_1, z_2, \ldots, z_N\} \subset \Omega$.
   b. Run the low-fidelity model $u_l(z_j)$ for each $z_j \in \Gamma_N$.
   c. Select M "important" points from Γ_N and denote it by $\gamma_M = \{z_{i_1}, \ldots z_{i_M}\} \subset \Gamma_N$. Construct the low-fidelity approximation space $U^L(\gamma_M)$.

2. **Online**

   a. Run high-fidelity simulations at each sample point of the selected sample set γ_M. Construct the high-fidelity approximation space $U^H(\gamma_M)$.
   b. For any given z, run the low-fidelity model to get the corresponding low-fidelity solution $u^L(z)$ and compute the low-fidelity coefficients by projection:

$$u^L(z) \approx \mathcal{P}_{U^L(\gamma_M)} u^L = \sum_{k=1}^{M} c_k(z) u^L(z_{i_k}).$$

   c. Construct the bi-fidelity approximation by applying the sample approximation rule learned from the low-fidelity model:

$$u^B(z) = \sum_{k=1}^{M} c_k(z) u^H(z_{i_k}).$$

4.2 Numerical Examples

We will apply the BFSC method to the Boltzmann equation, linear transport equation and epidemic kinetic models with multi-dimensional random parameters. The literature we based on is mainly from [2, 58, 59].

4.2.1 A BFSC Method for the Boltzmann Equation

In this section, we consider the application of the bi-fidelity algorithm to the study of the the Boltzmann equation (1) under the hydrodynamic scaling and with multi-dimensional random parameters. First, we consider a benchmark Sod shock tube test. In this test, the fluid regime with $\varepsilon = 10^{-4}$ is chosen. Assume the random cross section

$$b(z^b) = 1 + 0.5 \sum_{k=1}^{d_1+1} \frac{z_k^b}{2k},$$

and the uncertain initial distribution given by

$$f^0(x, v, z) = \frac{\rho^0}{2\pi T^0} e^{-\frac{|v-u^0|^2}{2T^0}},$$

where the initial data for ρ^0, u^0 and T^0 is

$$\begin{cases} \rho_l = 1, \quad u_l = (0, 0), \quad T_l(z^T) = 1 + 0.4 \sum_{k=1}^{d_1} \frac{z_k^T}{2k}, \quad x \le 0.5, \\ \rho_r = \frac{1}{8}, \quad u_r = (0, 0), \quad T_r(z^T) = \frac{1}{8}\left(1 + 0.4 \sum_{k=1}^{d_1} \frac{z_k^T}{2k}\right), \quad x > 0.5. \end{cases}$$

Here z^b and z^T represent the random variables in the collision kernel and initial temperature, with $\{z_i^b\}_{i=1}^{d_1+1}$ and $\{z_i^T\}_{i=1}^{d_1+1}$ following the uniform distribution on $[-1, 1]$. Set $d_1 = 7$, and the total dimension d_z of the random space to be 15. We solve the Boltzmann (high-fidelity) equation by using the asymptotic-preserving scheme developed in [23], with the second-order MUSCL scheme [51] for the spatial discretization, and fast spectral method [65] for the collision operator implementation. Here $\Delta x = 0.01$, $N_v^h = 24$, and the final time $t = 0.15$. We employ the Euler equation as the low-fidelity model, and solve it by using the same spatial and temporal resolution as in the high-fidelity Boltzmann equation.

Figure 3 suggests a fast convergence of L^2 errors between the high-fidelity and bi-fidelity solutions. With only 10 high-fidelity runs, the bi-fidelity approximation reaches an accuracy level of $O(10^{-3})$, which is remarkable for a 15-dimensional random variable problem. One observes that the high-fidelity and bi-fidelity solutions match really well in the case $M = 10$, whereas the low-fidelity solutions are quite

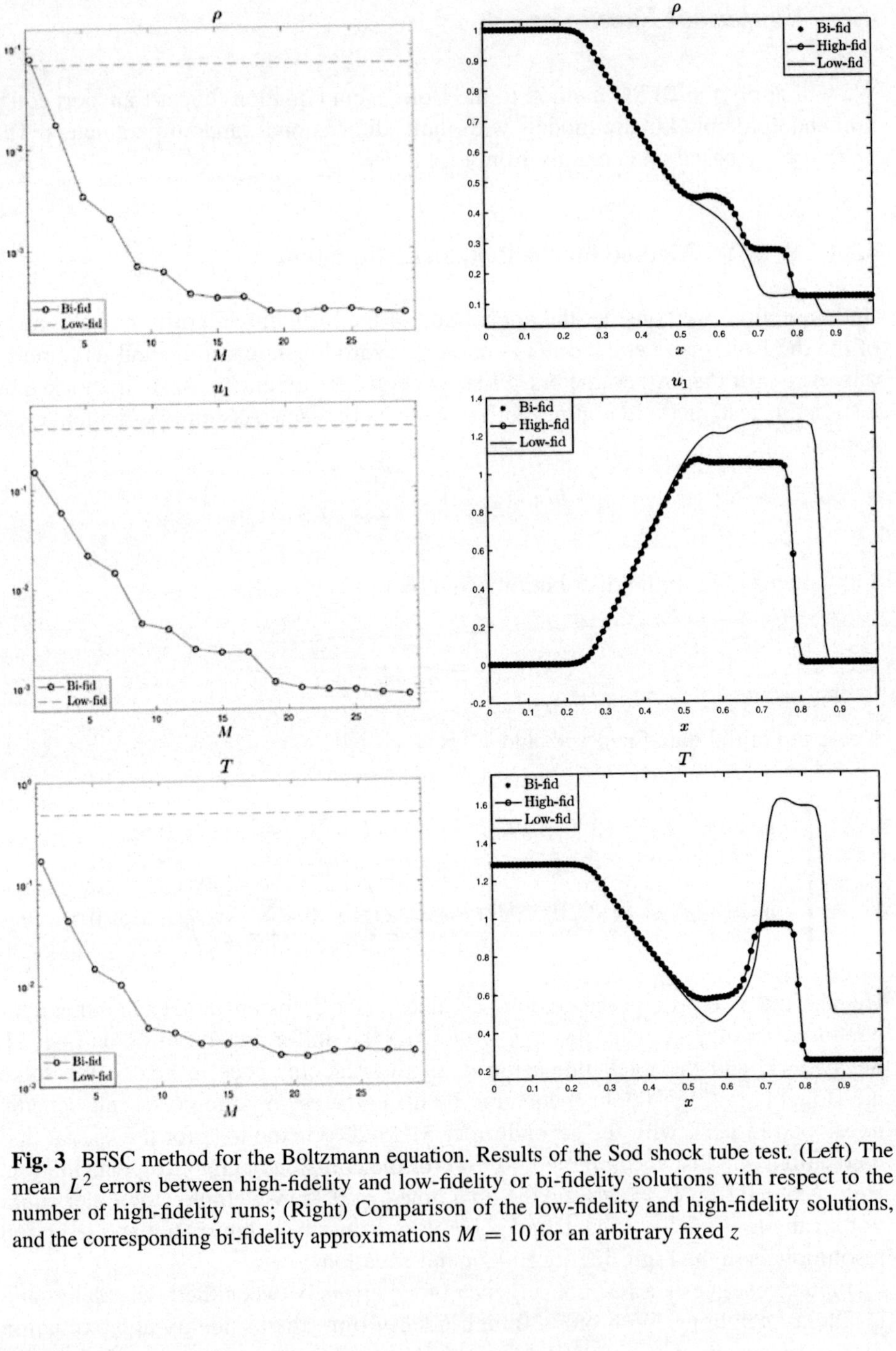

Fig. 3 BFSC method for the Boltzmann equation. Results of the Sod shock tube test. (Left) The mean L^2 errors between high-fidelity and low-fidelity or bi-fidelity solutions with respect to the number of high-fidelity runs; (Right) Comparison of the low-fidelity and high-fidelity solutions, and the corresponding bi-fidelity approximations $M = 10$ for an arbitrary fixed z

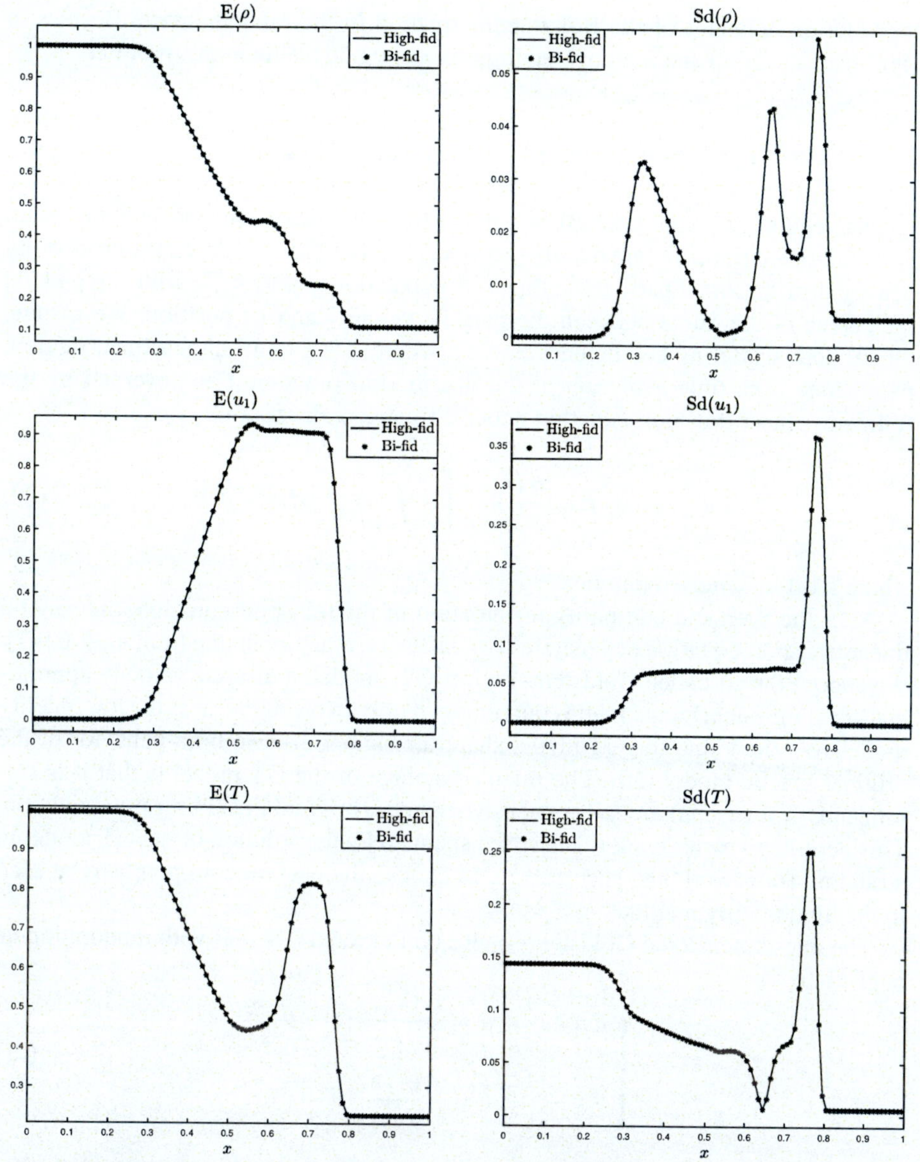

Fig. 4 BFSC method for the Boltzmann equation. Results of the Sod shock tube test. The mean and standard deviation of ρ, u_1, T of high-fidelity and bi-fidelity solutions with $M = 10$

inaccurate at some spatial points. Figure 4 shows that the mean and standard deviation of bi-fidelity approximations for ρ, u and T match well with the high-fidelity solutions under as few as 10 high-fidelity runs. These results indicate that even though the Euler model may not be accurate in the physical space, it still can capture well the solution behavior of the Boltzmann equation in the random space. Besides, a significant speedup and memory savings are quite noticeable in this problem.

4.2.2 A BFSC Method for the Linear Transport Equation

In this section, we apply the BFSC method to the linear transport equation with uncertainties. The results are mainly shown in [58]. Let $f(t, x, v, z)$ be the probability density distribution of particles at time $t > 0$, position $x \in \mathcal{D} \subseteq \mathbb{R}$, with $v \in [-1, 1]$ the cosine of the angle between the particle velocity and its position. We assume the random scattering coefficients $\sigma(x, z)$, where z is a multi-dimensional random parameters. The time evolution of the distribution function f is governed by the following linear transport equation under diffusive scaling:

$$\epsilon \partial_t f + v \partial_x f = \frac{\sigma(x, z)}{\varepsilon} \left[\frac{1}{2} \int_{-1}^{1} f(v')\, dv' - f \right], \tag{20}$$

where ε is the Knudsen number.

After the even and odd parity formulation of the transport equation, the general linear transport equation degenerates to a diffusion equation in the limit $\varepsilon \to 0$ [47]. We choose the so-called Goldstein-Taylor (GT) model, a discrete velocity approximation of the underlying kinetic equation with only two velocities, as the low-fidelity model. We show that the GT model share the same diffusive limit behavior of the original kinetic model (20). The main advantage of the GT model is that it is significantly less computationally expensive than the original linear transport equation. This permits to better explore the space spanned by the solution of the GT model in a random framework and successively to choose the best random points to be used in the high fidelity method.

The one-dimensional Goldstein-Taylor (GT) model [29, 84] with random inputs is given by

$$\begin{cases} \partial_t u + \dfrac{1}{\epsilon} \partial_x u = \dfrac{\sigma(x, z)}{2\epsilon^2} (v - u), \\[2ex] \partial_t v - \dfrac{1}{\epsilon} \partial_x v = \dfrac{\sigma(x, z)}{2\epsilon^2} (u - v). \end{cases} \tag{21}$$

Here we assume a random scattering coefficient $\sigma(x, z)$ as in the original high fidelity model (20). It is worth mentioning that the Goldstein-Taylor model (21) can be regarded as a discrete-velocity kinetic counterpart of the linear transport equation where u defines the density of particles traveling with velocity 1, whereas v that of particles traveling in the reverse direction with velocity -1. Besides, the GT model has significantly cheaper computational cost, yet shares the same limiting diffusion equations with the linear transport model as $\epsilon \to 0$. We refer to [54] for rigorous results concerning the diffusion limit of two-velocity models and extensions to nonlinear diffusion coefficients. This low-fidelity model is shown to be used in many interesting applications [31]. We introduce now the macroscopic variables: the mass density ρ and the flux s,

$$\rho = u + v, \qquad s = \frac{u - v}{\epsilon},$$

then the GT model (21) is equivalent to the following system:

$$\begin{cases} \partial_t \rho + \partial_x s = 0, \\ \partial_t s + \dfrac{1}{\epsilon^2} \partial_x \rho = -\dfrac{\sigma(x, z)}{\epsilon^2} s. \end{cases} \tag{22}$$

We employ the equivalent formulation of the GT equation (22) as our low-fidelity model. Assume a d_z-dimensional random variable $z = (z_1, \ldots, z_d)$ that follow the uniform distribution on $[-1, 1]^{d_z}$. In our tests of this section, we consider $d_z = 5$. To compute the reference solutions for the mean and standard deviation of the high-fidelity quantities of interests, we use the high-order stochastic collocation method over 5-dimensional sparse quadrature points with 5-level Clenshaw-Curtis rules [89]. For the high-fidelity solver, at each given sample we employ the AP scheme [47] developed for the deterministic linear transport equation (20) under the diffusive scaling. The standard 16-points Gauss-Legendre quadrature set is used for the velocity space to compute the density. For the low-fidelity (LF) solver, we use the deterministic AP method [46] to solve the linear Goldstein-Taylor model (22).

In this test, we consider a Riemann problem. Let us assume the random cross-section coefficient

$$\sigma(x, z) = 1 + 4 \sum_{i=1}^{d_z} \frac{1}{(i\pi)^2} \cos(2\pi i x) z_i, \tag{23}$$

with boundary conditions

$$f(x = 0, t, v, z) = 1 + 0.4 z_1, \quad \text{if } v \geq 0,$$
$$f(x = 1, t, v, z) = 0, \quad \text{if } v \leq 0,$$

with uncertain initial distribution given by

$$f(x, t = 0, v, z) = 1 + 0.4 z_1, \qquad 0 \leq x < 0.5,$$
$$f(x, t = 0, v, z) = 0, \qquad 0.5 \leq x \leq 1.$$

Let the output time $T = 0.01$ and $\epsilon = 10^{-8}$, set $\Delta t = 2 \times 10^{-4}$, $\Delta x = 0.04$ in the low-fidelity solver, and $\Delta t = 5 \times 10^{-5}$, $\Delta x = 0.0125$ in the high-fidelity solver. From Fig. 5, it is obvious that the bi-fidelity errors decay fast with respect to the selected high-fidelity runs. With only $M = 12$ high-fidelity simulation runs, the bi-

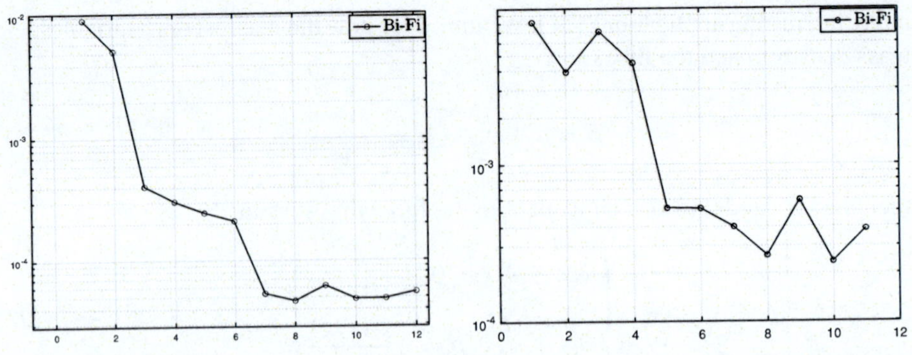

Fig. 5 BFSC method for the linear transport. Results of the Riemann problem in the diffusive scaling. Errors of the bi-fidelity approximation mean (left) and standard deviation (right) of the density ρ with respect to the number of high-fidelity runs

fidelity errors for the mean and standard deviation of the density ρ reach as small as $O(10^{-4})$. We mention that further increasing the high-fidelity samples after $M = 8$ will not help improving the quality of bi-fidelity approximation.

4.2.3 A BFSC Method for the Epidemic Kinetic Models

As an final example, we apply the BFSC method to study the UQ problem for the kinetic model of disease spread [2, 3]. Consider the epidemic transport system given by

$$\begin{aligned}
\frac{\partial f_S}{\partial t} + \frac{\partial (v_S f_S)}{\partial x} &= -F(f_S, I) + \frac{1}{\tau_S}\left(\frac{S}{2} - f_S\right) \\
\frac{\partial f_I}{\partial t} + \frac{\partial (v_I f_I)}{\partial x} &= F(f_S, I) - \gamma f_I + \frac{1}{\tau_I}\left(\frac{I}{2} - f_I\right) \\
\frac{\partial f_R}{\partial t} + \frac{\partial (v_R f_R)}{\partial x} &= \gamma f_I + \frac{1}{\tau_R}\left(\frac{R}{2} - f_R\right),
\end{aligned} \tag{24}$$

where S are the susceptible, I the infected and R the recovered individuals, with respectively kinetic densities f_S, f_I and f_R and $v_S = \lambda_S(x)v$, $v_I = \lambda_I(x)v$, $v_R = \lambda_R(x)v$ with $\lambda_S, \lambda_I, \lambda_R \geq 0$. The quantity $\gamma = \gamma(x, z)$ is the recovery rate of infected, and the transmission of infection is governed by an incidence function $F(\cdot, I)$ modeling the transmission of the disease [35]

$$F(g, I) = \beta \frac{g I^p}{1 + \kappa I}, \tag{25}$$

with the classic bi-linear case corresponding to $p = 1, \kappa = 0$. Here τ_S, τ_I, τ_R represent the relaxation frequencies playing the role of the Knudsen number in (1).

The low-fidelity model, is based on considering individuals moving in two opposite directions (indicated by signs "+" and "−"), with velocities $\pm\lambda_S$ for susceptible, $\pm\lambda_I$ for infected and $\pm\lambda_R$ for removed. This dynamics of the population through the two-velocity epidemic model [3] is given by

$$\begin{aligned}\frac{\partial S^{\pm}}{\partial t}+\lambda_S\frac{\partial S^{\pm}}{\partial x}&=-F(S^{\pm},I)\mp\frac{1}{2\tau_S}\left(S^{+}-S^{-}\right)\\ \frac{\partial I^{\pm}}{\partial t}+\lambda_I\frac{\partial I^{\pm}}{\partial x}&=F(S^{\pm},I)-\gamma I^{\pm}\mp\frac{1}{2\tau_I}\left(I^{+}-I^{-}\right)\\ \frac{\partial R^{\pm}}{\partial t}+\lambda_R\frac{\partial R^{\pm}}{\partial x}&=\gamma I^{\pm}\mp\frac{1}{2\tau_R}\left(R^{+}-R^{-}\right).\end{aligned} \tag{26}$$

In the above system, individuals $S(x,t,z)$, $I(x,t,z)$ and $R(x,t,z)$ are defined as

$$S=S^{+}+S^{-},\quad I=I^{+}+I^{-},\quad R=R^{+}+R^{-},$$

with the fluxes

$$J_S=\lambda_S\left(S^{+}-S^{-}\right),\quad J_I=\lambda_I\left(I^{+}-I^{-}\right),\quad J_R=\lambda_R\left(R^{+}-R^{-}\right). \tag{27}$$

Then we derive a hyperbolic model equivalent to (26), while presenting a macroscopic description of propagation of the epidemic at finite speeds. In this system the fluxes satisfy

$$\begin{aligned}\frac{\partial J_S}{\partial t}+\lambda_S^2\frac{\partial S}{\partial x}&=-F(J_S,I)-\frac{J_S}{\tau_S}\\ \frac{\partial J_I}{\partial t}+\lambda_I^2\frac{\partial I}{\partial x}&=\frac{\lambda_I}{\lambda_S}F(J_S,I)-\gamma J_I-\frac{J_I}{\tau_I}\\ \frac{\partial J_R}{\partial t}+\lambda_R^2\frac{\partial R}{\partial x}&=\frac{\lambda_R}{\lambda_I}\gamma J_I-\frac{J_R}{\tau_R}.\end{aligned} \tag{28}$$

Now consider the behavior of the low-fidelity model in the diffusive regime. To this aim, we define the diffusion coefficients

$$D_S=\lambda_S^2\tau_S,\quad D_I=\lambda_I^2\tau_I,\quad D_R=\lambda_R^2\tau_R\,. \tag{29}$$

Its diffusion limit can be formally recovered by letting the relaxation times $\tau_{S,I,R}\to 0$. We remark that the motivation of choosing (26) as our low-fidelity model in the bi-fidelity approximation is that it shares the same diffusion limit as the high-fidelity model. The only difference lies in the definition of two diffusion coefficients, see [2] for details.

In this test, consider a 2-dimensional random vector $\mathbf{z}=(z_1,z_2)^T$, with z_1 and z_2 following a uniform distribution on $[-1,1]$. To compute the reference solutions, we adopt a 3-rd level sparse grid quadrature based on Clenshaw-Curtis rules for the choice of the stochastic collocation nodes, for both the high-fidelity and low-fidelity models. We assume the initial distributions of the high-fidelity kinetic SIR model (24) as follows:

$$f_i(x, v, 0) = c\, i(x, 0)\, e^{-\frac{v^2}{2}}, \qquad i \in \{S, I, R\} \tag{30}$$

where $c = \frac{1}{2}\sum_{i=1}^{N_v} w_i\, e^{-\frac{\zeta_i^2}{2}}$ is a re-normalization constant, with N_v the number of Gauss-Legendre quadrature points used in velocity discretization. Let the initial densities be

$$S(x, 0) = 1 - I(x, 0), \qquad I(x, 0) = 0.01e^{-(x-10)^2}, \qquad R(x, 0) = 0,$$

with a physical domain $L = [0, 20]$. Let the initial fluxes $J_S(x, 0)$, $J_I(x, 0)$ and $J_R(x, 0)$ be zeros and consider periodic boundary conditions. The same initial conditions for S, I, R and J_S, J_I, J_R are imposed in the low-fidelity SIR model to make the two models consistent. Concerning spatially heterogeneous environments, we assume the contact rate [3, 87] be spatially dependent and contains uncertainties

$$\beta(x, z) = \beta^0(z)\left(1 + 0.05 \sin\left(\frac{13\pi x}{20}\right)\right),$$

where $\beta^0(z) = 11(1 + 0.6z_1)$. The uncertain recovery rate is assumed $\gamma(z) = 10(1 + 0.4z_2)$. In the incidence function, let $\kappa = 0$ and $p = 1$. In the first test, a parabolic configuration of speeds and relaxation parameters is considered, by letting $\lambda_i^2 = 10^5, i \in \{S, I, R\}$. Here $\tau_i = 10^{-5}$ in the low-fidelity model and $\tau_i = 3 \times 10^{-5}$ in the high-fidelity model, to maintain consistency of the two simulations. For spatial and velocity discretizations, we use $N_x = 150$ in both the high-fidelity and low-fidelity models and $N_v = 8$ for the high-fidelity model.

In the first row of Fig. 6, the mean and standard deviation of the solution of compartment I for the high-fidelity model, the low-fidelity model and the bi-fidelity approximation at time $t = 5$ are shown. Since the high-fidelity and low-fidelity model share the same diffusive limit, a perfect agreement of the solutions are observed. In the second row, we plot the relative L^2 errors of the mean and standard deviation between the bi-fidelity and high-fidelity solutions at $t = 5$. A fast error decay with respect to the number of selected important points in the bi-fidelity algorithm is clearly observed. One can conclude that with only 8 hi-fidelity sample points, bi-fidelity approximation is able to achieve a relative error of $O(10^{-6})$ for both the mean and standard deviation. In the second test, we consider the hyperbolic regime for $\lambda_i = 1$, $i \in \{S, I, R\}$, with $\tau_i = 1$ in the low-fidelity model and $\tau_i = 3$ in the high-fidelity model. Results are shown in Fig. 7. We refer [2] for details.

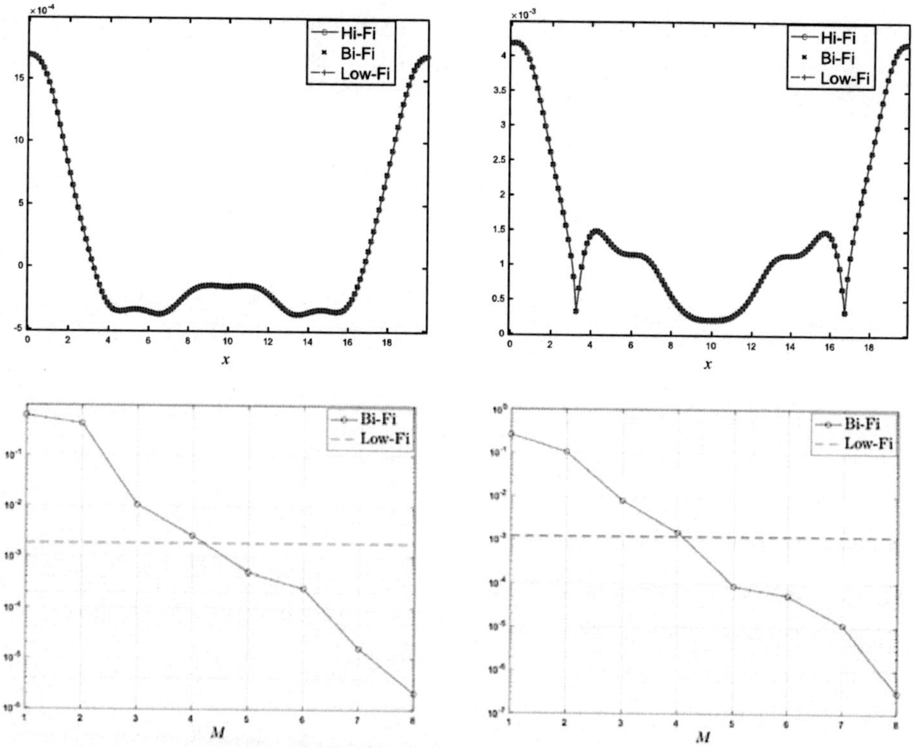

Fig. 6 BFSC method for the kinetic SIR model. Results in the diffusive regime. First row: mean (left) and standard deviation (right) for the density I, by using $M = 8$ high-fidelity runs. Second row: relative L^2 errors of the bi-fidelity approximation

5 Conclusion

In this article, we reviewed some of the recent progress on UQ for multiscale kinetic problems with uncertainties. We address from both aspects of mathematical analysis and numerical computation. A general framework on the regularity and long-time behavior of the solution to collisional multiscale kinetic models with random inputs is given, by using the hypocoercivity of kinetic operators. For numerical approximations, we study both the non-intrusive and intrusive methods, in particular the SG scheme that satisfies the s-AP property and SC method in a multi-fidelity framework. We study several important kinetic equations, with applications in various fields such as rarefied gas, semiconductor device modeling, social and biological sciences. There are many open problems on this topic, e.g., analysis for boundary value problems, study of efficient, high-accuracy sampling based methods for kinetic problems with high-dimensional random parameters, which worth further exploring.

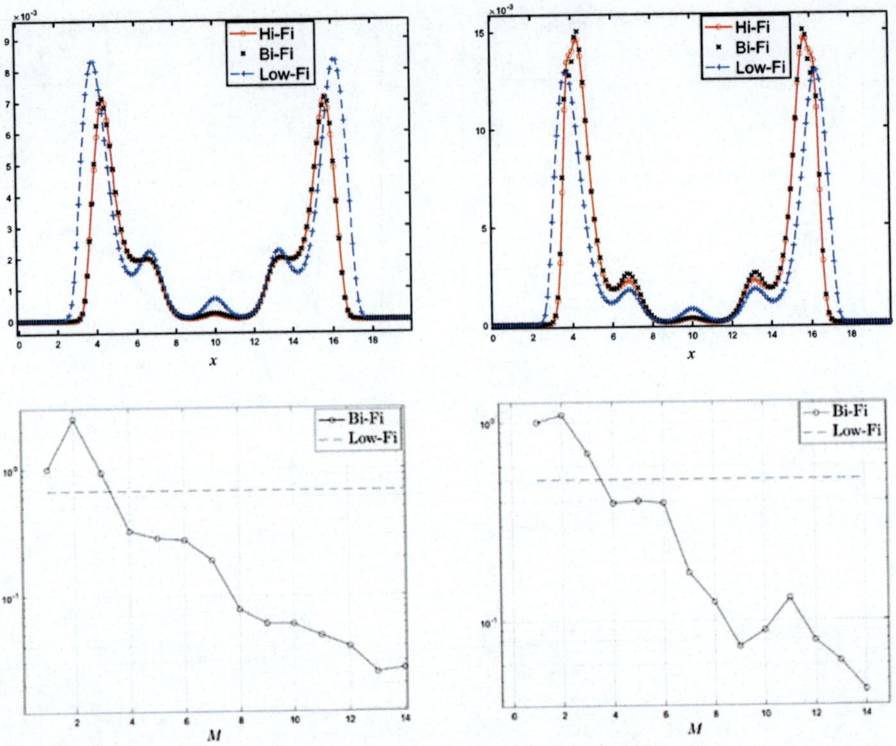

Fig. 7 BFSC method for the kinetic SIR model. Results in the hyperbolic regime. First row: mean (left) and standard deviation (right) for the density I, by using $M = 8$ high-fidelity runs. Second row: relative L^2 errors of the bi-fidelity approximation

References

1. Bardos, C., Golse, F., Levermore, D.: Fluid dynamic limits of kinetic equations. I. Formal derivations. J. Statist. Phys. **63**, 323–344 (1991)
2. Bertaglia, G., Liu, L., Pareschi, L., Zhu, X.: Bi-fidelity stochastic collocation methods for epidemic transport models with uncertainties. Netw. Heterog. Media **17**, 401–425 (2022)
3. Bertaglia, G., Pareschi, L.: Hyperbolic models for the spread of epidemics on networks: kinetic description and numerical methods. ESAIM Math. Model. Numer. Anal **55**, 381–407 (2021)
4. Bird, G.: Direct simulation and the Boltzmann equation. Phys. Fluids **13**, 2676–2681 (1970)
5. Bouchut, F., Golse, F., Pulvirenti, M.: Kinetic Equations and Asymptotic Theory. Elsevier (2000)
6. Briant, M.: From the Boltzmann equation to the incompressible Navier-Stokes equations on the torus: a quantitative error estimate. J. Differ. Equ. **259**, 6072–6141 (2015)
7. Carrillo, J.A., Hu, J., Wang, L., Wu, J.: A particle method for the homogeneous Landau equation. J. Comput. Phys. X **7**, 100066, 24 (2020)
8. Cercignani, C.: The Boltzmann Equation and Its Applications. Springer, New York (1988)
9. Cercignani, C.: Rarefied Gas Dynamics: From Basic Concepts to Actual Calculations. Cambridge Texts in Applied Mathematics, Cambridge University Press, Cambridge (2000)

10. Daus, E.S., Jin, S., Liu, L.: Spectral convergence of the stochastic Galerkin approximation to the Boltzmann equation with multiple scales and large random perturbation in the collision kernel. Kinet. Relat. Models **12**, 909–922 (2019)
11. Daus, E.S., Jin, S., Liu, L.: On the multi-species Boltzmann equation with uncertainty and its stochastic Galerkin approximation. ESAIM Math. Model. Numer. Anal. **55**, 1323–1345 (2021)
12. Degond, P., Deluzet, F.: Asymptotic-preserving methods and multiscale models for plasma physics. J. Comput. Phys. **336**, 429–457 (2017)
13. Degond, P., Pareschi, L., Russo, G. (eds.): Birkhäuser Boston Inc., Boston, MA (2004)
14. Dimarco, G., Loubére, R., Narski, J., Rey, T.: An efficient numerical method for solving the Boltzmann equation in multidimensions. J. Comp. Phys. **353**, 46–81 (2018)
15. Dimarco, G., Pareschi, L.: Numerical methods for kinetic equations. Acta Numerica **23**, 369–520 (2014)
16. Dimarco, G., Pareschi, L.: Multi-scale control variate methods for uncertainty quantification in kinetic equations. J. Comput. Phys. **388**, 63–89 (2019)
17. Dimarco, G., Pareschi, L.: Multi-scale variance reduction methods based on multiple control variates for kinetic equations with uncertainties. Multiscale Model Simul. **18**, 351–382 (2020)
18. Dimarco, G., Pareschi, L., Zanella, M.: Uncertainty quantification for kinetic models in socio-economic and life sciences. In: Uncertainty Quantification for Hyperbolic and Kinetic Equations. SEMA SIMAI Springer Series, vol. 14, pp. 151–191. Springer, Cham (2017)
19. DiPerna, R.J., Lions, P.-L.: On the Cauchy problem for Boltzmann equations: global existence and weak stability. Ann. Math. **130**(2), 321–366 (1989)
20. Dolbeault, J., Mouhot, C., Schmeiser, C.: Hypocoercivity for linear kinetic equations conserving mass. Trans. Am. Math. Soc. **367**, 3807–3828 (2015)
21. Dolbeault, J., Mouhot, C., Schmeiser, C.: Hypocoercivity for linear kinetic equations conserving mass. Trans. Amer. Math. Soc. **367**, 3807–3828 (2015)
22. Fernández-Godino, M.G., Park, C., Kim, N.-H., Haftka, R.T.: Review of multi-fidelity models (2016). arXiv:1609.07196
23. Filbet, F., Jin, S.: A class of asymptotic-preserving schemes for kinetic equations and related problems with stiff sources. J. Comput. Phys. **229**, 7625–7648 (2010)
24. Gamba, I.M., Tharkabhushanam, S.H.: Spectral-Lagrangian methods for collisional models of non-equilibrium statistical states. J. Comput. Phys. **228**, 2012–2036 (2009)
25. Gerritsma, M., van der Steen, J.-B., Vos, P., Karniadakis, G.: Time-dependent generalized polynomial chaos. J. Comput. Phys. **229**, 8333–8363 (2010)
26. Ghanem, R.G., Spanos, P.D.: Stochastic Finite Elements: A Spectral Approach. Springer, New York (1991)
27. Giacomo, D., Lorenzo, P., Mattia, Z.: Micro-macro stochastic Galerkin methods for nonlinear Fokker-Plank equations with random inputs, preprint
28. Giles, M.B.: Multilevel Monte Carlo methods. Acta Numerica **24**, 259–328 (2015)
29. Goldstein, S.: On diffusion by discontinuous movements, and on the telegraph equation. Quart. J. Mech. Appl. Math. **4**, 129–156 (1951)
30. Golse, F., Saint-Raymond, L.: The Navier-Stokes limit of the Boltzmann equation for bounded collision kernels. Invent. Math. **155**, 81–161 (2004)
31. Gottlieb, D., Xiu, D.: Galerkin method for wave equations with uncertain coefficients. Commun. Comput. Phys. **3**, 505–518 (2008)
32. Gunzburger, M.D., Webster, C.G., Zhang, G.: Stochastic finite element methods for partial differential equations with random input data. Acta Numer. **23**, 521–650 (2014)
33. Guo, Y.: Boltzmann diffusive limit beyond the Navier-Stokes approximation. Comm. Pure Appl. Math. **59**, 626–687 (2006)
34. Hérau, F., Nier, F.: Isotropic hypoellipticity and trend to equilibrium for the Fokker-Planck equation with a high-degree potential. Arch. Ration. Mech. Anal. **171**, 151–218 (2004)
35. Hethcote, H.W.: The mathematics of infectious diseases. SIAM Rev. **42**, 599–653 (2000)
36. Hu, J., Jin, S.: A stochastic Galerkin method for the Boltzmann equation with uncertainty. J. Comput. Phys. **315**, 150–168 (2016)

37. Hu, J., Jin, S.: Uncertainty quantification for kinetic equations. In: Uncertainty Quantification for Hyperbolic and Kinetic Equations. SEMA SIMAI Springer Series, vol. 14, pp. 193–229. Springer, Cham (2017)
38. Hu, J., Jin, S., Li, Q.: Asymptotic-preserving schemes for multiscale hyperbolic and kinetic equations. In: Handbook of Numerical Methods for Hyperbolic Problems. Handbook of Numerical Analysis, vol. 18, pp. 103–129. Elsevier/North-Holland, Amsterdam (2017)
39. Hu, J., Pareschi, L., Wang, Y.: Uncertainty quantification for the BGK model of the Boltzmann equation using multilevel variance reduced Monte Carlo methods. SIAM/ASA J. Uncertain. Quantif. **9**, 650–680 (2021)
40. Hu, J., Qi, K.: A fast Fourier spectral method for the homogeneous Boltzmann equation with non-cutoff collision kernels. J. Comput. Phys. **423**, 109806, 21 (2020)
41. Jin, S.: Efficient asymptotic-preserving (AP) schemes for some multiscale kinetic equations. SIAM J. Sci. Comput. **21**, 441–454 (1999)
42. Jin, S.: Mathematical analysis and numerical methods for multiscale kinetic equations with uncertainties. In: Proceedings of the International Congress of Mathematicians—Rio de Janeiro 2018. Invited lectures, vol. IV, pp. 3611–3639. World Sci. Publ., Hackensack, NJ (2018)
43. Jin, S.: Asymptotic-preserving schemes for multiscale physical problems. Acta Numer. **31**, 415–489 (2022)
44. Jin, S., Liu, J., Ma, Z.: Uniform spectral convergence of the stochastic Galerkin method for the linear transport equations with random inputs in diffusive regime and a micro–macro decomposition-based asymptotic-preserving method. Res. Math. Sci. **4**, Paper No. 15, 25 (2017)
45. Jin, S., Pareschi, L. (eds.): SEMA-SIMAI Springer Series, vol. 14. Springer (2017)
46. Jin, S., Pareschi, L., Toscani, G.: Diffusive relaxation schemes for multiscale discrete-velocity kinetic equations. SIAM J. Numer. Anal. **35**, 2405–2439 (1998)
47. Jin, S., Pareschi, L., Toscani, G.: Uniformly accurate diffusive relaxation schemes for multiscale transport equations. SIAM J. Numer. Anal. **38**, 913–936 (2000)
48. Jin, S., Xiu, D., Zhu, X.: Asymptotic-preserving methods for hyperbolic and transport equations with random inputs and diffusive scalings. J. Comp. Phys. **289**, 25–52 (2015)
49. Jin, S., Xiu, D., Zhu, X.: Asymptotic-preserving methods for hyperbolic and transport equations with random inputs and diffusive scalings. J. Comput. Phys. **289**, 35–52 (2015)
50. Jüngel, A.: Transport Equations for Semiconductors. Lecture Notes in Physics, vol. 773. Springer, Berlin (2009)
51. LeVeque, R.J.: Numerical Methods for Conservation Laws. Birkhäuser (1992)
52. Levermore, C.D., Masmoudi, N.: From the Boltzmann equation to an incompressible Navier-Stokes-Fourier system. Arch. Ration. Mech. Anal. **196**, 753–809 (2010)
53. Li, Q., Wang, L.: Uniform regularity for linear kinetic equations with random input based on hypocoercivity. SIAM/ASA J. Uncertain. Quantif. **5**, 1193–1219 (2017)
54. Lions, P.L., Toscani, G.: Diffusive limit for finite velocity Boltzmann kinetic models. Rev. Mat. Iberoamericana **13**, 473–513 (1997)
55. Liu, L.: Uniform spectral convergence of the stochastic Galerkin method for the linear semiconductor Boltzmann equation with random inputs and diffusive scaling. Kinet. Relat. Models **11**, 1139–1156 (2018)
56. Liu, L.: A stochastic asymptotic-preserving scheme for the bipolar semiconductor Boltzmann-Poisson system with random inputs and diffusive scalings. J. Comput. Phys. **376**, 634–659 (2019)
57. Liu, L., Jin, S.: Hypocoercivity based sensitivity analysis and spectral convergence of the stochastic Galerkin approximation to collisional kinetic equations with multiple scales and random inputs. Multiscale Model. Simul. **16**, 1085–1114 (2018)
58. Liu, L., Pareschi, L., Zhu, X.: A bi-fidelity stochastic collocation method for transport equations with diffusive scaling and multi-dimensional random inputs. J. Comput. Phys. **462**, Paper No. 111252, 16 (2022)
59. Liu, L., Zhu, X.: A bi-fidelity method for the multiscale Boltzmann equation with random parameters. J. Comput. Phys. **402**, 108914, 23 (2020)

60. Maitre, O.L., Knio, O.M.: Spectral methods for uncertainty quantification: with applications to computational fluid dynamics, Scientific Computation. Springer, Netherlands (2010)
61. Markowich, P.A., Ringhofer, C., Schmeiser, C.: Semiconductor Equations. Springer, Vienna (1990)
62. Medaglia, A., Colelli, G., Farina, L., Bacila, A., Bini, P., Marchioni, E., Figini, S., Pichiecchio, A., Zanella, M.: Uncertainty quantification and control of kinetic models of tumour growth under clinical uncertainties. Int. J. Non-Linear Mech. (2022)
63. Mishra, S., Schwab, C., Sukys, J.: Multi-level Monte Carlo finite volume methods for nonlinear systems of conservation laws in multi-dimensions. J. Comp. Phys. **231**, 3365–3388 (2012)
64. Mouhot, C., Neumann, L.: Quantitative perturbative study of convergence to equilibrium for collisional kinetic models in the torus. Nonlinearity **19**, 969–998 (2006)
65. Mouhot, C., Pareschi, L.: Fast algorithms for computing the Boltzmann collision operator. Math. Comp. **75**, 1833–1852 (2006)
66. Naldi, G., Pareschi, L., Toscani, G. (eds.): Modeling and Simulation in Science Engineering and Technology. Birkhäuser Boston Inc, Boston, MA (2010)
67. Naldi, G., Pareschi, L., Toscani, G. (eds.): Mathematical Modeling of Collective Behavior in Socio-Economic and Life Sciences. Modeling and Simulation in Science, Engineering and Technology. Springer (2010)
68. Nanbu, K.: Direct simulation scheme derived from the Boltzmann equation I: Monocomponent gases. J. Phys. Soc. Jpn. **49**, 2042–2049 (1980)
69. Nobile, F., Tempone, R., Webster, C.G.: A sparse grid stochastic collocation method for partial differential equations with random input data. SIAM J. Num. Anal. **46**, 2309–2345 (2008)
70. Pareschi, L.: An introduction to uncertainty quantification for kinetic equations and related problems. In: Trails in Kinetic Theory. SEMA SIMAI Springer Series, vol. 25, pp. 141–181. Springer, Cham (2021)
71. Pareschi, L., Russo, G.: An introduction to Monte Carlo method for the Boltzmann equation. ESAIM: Proc. EDP Sci. **10**, 35–75 (2001)
72. Pareschi, L., Russo, G.: An introduction to the numerical analysis of the Boltzmann equation, Riv. Mat. Univ. Parma (7), 4**, 145–250 (2005)
73. Pareschi, L., Toscani, G.: Interacting Multiagent Systems: Kinetic Equations and Monte Carlo Methods. Oxford University Press (2013)
74. Pareschi, L., Zanella, M.: Monte Carlo stochastic Galerkin methods for the Boltzmann equation with uncertainties: space-homogeneous case. J. Comput. Phys. 423:109822 (2020)
75. Park, C., Haftka, R.T., Kim, N.H.: Remarks on multi-fidelity surrogates. Struct. Multidiscip. Optim. **55**, 1029–1050 (2017)
76. Peherstorfer, B., Willcox, K., Gunzburger, M.: Survey of multifidelity methods in uncertainty propagation, inference, and optimization. SIAM Rev. **60**, 550–591 (2018)
77. Pettersson, P., Iaccarino, G., Nordström, J.: Polynomial Chaos Methods for Hyperbolic Partial Differential Equations: Numerical Techniques for Fluid Dynamics Problems in the Presence of Uncertainties, Mathematical Engineering. Springer (2015)
78. Poëtte, G.: A gPC-intrusive Monte-Carlo scheme for the resolution of the uncertain linear Boltzmann equation. J. Comput. Phys. **385**, 135–162 (2019)
79. Poëtte, G., Després, B., Lucor, D.: Uncertainty quantification for systems of conservation laws. J. Comp. Phys. **228**, 2443–2467 (2009)
80. Poupaud, F.: On a system of nonlinear Boltzmann equations of semiconductor physics. SIAM J. Appl. Math. **50**, 1593–1606 (1990)
81. Rjasanow, S., Wagner, W.: Stochastic Numerics for the Boltzmann Equation. Computational Mathematics, , vol. 37. Springer (2005)
82. Shu, R., Hu, J., Jin, S.: A stochastic Galerkin method for the Boltzmann equation with multi-dimensional random inputs using sparse wavelet bases. Num. Math. Theory, Methods Appl. (NMTMA) **10**, 465–488 (2017)
83. Strain, R.M., Guo, Y.: Almost exponential decay near Maxwellian. Comm. Partial Differ. Equ. **31**, 417–429 (2006)
84. Taylor, G.I.: Diffusion by continuous movements. Proc. London Math. **20**, 196–212 (1921)

85. Villani, C.: Hypocoercivity. Mem. Amer. Math. Soc. (2009)
86. Sukys, J., Rasthofer, U., Wermelinger, F., Hadjidoukas, P., Koumoutsakos, P.: Multilevel control variates for uncertainty quantification in simulations of cloud cavitation. SIAM J. Sci. Comput. **40**, B1361–B1390 (2018)
87. Wang, J., Xie, F., Kuniya, T.: Analysis of a reaction-diffusion cholera epidemic model in a spatially heterogeneous environment. Commun. Nonlinear Sci. Numer. Simul. **80**, 104951 (2020)
88. Xiu, D.: Numerical Methods for Stochastic Computations. Princeton University Press, Princeton, NJ (2010). A spectral method approach
89. Xiu, D., Hesthaven, J.S.: High-order collocation methods for differential equations with random inputs. SIAM J. Sci. Comput. **27** (2005)
90. Zhu, X., Narayan, A., Xiu, D.: Computational aspects of stochastic collocation with multifidelity models. SIAM/ASA J. Uncertain. Quantif. **2**, 444–463 (2014)

On the Shock Wave Discontinuities in Grad Hierarchy for a Binary Mixture of Inert Gases

Fiammetta Conforto and Giorgio Martalò

Abstract The shock wave problem is investigated for a binary mixture of inert gases described at each closure level within the hierarchy of Grad 13–moment system. The analysis focuses on the occurrence of singularities as the Mach number increases: their compatibility with jump discontinuities is examined through a geometric approach, and confirmed by stability arguments.

Keywords Hyperbolic systems · Balance laws · Grad hierarchy · Subshock formation

1 Introduction

As well known [1], hyperbolic systems of balance laws do not admit continuous travelling wave solutions if the front speed exceeds the maximum characteristic velocity evaluated at the unperturbed equilibrium state.

Recently [2–6], the occurrence of multiple sub–shocks has been investigated to conclude that different jump discontinuities may arise along the wave profile, when the front speed overcomes any characteristic velocity of the system evaluated at either upstream or downstream equilibrium states.

In such framework, fluid mixtures are of particular interest [5, 7–18] since all constituents, although characterized by their own distinctive features, play analogous roles in the spectrum of the governing system, and the coupling effects are confined only in the source terms. As a consequence, each species may create a subshock

F. Conforto
Department of ChiBioFarAm, University of Messina, V.le F. Stagno d'Alcontres 31, 98166 Messina, Italy
e-mail: fiammetta.conforto@unime.it

G. Martalò (✉)
Department of SMFI, University of Parma, Parco Area delle Scienze 53/A, 43124 Parma, Italy
e-mail: giorgio.martalo@unipr.it

G. Albi et al. (eds.), *Advances in Numerical Methods for Hyperbolic Balance Laws and Related Problems*, SEMA SIMAI Springer Series 32,
https://doi.org/10.1007/978-3-031-29875-2_8

front, at different times and positions, along the solution profile, corresponding to a stable jump discontinuity in the travelling wave.

In [14–18], multi–velocity and multi–temperature binary gas mixtures have been considered, whose macroscopic description is deduced by suitable closures at Euler [14, 15], 10– [16, 17] and 13–moment [18] levels of the Boltzmann equations. The results confirm that the light gas component always generates a sub–shock for wave speeds greater than the maximum eigenvalue evaluated at the unperturbed upstream equilibrium state; even the heavy species may induce sub–shock formation if its equilibrium concentration belongs to a proper range defined for any fixed mass ratio, at any closure level. Moreover, at Euler [14, 15] and 10–moment [16, 17] levels, up to two sub–shocks may occur, up to one in each component; whereas, up to four sub–shocks, up to two in each species, may arise for 13–moment equations [18], since four different Lax–admissible jump discontinuities exist [19]. It is worth to be underlined that multiple sub–shocks are not allowed in single–velocity formulations, where at most an unique jump discontinuity involving all species and global fields may appear [20, 21].

This paper deals with the shock wave structure problem for the whole hierarchy of subsystems which can be built on the Grad 13–moment equations, deduced in [9]. The analysis developed in [16–18] is extended to any subsystem of the hierarchy (Euler, 8-, 10- and 13-moments); some interesting features of the singularities, corresponding to the upstream and downstream equilibrium values of the characteristic velocities, are pointed out by deepening the geometric approach introduced in the previous papers and by discussing some stability arguments. The results clearly show the strong similarity between Euler and 10–moment descriptions, as well as between 8–and 13–moment ones, which both involve third order moments of the distribution functions.

The paper is organized as follows. The hierarchy is briefly presented in Sect. 2. In Sect. 3, the shock wave structure problem is stated for the whole hierarchy, and specified in terms of the singularity manifolds and the related critical values of the Mach number at each closure level. Finally, in Sect. 4, for a mixture of Helium and Argon (He–Ar), the influence of the concentrations of the two constituents on the role of the different critical values of the Mach number in the sub–shock admissibility is shown at each closure level. Then, for a fixed concentration, the compatibility of jump discontinuities, corresponding to the singularities admitted by each subsystem in the hierarchy, is discussed through geometric arguments and confirmed by a stability analysis.

2 13–Moment Equations and Principal Subsystems

In [9] a Grad 13–moment expansion is proposed for studying the evolution of a rarefied gas mixture of four monatomic species undergoing a bimolecular reversible chemical reaction, along with all possible elastic collisions.

In this paper, the macroscopic equations deduced in [9] are rewritten in the simpler case of a mixture of two components, indexed by $i = 1, 2$, where no chemical interaction occurs. In one space dimension (i.e. spherical symmetry around the x-axis is assumed for the distribution functions, so that the vector and tensor variables reduce to their only significant x–component), the field variables are number densities n_i, mean velocities v_i, temperatures T_i, viscous stress tensors σ_i and heat fluxes q_i of each species, and the governing system writes as

$$\partial_t n_i + \partial_x (n_i v_i) = 0, \tag{1}$$

$$m_i n_i (\partial_t v_i + v_i \partial_x v_i) + \partial_x (n_i k_B T_i + \sigma_i) = R_i, \tag{2}$$

$$\frac{3}{2} n_i [\partial_t (k_B T_i) + v_i \partial_x (k_B T_i)] + (n_i k_B T_i + \sigma_i) \partial_x v_i + \partial_x q_i = S_i, \tag{3}$$

$$\partial_t \sigma_i + \partial_x (v_i \sigma_i) + \frac{4}{3} (n_i k_B T_i + \sigma_i) \partial_x v_i + \frac{8}{15} \partial_x q_i = V_i, \tag{4}$$

$$\begin{aligned} &\partial_t q_i + \partial_x (v_i q_i) + \frac{11}{5} q_i \partial_x v_i + \frac{5}{2} (n_i k_B T_i + \sigma_i) \partial_x \left(\frac{k_B T_i}{m_i} \right) \\ &\quad + n_i k_B T_i \partial_x \left(\frac{\sigma_i}{m_i n_i} \right) - \frac{\sigma_i}{m_i n_i} \partial_x \sigma_i = W_i, \end{aligned} \tag{5}$$

In what follows, any quantity relevant to each component of the gas mixture will be denoted by the subscripts i, where $i = 1, 2$.
The source terms at the right hand side of (1)–(5) are given by

$$R_i = n_i \sum_{j=1}^{2} \nu_{1ij} \mu_{ij} n_j (v_j - v_i), \tag{6}$$

$$S_i = n_i \sum_{j=1}^{2} \frac{\nu_{1ij} \mu_{ij}}{m_i + m_j} n_j \left[3 k_B (T_j - T_i) + m_j (v_j - v_i)^2 \right], \tag{7}$$

$$\begin{aligned} V_i = &\sum_{j=1}^{2} \frac{2 \nu_{1ij} \mu_{ij}}{m_i + m_j} \left[\frac{2}{3} m_j n_j n_i (v_j - v_i)^2 + n_i \sigma_j - n_j \sigma_i \right] \\ &- \sum_{j=1}^{2} \frac{3 \nu_{2ij} m_j}{2 (m_i + m_j)^2} \left[\frac{2}{3} m_i m_j n_i n_j (v_j - v_i)^2 + m_i n_i \sigma_j + m_j n_j \sigma_i \right], \end{aligned} \tag{8}$$

$$\begin{aligned} W_i = &- \left(\frac{5}{2} n_i k_B T_i + \sigma_i \right) \frac{1}{m_i} \sum_{j=1}^{2} \nu_{1ij} \mu_{ij} n_j (v_j - v_i) + \sum_{j=1}^{2} \frac{m_j n_j}{(m_i + m_j)^3} \\ &\times \left\{ \beta_{1ij} q_i + \frac{m_i n_i}{m_j n_j} \beta_{4ij} q_j + (v_j - v_i) \left(\frac{1}{2} \beta_{2ij} \sigma_i + \frac{m_i n_i}{m_j n_j} \beta_{4ij} \sigma_j \right) \right. \\ &+ \frac{1}{2} (v_j - v_i) \left[(\beta_{2ij} + 3 \beta_{3ij}) n_i k_B T_i + 5 \frac{m_i n_i}{m_j n_j} \beta_{4ij} n_j k_B T_j \right. \\ &\left. \left. + m_i n_i \beta_{4ij} (v_j - v_i)^2 \right] \right\}, \end{aligned} \tag{9}$$

where m_i are the particle masses (from here on, it is assumed that $m_1 < m_2$), $\mu_{ij} = m_i m_j/(m_i + m_j)$ are the reduced masses, k_B is the Boltzmann constant, $p_i = n_i k_B T_i + \sigma_i$ are the pressures, ν_{kij} $(i, j, k = 1, 2)$ are the weighted elastic collision frequencies and the coefficients

$$\begin{aligned}
\beta_{1ij} &= -\left(3m_i^2 + m_j^2\right)\nu_{1ij} - 2m_i m_j \nu_{2ij},\\
\beta_{2ij} &= 2(m_i - m_j)^2 \nu_{1ij} + m_j(m_i - 3m_j)\nu_{2ij},\\
\beta_{3ij} &= (m_i - m_j)^2 \nu_{1ij} + m_j(3m_i + m_j)\nu_{2ij},\\
\beta_{4ij} &= 2m_j^2\left(2\nu_{1ij} - \nu_{2ij}\right),
\end{aligned}$$

are proper weighted (by masses) combinations of moments of the collision frequencies.
Equations (1)–(5) can be set in conservative form

$$\partial_t \bar{\mathbf{F}}_i^0(\bar{\boldsymbol{\omega}}_i) + \partial_x \bar{\mathbf{F}}_i^1(\bar{\boldsymbol{\omega}}_i) = \bar{\mathbf{G}}_i(\bar{\boldsymbol{\omega}}), \tag{10}$$

where $\bar{\boldsymbol{\omega}}(t, x) = (\bar{\boldsymbol{\omega}}_1, \bar{\boldsymbol{\omega}}_2)$ and $\bar{\boldsymbol{\omega}}_i(t, x) = (n_i, v_i, T_i, \sigma_i, q_i)$; moreover, the conservation laws of total mass, momentum and energy hold

$$\partial_t \rho + \partial_x(\rho v) = 0, \tag{11}$$

$$\partial_t(\rho v) + \partial_x\left(\rho v^2 + nk_B T + \sigma\right) = 0, \tag{12}$$

$$\partial_t\left(\frac{1}{2}\rho v^2 + \frac{3}{2}nk_B T\right) + \partial_x\left[\left(\frac{1}{2}\rho v^2 + \frac{5}{2}nk_B T + \sigma\right)v + q\right] = 0, \tag{13}$$

where the global variables are defined by

$$n = \sum_{i=1}^{2} n_i, \qquad \rho = \sum_{i=1}^{2}\rho_i = \sum_{i=1}^{2} m_i n_i, \qquad v = \frac{1}{\rho}\sum_{i=1}^{2}\rho_i v_i,$$

$$T = \frac{1}{nk_B}\sum_{i=1}^{2}\left[n_i k_B T_i + \frac{1}{3}\rho_i\left(v_i^2 - v^2\right)\right], \quad \sigma = \sum_{i=1}^{2}\left[\sigma_i + \frac{2}{3}\rho_i\left(v_i^2 - v^2\right)\right],$$

$$q = \sum_{i=1}^{2}\left[q_i + \sigma_i(v_i - v) + \frac{5}{2}n_i k_B T_i(v_i - v) + \frac{1}{2}\rho_i(v_i - v)^3\right],$$

and global pressure is given by $p = nk_B T + \sigma$.

The subsystems of the hierarchy, ordered by decreasing closure level, are: the 8–moment description, consisting of Eqs. (1)–(3) and (5) where $\sigma_i = 0$; the 10–moment system, consisting of Eqs. (1)–(4) where $q_i = 0$; the Euler equations, consisting of Eqs. (1)–(3) where $\sigma_i = 0$ and $q_i = 0$. The equilibrium subsystem of the whole hierarchy has the same structure of the Euler equations for a single gas (conservation laws of global mass, momentum and energy), with $v_i = v$, $T_i = T$, $\sigma = \sigma_i = 0$ and

$q = q_i = 0$, to which one of the species mass conservation Eq. (1) has to be added, i.e.

$$\partial_t n_1 + \partial_x (n_1 v) = 0, \tag{14}$$

$$\partial_t \rho + \partial_x (\rho v) = 0, \tag{15}$$

$$\partial_t (\rho v) + \partial_x \left(\rho v^2 + n k_B T\right) = 0, \tag{16}$$

$$\partial_t \left(\frac{1}{2}\rho v^2 + \frac{3}{2} n k_B T\right) + \partial_x \left[\left(\frac{1}{2}\rho v^2 + \frac{5}{2} n k_B T\right) v\right] = 0. \tag{17}$$

3 The Shock Wave Problem

By introducing the variable

$$\varphi = x - st\,,$$

where $s > 0$ is the velocity of propagation of the wave front, (1)–(5) rewrites in terms of φ as

$$\left(\tilde{\mathbf{A}}_i\left(\tilde{\boldsymbol{\omega}}_i\right) - s\mathbf{I}\right)\frac{d\tilde{\boldsymbol{\omega}}_i}{d\varphi} = \tilde{\mathbf{B}}_i\left(\tilde{\boldsymbol{\omega}}\right)\,, \tag{18}$$

where $\tilde{\boldsymbol{\omega}} = (\tilde{\boldsymbol{\omega}}_1, \tilde{\boldsymbol{\omega}}_2)$ is the field vector $\bar{\boldsymbol{\omega}}$ expressed as function of φ.
A shock wave solution is a function $\tilde{\boldsymbol{\omega}} = \tilde{\boldsymbol{\omega}}(\varphi)$ solving Eq. (18) and fulfilling the asymptotic conditions

$$\lim_{\varphi \to \pm\infty} \tilde{\boldsymbol{\omega}} = \tilde{\boldsymbol{\omega}}_{\pm}\,, \qquad \lim_{\varphi \to \pm\infty} \frac{d\tilde{\boldsymbol{\omega}}}{d\varphi} = \mathbf{0}\,, \tag{19}$$

where equilibrium states are $\tilde{\boldsymbol{\omega}}_{i\pm} = (n_{i\pm}, v_{\pm}, T_{\pm}, 0, 0)$. From now on, the subscripts $+$ and $-$ will denote quantities evaluated at the unperturbed and perturbed equilibrium state, respectively.
By introducing the species relative velocities u_i, and the global relative velocity u defined by

$$u_i = s - v_i\,, \qquad u = \frac{1}{\rho}\sum_{i=1}^{2} \rho_i u_i, \tag{20}$$

system (18) reads as

$$\mathbf{A}_i(\boldsymbol{\omega}_i)\frac{d\boldsymbol{\omega}_i}{d\varphi} = \mathbf{B}_i\left(\boldsymbol{\omega}\right)\,, \tag{21}$$

where $\boldsymbol{\omega}_i = \boldsymbol{\omega}_i(\varphi) = (n_i, u_i, T_i, \sigma_i, q_i)$, the matrices $\mathbf{A}_i$ are given by

$$\mathbf{A}_i(\boldsymbol{\omega}_i) = \begin{pmatrix} u_i & n_i & 0 & 0 & 0 \\ \frac{k_B T_i}{m_i n_i} & u_i & \frac{k_B}{m_i} & \frac{1}{m_i n_i} & 0 \\ 0 & \frac{2(n_i k_B T_i + \sigma_i)}{3 k_B n_i} & u_i & 0 & -\frac{2}{3 k_B n_i} \\ 0 & \frac{4 n_i k_B T_i + 7\sigma_i}{3} & 0 & u_i & -\frac{8}{15} \\ \frac{k_B T_i \sigma_i}{m_i n_i} & \frac{16}{5} q_i & -\frac{5 k_B (k_B n_i T_i + \sigma_i)}{2 m_i} & \frac{\sigma_i - n_i k_B T_i}{m_i n_i} & u_i \end{pmatrix},$$

and the source terms are

$$\mathbf{B}_i\left(\boldsymbol{\omega}\right) = \left(0, \frac{R_i}{m_i n_i}, \frac{2 S_i}{3 k_B n_i}, V_i, W_i\right)^T, \tag{22}$$

expressed in terms of relative velocities u_i, the equilibrium states being $\boldsymbol{\omega}_{i\pm} = (n_{i\pm}, u_\pm, T_\pm, 0, 0)$.

Finally, the asymptotic equilibria at $\pm\infty$ defined in (19) are assumed to satisfy the Rankine–Hugoniot (R–H) conditions of the equilibrium subsystem (14)–(17), i.e. $\boldsymbol{\omega}_- = \boldsymbol{\omega}_-\left(\boldsymbol{\omega}_+; M_+\right)$, yielding the relations

$$n_{i-} = n_{i+} \frac{4 M_+^2}{M_+^2 + 3}, \qquad i = 1, 2, \tag{23}$$

$$v_- = v_+ + \frac{3}{4} c_+ \frac{M_+^2 - 1}{M_+}, \tag{24}$$

$$T_- = T_+ \frac{\left(M_+^2 + 3\right)\left(5 M_+^2 - 1\right)}{16 M_+^2}, \tag{25}$$

and, consequently,

$$n_- = n_+ \frac{4 M_+^2}{M_+^2 + 3}, \quad \rho_- = \rho_+ \frac{4 M_+^2}{M_+^2 + 3}, \quad u_- = c_+ \frac{M_+^2 + 3}{4 M_+}, \quad p_- = p_+ \frac{5 M_+^2 - 1}{4},$$

where the Mach number M_+ is defined, in terms of the sound speed of the equilibrium subsystem evaluated at the unperturbed state $\boldsymbol{\omega}_+$, denoted by c_+, as

$$M_+ = \frac{u_+}{c_+} = \frac{s - v_+}{c_+}, \qquad c_+^2 = \frac{5 n_+ k_B T_+}{3 \rho_+}. \tag{26}$$

3.1 Singularity Manifolds and Critical Mach Numbers

A crucial role in our analysis is played by the eigenvalues of the hierarchy of hyperbolic systems

$$\partial_t \tilde{\boldsymbol{\omega}}_i + \tilde{\mathbf{A}}_i \left(\tilde{\boldsymbol{\omega}}_i\right) \partial_x \tilde{\boldsymbol{\omega}}_i = \tilde{\mathbf{B}}_i \left(\tilde{\boldsymbol{\omega}}\right), \qquad i = 1, 2,$$

which obviously coincide with the eigenvalues of the matrices $\tilde{\mathbf{A}}_i \left(\tilde{\boldsymbol{\omega}}_i\right)$ for each closure; the eigenvalues, evaluated at equilibrium, for the whole hierarchy are listed below:
at 13–moment level,

$$v_\pm - \sqrt{\frac{13 \pm \sqrt{94}}{5} \frac{k_B T_\pm}{m_i}}, \qquad v_\pm, \qquad v_\pm + \sqrt{\frac{13 \pm \sqrt{94}}{5} \frac{k_B T_\pm}{m_i}}; \tag{27}$$

at 8–moment level,

$$v_\pm - \sqrt{\frac{5 \pm \sqrt{10}}{3} \frac{k_B T_\pm}{m_i}}, \qquad v_\pm, \qquad v_\pm + \sqrt{\frac{5 \pm \sqrt{10}}{3} \frac{k_B T_\pm}{m_i}}; \tag{28}$$

at 10–moment level,

$$v_\pm - \sqrt{3 \frac{k_B T_\pm}{m_i}}, \qquad v_\pm, \qquad v_\pm + \sqrt{3 \frac{k_B T_\pm}{m_i}}; \tag{29}$$

at Euler level,

$$v_\pm - \sqrt{\frac{5}{3} \frac{k_B T_\pm}{m_i}}, \qquad v_\pm, \qquad v_\pm + \sqrt{\frac{5}{3} \frac{k_B T_\pm}{m_i}}; \tag{30}$$

finally, the characteristic velocities of the equilibrium subsystem are

$$v_\pm - \sqrt{\frac{5}{3} \frac{n_\pm k_B T_\pm}{\rho_\pm}}, \qquad v_\pm, \qquad v_\pm + \sqrt{\frac{5}{3} \frac{n_\pm k_B T_\pm}{\rho_\pm}}. \tag{31}$$

As well–known [1, 22], since we are interested in discontinuous solutions, only the higher eigenvalues evaluated at the equilibrium states $\boldsymbol{\omega}_\pm$ are involved in the problem, and from now on they will be definitively denoted by

$$\lambda_{i\pm}^{(13)\pm} = v_\pm + \sqrt{\frac{13 \pm \sqrt{94}}{5} \frac{k_B T_\pm}{m_i}}, \tag{32}$$

$$\lambda_{i\pm}^{(8)\pm} = v_\pm + \sqrt{\frac{5 \pm \sqrt{10}}{3} \frac{k_B T_\pm}{m_i}}, \tag{33}$$

$$\lambda_{i\pm}^{(10)} = v_\pm + \sqrt{3 \frac{k_B T_\pm}{m_i}}, \tag{34}$$

$$\lambda_{i\pm}^{(E)} = v_\pm + \sqrt{\frac{5}{3} \frac{k_B T_\pm}{m_i}}, \tag{35}$$

together with the unperturbed maximum eigenvalue of the equilibrium subsystem

$$\mu_+ = v_+ + c_+ = v_+ + \sqrt{\frac{5}{3} \frac{n_+ k_B T_+}{\rho_+}}. \tag{36}$$

The ODEs (21) allows to define the singularity manifolds of the system, characterized by $s = \lambda_i\,(\boldsymbol{\omega}_i)$, the so–called singular barriers [23], in the phase space through the equations

$$|\mathbf{A}_i\,(\boldsymbol{\omega}_i)| = 0\,, \tag{37}$$

which, evaluated at equilibrium ($u_i = u$, with $u \neq 0$, $T_i = T$, $\sigma_i = 0$ and $q_i = 0$) for each closure level, specify as follows:
at 13–moments

$$B_i^{(13)} := u^4 - \frac{26}{5} \frac{k_B T}{m_i} u^2 + 3 \frac{k_B^2 T^2}{m_i^2} = 0\,; \tag{38}$$

at 8–moments

$$B_i^{(8)} := u^4 - \frac{10}{3} \frac{k_B T}{m_i} u^2 + \frac{5}{3} \left(\frac{k_B T}{m_i} \right)^2 = 0\,; \tag{39}$$

at 10–moments

$$B_i^{(10)} := u^2 - 3 \frac{k_B T}{m_i} = 0\,; \tag{40}$$

at Euler level

$$B_i^{(E)} := u^2 - \frac{5}{3} \frac{k_B T}{m_i} = 0\,. \tag{41}$$

The roots of Eqs. (38), (39), (40) and (41), evaluated at the equilibrium states $\boldsymbol{\omega}_\pm$ related by R–H conditions (23)–(25), i.e. $B_{i\pm} := B_i\,(\boldsymbol{\omega}_\pm) = 0$, yields the following critical values of the Mach number for each closure, respectively:
at 13–moments

$$M_{i+}^{(13)\pm} := \sqrt{\frac{3\left(13 \pm \sqrt{94}\right)}{25} \frac{1-(1-\alpha)\,\chi}{\alpha^{2-i}}}, \tag{42}$$

$$M_{i-}^{(13)\pm} := \sqrt{\frac{3}{5} \frac{\left(13 \pm \sqrt{94}\right)[1-(1-\alpha)\,\chi] + 25\alpha^{2-i}}{3\left(13 \pm \sqrt{94}\right)[1-(1-\alpha)\,\chi] - 5\alpha^{2-i}}}; \tag{43}$$

at 8–moments

$$M_{i+}^{(8)\pm} := \sqrt{\frac{5 \pm \sqrt{10}}{5} \frac{1-(1-\alpha)\,\chi}{\alpha^{2-i}}}, \tag{44}$$

$$M_{i-}^{(8)\pm} := \sqrt{\frac{1}{5} \frac{\left(5 \pm \sqrt{10}\right)[1-(1-\alpha)\,\chi] + 15\alpha^{2-i}}{\left(5 \pm \sqrt{10}\right)[1-(1-\alpha)\,\chi] - \alpha^{2-i}}}; \tag{45}$$

at 10–moments

$$M_{i+}^{(10)} := \sqrt{\frac{9}{5} \frac{1-(1-\alpha)\,\chi}{\alpha^{2-i}}}, \tag{46}$$

$$M_{i-}^{(10)} := \sqrt{\frac{3}{5} \frac{3[1-(1-\alpha)\,\chi] + 5\alpha^{2-i}}{9[1-(1-\alpha)\,\chi] - \alpha^{2-i}}}; \tag{47}$$

at Euler level

$$M_{i+}^{(E)} := \sqrt{\frac{1-(1-\alpha)\,\chi}{\alpha^{2-i}}}, \tag{48}$$

$$M_{i-}^{(E)} := \sqrt{\frac{1-(1-\alpha)\,\chi + 3\alpha^{2-i}}{5[1-(1-\alpha)\,\chi] - \alpha^{2-i}}}. \tag{49}$$

In (43)–(49), the mass ratio $\alpha = m_1/m_2 \in (0,1)$ and species concentrations χ_i have been introduced, defined by

$$\chi_i = \frac{n_i}{n} \in (0,1), \qquad \chi_1 + \chi_2 = 1, \qquad \chi_{i+} = \chi_{i-}, \quad i = 1, 2, \tag{50}$$

denoting $\chi = \chi_{1\pm}$, so that $\chi_{2\pm} = 1 - \chi$; in fact, the relations

$$\frac{1}{m_i} = \frac{n_+}{\rho_+} \frac{[1-(1-\alpha)\,\chi]}{\alpha^{2-i}}, \qquad i = 1, 2, \tag{51}$$

allow to express the critical values of Mach number only in terms of the mass ratio α and the equilibrium concentration χ.

Each compatible jump discontinuity along the wave front satisfies Lax admissibility conditions [19]

$$\lambda_{i+} < s < \lambda_{i-}\,, \tag{52}$$

provided that $s > \mu$, for each pair of unperturbed and perturbed equilibrium values of the characteristic velocities of the system, defined by (32)–(35) at each closure level, which, in terms of the critical values of Mach number, write as

$$\begin{cases} M_+ > M_{i+} \\ M_+ > M_{i-} \end{cases} \tag{53}$$

with $M_+ > 1$, and correspond to a change in sign of $|\mathbf{A}_i\,(\boldsymbol{\omega}_i)|$, allowing the trajectory to cross the relevant singularity manifold defined by (37).

4 Singularity Analysis

In this section we want to discuss mainly three aspects of the problem, which allow to point out some interesting features characterizing the singularities of each closure in the hierarchy; the comparison between the different levels of description clearly shows the strong analogies (Euler and 10–moment levels, 8– and 13–moment levels), as well as the obvious differences (due to the presence of third order moments of the distribution functions).

The investigation is carried out in the test case of a mixture of Helium and Argon (characterized by the mass ratio $\alpha = 0.1$) defined by fixing the unperturbed equilibrium mean velocity $v_+ = 0$ and temperature $T_+ = 2$ (the same values used in the numerical tests performed in the previous papers [16–18]), for any arbitrary (physically consistent) values of the species number densities set at the unperturbed equilibrium state, n_{1+} and n_{2+}.

At first, it is worth to be deepened the relation between the jump admissibility and the mass ratio and species concentrations (α and χ). In fact, the critical values $M_{i\pm}^{(\cdot)}$, given at each closure level by (43)–(49), allow to define the ranges of Mach number compatible with any admissible sub–shock characterized by the Lax conditions (53), varying $\alpha \in (0, 1)$ and $\chi \in (0, 1)$.
For any fixed $\alpha \in (0, 1)$, the ranges correspond to the regions of the (χ, M_+)–plane bounded by the curves $M_+ = M_{i\pm}^{(\cdot)}$; each region is characterised by the admissible sub–shocks in each gas component, corresponding to the critical values $M_{i\pm}^{(\cdot)} > 1$ below M_+, for any fixed $\chi \in (0, 1)$.

In Figs. 1, 2, 3 and 4, the plots show all the sub–shock admissibility regions on the (χ, M_+)–plane, being $\chi \in (0, 1)$ and $M_+ > 1$, at each closure level, for $\alpha = 0.1$. Above each plot, the notation $(s_1, s_2) = (h, k)$, with $h, k = 0, 1, 2$, is used to specify the number s_i (up to two) of compatible sub–shocks in each species, $i = 1, 2$, relevant to each region.

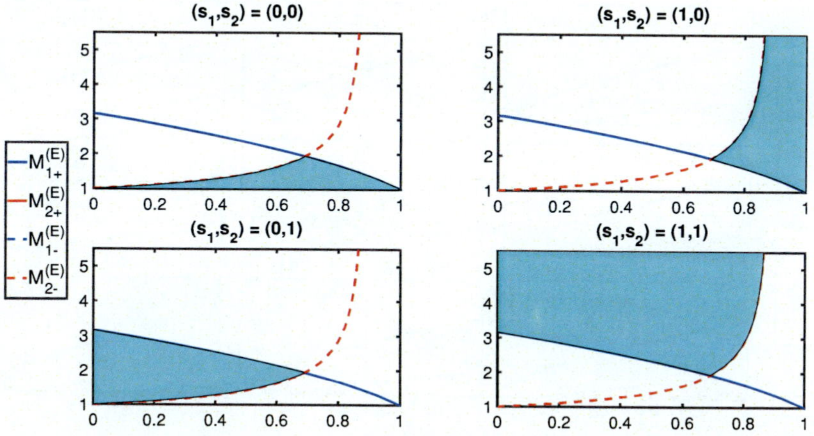

Fig. 1 Plots of the critical Mach numbers versus χ, set $\alpha = 0.1$, at Euler level. The coloured regions represent the different ranges of admissibility of sub–shocks: continuous range (up–left); one sub–shock in species $i = 1$ (up–right), or one sub–shock in species $i = 2$ (down–left); one sub–shock in each species (down–right)

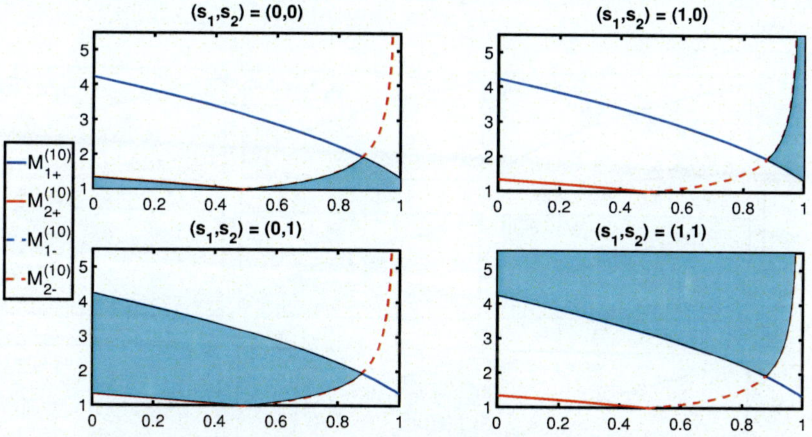

Fig. 2 Plots of the critical Mach numbers versus χ, set $\alpha = 0.1$, at 10–moment level. The coloured regions represent the different ranges of admissibility of sub–shocks: continuous range (up–left); one sub–shock in species $i = 1$ (up–right), or one sub–shock in species $i = 2$ (down–left); one sub–shock in each species (down–right)

Figures 1 and 2, relevant to Euler and 10–moment level descriptions, respectively, show the four regions bounded by the curves $M_+ = M_{i\pm}^{(\cdot)}$; in particular, the two curves $M_+ = M_{1+}^{(E)}$ and $M_+ = M_{2-}^{(E)}$ defined by (48) and (49) in Fig. 1, and the three curves $M_+ = M_{1+}^{(10)}$, $M_+ = M_{2+}^{(10)}$ and $M_+ = M_{2-}^{(10)}$ defined by (46) and (47) in Fig. 2. At both levels, in the region below any critical Mach curve, denoted by $(s_1, s_2) = (0, 0)$ (up–left), there are no admissible sub–shocks in Lax sense (52); for sufficiently

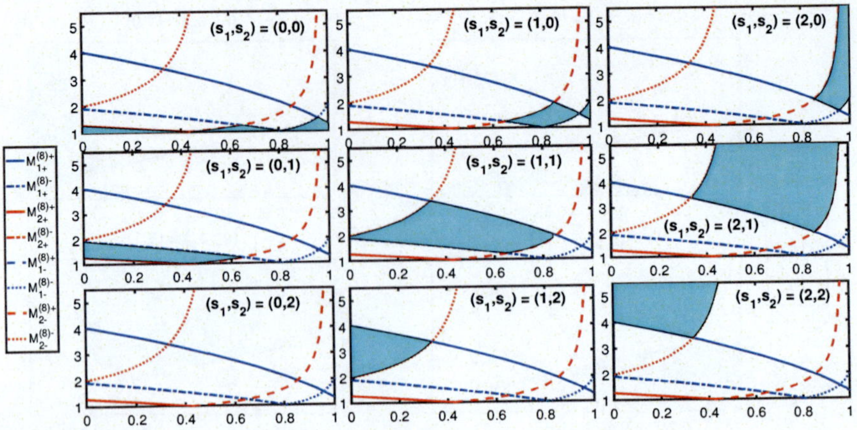

Fig. 3 Plots of the critical Mach numbers versus χ, set $\alpha = 0.1$, at 8–moment level. The coloured regions represent the different ranges of admissibility of sub–shocks: continuous range (up–left); one sub–shock in species $i = 1$ (up–middle), or in species $i = 2$ (middle–left); two sub–shocks in species $i = 1$ (up–right), or in species $i = 2$ (down–left); one sub–shock in each species (middle–middle); two sub–shocks in species $i = 1$ and one in species $i = 2$ (middle–right), or two sub–shocks in species $i = 2$ and one in species $i = 1$ (down–middle); two sub–shocks in each species (down–right)

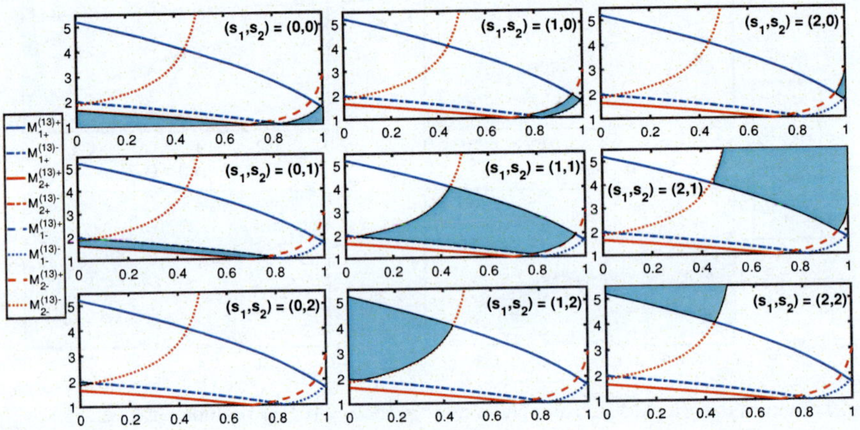

Fig. 4 Plots of the critical Mach numbers versus χ, set $\alpha = 0.1$, at 13–moment level. The coloured regions represent the different ranges of admissibility of sub–shocks: continuous range (up–left); one sub–shock in species $i = 1$ (up–middle), or in species $i = 2$ (middle–left); two sub–shocks in species $i = 1$ (up–right), or in species $i = 2$ (down–left); one sub–shock in each species (middle–middle); two sub–shocks in species $i = 1$ and one in species $i = 2$ (middle–right), or two sub–shocks in species $i = 2$ and one in species $i = 1$ (down–middle); two sub–shocks in each species (down–right)

large χ, in the region denoted by $(s_1, s_2) = (1, 0)$ (up–right plot), satisfying the Lax conditions

$$M_+ > M_{1+}^{(E)} > 1 > M_{1-}^{(E)}$$

in Fig. 1, and

$$M_+ > M_{1+}^{(10)} > 1 > M_{1-}^{(10)}$$

in Fig. 2, only one sub–shock in species $i = 1$ is compatible due to the unperturbed critical value $M_{1+}^{(\cdot)}$; for sufficiently small χ, in the region denoted by $(s_1, s_2) = (0, 1)$ (down–left plot), satisfying the Lax conditions

$$M_+ > M_{2+}^{(E)} > 1 > M_{2-}^{(E)}$$

in Fig. 1, and

$$M_+ > M_{2+}^{(10)} > 1 > M_{2-}^{(10)} \quad \text{or} \quad M_+ > M_{2-}^{(10)} > 1 > M_{2+}^{(10)}$$

in Fig. 2, only one sub–shock in species $i = 2$ is compatible due only to the perturbed critical value $M_{2-}^{(E)}$ at Euler level, and to the critical values $M_{2+}^{(10)}$ for lower χ, or $M_{2-}^{(10)}$ for higher χ at 10–moment level; finally, in the region denoted by $(s_1, s_2) = (1, 1)$ (down–right plot), satisfying the Lax conditions

$$M_+ > M_{1+}^{(E)} > M_{2-}^{(E)} > 1 > M_{2+}^{(E)} > M_{1-}^{(E)}$$

in Fig. 1, and

$$M_+ > M_{1+}^{(10)} > M_{2+}^{(10)} > 1 > M_{2-}^{(10)} > M_{1-}^{(10)}$$

or

$$M_+ > M_{1+}^{(10)} > M_{2-}^{(10)} > 1 > M_{2+}^{(10)} > M_{1-}^{(10)}$$

in Fig. 2, two sub–shocks are admissible, in species $i = 1$ always due to $M_+ = M_{1+}$ and in species $i = 2$ due only to $M_+ = M_{2-}^{(E)}$ at Euler level, and again to the critical values $M_{2+}^{(10)}$ for lower χ, or $M_{2-}^{(10)}$ for higher χ at 10–moment level.

Analogously, Figs. 3 and 4 show the nine regions at 8–and 13–moment level, respectively, bounded by the six curves defined by the critical values greater than one, three relevant to species $i = 1$ and three to species $i = 2$, being $M_{1+}^{(\cdot)+}$, $M_{1+}^{(\cdot)-}$, $M_{1-}^{(\cdot)-}$, $M_{2+}^{(\cdot)+}$, $M_{2-}^{(\cdot)+}$, $M_{2-}^{(\cdot)-}$ for both closure levels, given by (44) and (45) in Fig. 3, and by (42) and (43) in Fig. 4. Obviously, the larger number of regions is due to the characteristic velocities of the governing system, which passes from two (and then four critical Mach numbers) at both Euler and 10–moment levels, to four (and then eight critical Mach numbers) for both 8–and 13–moment descriptions. Therefore, the nine regions are characterized by all possible combinations $(s_1, s_2) = (h, k)$, with $h, k = 0, 1, 2$, from the continuous region denoted by $(s_1, s_2) = (0, 0)$ (up–left plot), to the region denoted by $(s_1, s_2) = (2, 2)$ (down–right plot), for sufficiently small χ, where four compatible sub–shocks are admitted: two in the lightest species $i = 1$, one always due to the maximum unperturbed critical Mach number $M_{1+}^{(\cdot)+}$ and the other one due to $M_{1+}^{(\cdot)-}$ for lower χ, or to $M_{1-}^{(\cdot)-}$ for higher χ, and two in the heaviest species $i = 2$, one due to $M_{2-}^{(\cdot)-}$ and the other due to $M_{2+}^{(\cdot)+}$ for lower χ, or to $M_{2-}^{(\cdot)+}$ for higher χ.

A second issue that, in authors' opinion, has to be deepened is the geometric interpretation of the Lax conditions in the phase space, where the Lax–admissible sub–shocks can be exploited by studying the positions of the upstream and downstream equilibrium configurations with respect to the singularity manifolds, by means of proper projections on the (u_i, T_i)–planes. Lax admissible sub–shocks relevant to each pair of unperturbed and perturbed critical values of the Mach number, $M_{i+}^{(\cdot)}$ and $M_{i-}^{(\cdot)}$, $i = 1, 2$, defined for each subsystem by (42), (43)–(48), (49), correspond to solutions connecting two equilibrium configurations separated by the species singular barrier, defined by the equation $B_i\,(u, T) = 0$ as the locus of the states characterized by an unique relative velocity $u_i = u$, an unique temperature $T_i = T$ and vanishing viscous stress and heat flux $\sigma_i = q_i = 0$, for $i = 1, 2$, relevant to each species; in fact, the components are distinguished only by the different masses ($m_1 < m_2$) in the Eqs. (38)–(41) of the two equilibrium singular barriers, for $i = 1, 2$, admitted at any order closure. Moreover, since Eqs. (38)–(41) do not explicitly depend on the species number densities n_i, the singularity manifolds (38)–(41) of the two species, at each closure level, intercept two curves on the (u_i, T_i)–planes of the phase space defined by $\sigma_i = 0$ and $q_i = 0$, $i = 1, 2$, which do not depend on the equilibrium values assigned to the species number densities $n_{i\pm}$. Therefore, the positions on the (u_i, T_i)–planes of the two equilibrium configurations $\boldsymbol{\omega}_{\pm} = (u_{\pm}, T_{\pm})$ with respect to the curve defined by Eqs. (38)–(41) for each subsystem, allow to determine their positions with respect the whole singularity manifolds in the phase space; in fact, the equilibrium configurations $\boldsymbol{\omega}_{\pm} = (u_{\pm}, T_{\pm})$ define two curves parametrized by the Mach number M_+: the locus of the unperturbed equilibrium states $\boldsymbol{\omega}_+ = (u_+, T_+)$ is characterized by

$$u_+ = c_+ M_+ \quad , \quad T_+ = \text{ constant} \, ,$$

whereas, the dependence of the perturbed equilibrium configurations $\boldsymbol{\omega}_- = (u_-, T_-)$ on the parameter M_+ is determined by assuming that the upstream and downstream equilibrium are related by the R–H conditions (23)–(25) of the equilibrium subsystem (14)–(17), yielding the relations

$$u_- = c_+ \frac{M_+^2 + 3}{4M_+} \quad , \quad T_- = T_+ \frac{\left(M_+^2 + 3\right)\left(5M_+^2 - 1\right)}{16M_+^2} \, .$$

In order to deal with numerical values of critical Mach numbers (43)–(49), also the equilibrium value of the lightest species concentration χ needs to be fixed in the considered test case, and it is set $\chi = 0.25$, allowing to represent such geometric arguments on the (u_i, T_i)–planes of the phase space.

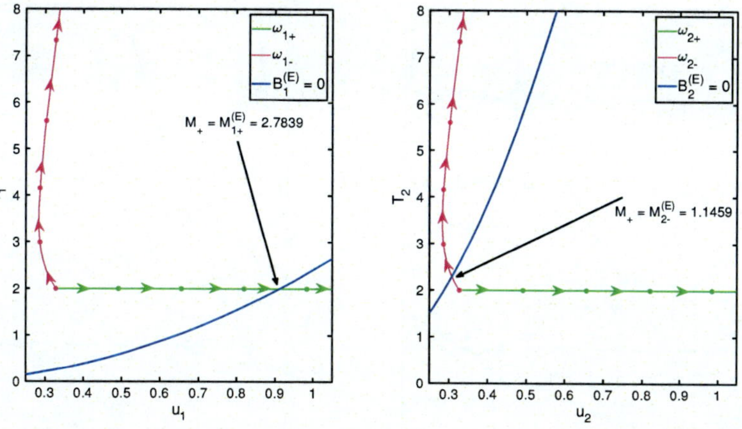

Fig. 5 Plots of the barrier $B_i^{(E)} = 0$, for $i = 1$ (left) and $i = 2$ (right), and the equilibrium states $\boldsymbol{\omega}_{i\pm} = (u_\pm, T_\pm)$ on the (u_i, T_i)-plane of the phase space, set $\alpha = 0.1$ and $\chi = 0.25$

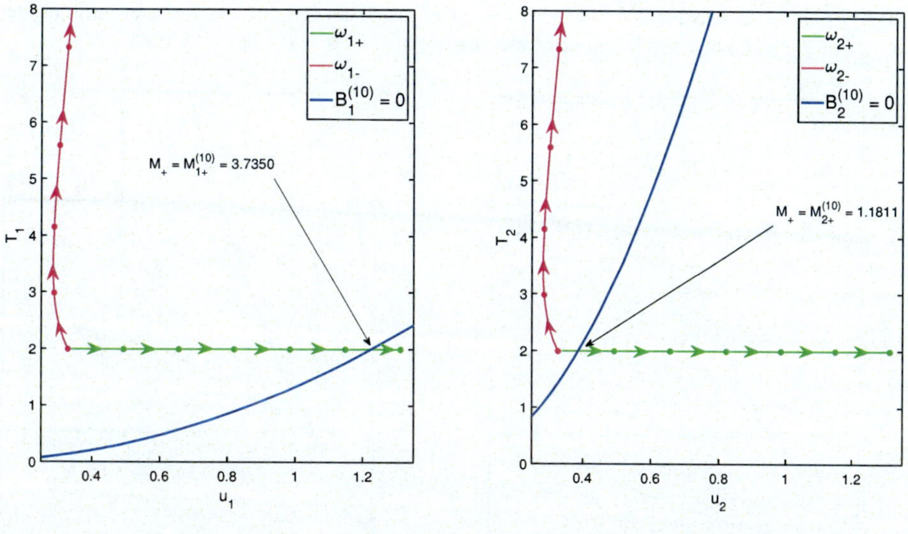

Fig. 6 Plots of the barrier $B_i^{(10)} = 0$ (down), for $i = 1$ (left) and $i = 2$ (right), and the equilibrium states $\boldsymbol{\omega}_{i\pm} = (u_\pm, T_\pm)$ on the (u_i, T_i)-plane of the phase space, set $\alpha = 0.1$ and $\chi = 0.25$

Figures 5, 6, 7 and 8 show the plots on the (u_i, T_i)–planes of the equilibrium singular barriers of species $i = 1$ on the left and species $i = 2$ on the right, together with the loci of the equilibrium configurations (u_+, T_+) and (u_-, T_-), which are obviously the same for the two species; the arrows on both the equilibrium state loci points the direction of increasing Mach numbers and, obviously, the equilibrium configurations coincide for $M_+ = 1$, i.e. $\boldsymbol{\omega}_{i-} = \boldsymbol{\omega}_{i+} = (u_+, T_+)$.

It can be observed that, at each closure level (Euler in Fig. 5; 10–moment in Fig. 6; 8–moment in Fig. 7; 13–moment in Fig. 8), the two equilibria lie on the same region

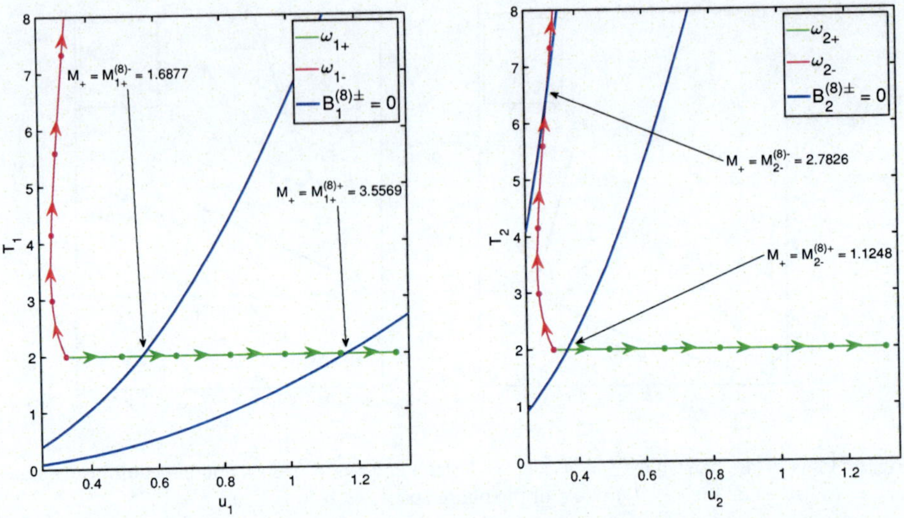

Fig. 7 Plots of the barriers $B_i^{(8)\pm} = 0$ for $i = 1$ (left) and $i = 2$ (right), and the equilibrium states $\boldsymbol{\omega}_{i\pm} = (u_\pm, T_\pm)$ on the (u_i, T_i)-plane of the phase space, set $\alpha = 0.1$ and $\chi = 0.25$

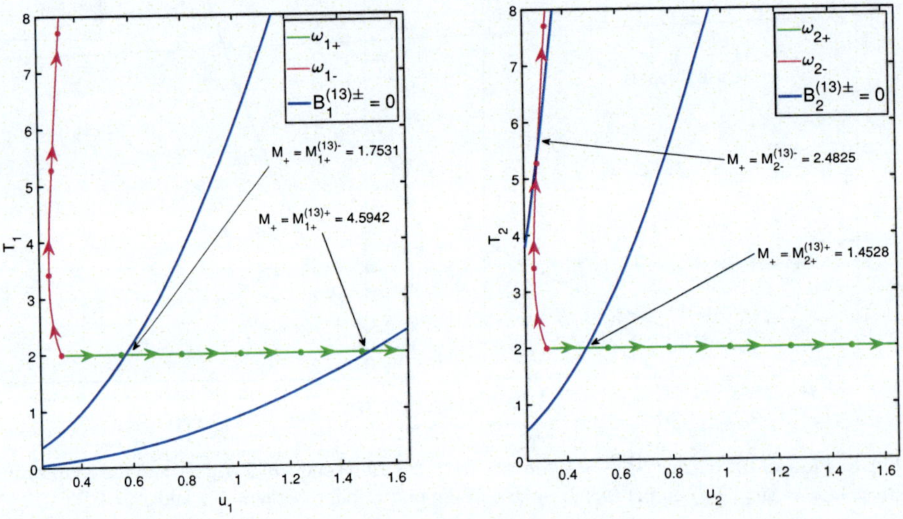

Fig. 8 Plots of the barriers $B_i^{(13)\pm} = 0$ for $i = 1$ (left) and $i = 2$ (right), and the equilibrium states $\boldsymbol{\omega}_{i\pm} = (u_\pm, T_\pm)$ on the (u_i, T_i)-plane of the phase space, set $\alpha = 0.1$ and $\chi = 0.25$

with respect to the singular barrier $B_i\,(u,T)=0$ until the Mach number is lower than its minimum critical value. At each critical value $M_{i\pm}^{(\cdot)}$, the corresponding equilibrium state $\boldsymbol{\omega}_{i\pm}=(u_\pm,T_\pm)$ lies at the intersection with the singular barrier relevant to the species i. When M_+ overcomes a critical value relevant to the species i and to an equilibrium state $\boldsymbol{\omega}_{i\pm}$, a singularity manifold corresponding to the same component and equilibrium state, $B_{i\pm}=0$, separates the two equilibrium configurations.

This means that any solution connecting two equilibrium states has to cross as many singular barriers as many critical values of the Mach number locate the distinct solutions of the four equations $B_{i\pm}=0$, corresponding to the intersection points between the equilibrium singular barrier of each species with each locus of unperturbed and perturbed equilibrium states on the (u_i,T_i)–planes.

As shown in Fig. 5 for the Euler level closure and in Fig. 6 for 10–moment description, each species barrier $B_i\,(u,T)=0$, respectively defined by (41) and (40), intersect only one of the two equilibrium loci $\boldsymbol{\omega}_{i+}=(u_+,T_+)$ and $\boldsymbol{\omega}_{i-}=(u_-,T_-)$, at the unique point on the (u_i,T_i)–planes corresponding to each species critical value of the Mach number, confirming that up to two sub–shocks, one for each constituent, may arise at both levels.

More precisely, at Euler level, the intersection relevant to species $i=1$ lies on the unperturbed equilibrium locus $\boldsymbol{\omega}_+$ for $M_+=M_{1+}^{(E)}$ (left plot in Fig. 5); whereas, the intersection relevant to species $i=2$ lies on the perturbed equilibrium locus $\boldsymbol{\omega}_-$ for $M_+=M_{2-}^{(E)}$ (right plot in Fig. 5).

At 10–moment level, both intersections lie on the unperturbed equilibrium curve $\boldsymbol{\omega}_+$ for $M_+=M_{1+}^{(10)}$ (left plot in Fig. 6) and $M_+=M_{2+}^{(10)}$ (right plot in Fig. 6).

At both 8–and 13–moment levels, the main difference with respect to the previous ones consists in the number of intersection points between each species barrier, respectively defined by (39) and (38) and the equilibrium loci $\boldsymbol{\omega}_{i+}=(u_+,T_+)$ and $\boldsymbol{\omega}_{i-}=(u_-,T_-)$, which can be two for each species.

At 8–moment level, Fig. 7 exhibits the two intersections relevant to species $i=1$, both with the unperturbed equilibrium locus $\boldsymbol{\omega}_+$ for $M_+=M_{1+}^{(8)-}$ and $M_+=M_{1+}^{(8)+}$ (left plot), and the two intersections relevant to species $i=2$, one lying on the perturbed equilibrium locus $\boldsymbol{\omega}_-$ corresponding to $M_+=M_{2-}^{(8)-}$ and the other on the unperturbed equilibrium locus $\boldsymbol{\omega}_+$ corresponding to $M_+=M_{2+}^{(8)+}$ (right plot).

Analogously to the previous closure, as shown in Fig. 8, at 13–moment level, both the intersections relevant to species $i=1$ lie on the unperturbed equilibrium locus $\boldsymbol{\omega}_+$ for $M_+=M_{1+}^{(13)-}$ and $M_+=M_{1+}^{(13)+}$ (left plot); whereas, one of the two intersections relevant to species $i=2$ lies on the perturbed equilibrium locus $\boldsymbol{\omega}_-$ corresponding to $M_+=M_{2-}^{(13)-}$ and the other one on the unperturbed equilibrium locus $\boldsymbol{\omega}_+$ corresponding to $M_+=M_{2+}^{(13)+}$ (right plot).

The last feature to be pointed out concerns the stability of the asymptotic configurations, analyzed by computing the eigenvalues of the Jacobian matrix of system (21) rewritten in normal form. Figures. 9, 10, 11 and 12 depict the behaviour of the real part of the eigenvalues versus M_+ at both equilibria ($\boldsymbol{\omega}_-$ on the left, $\boldsymbol{\omega}_+$ on the right) for each closure; vertical lines are plotted in correspondence of the critical values of the Mach number.

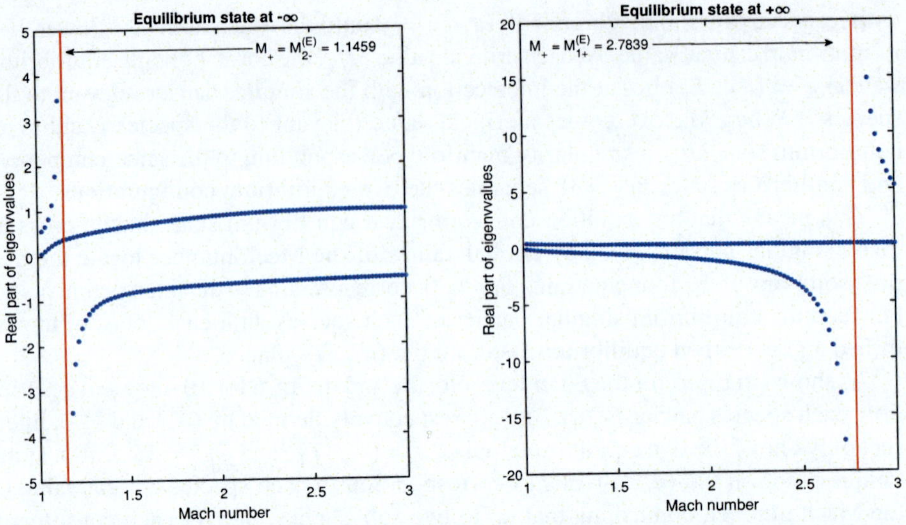

Fig. 9 Stability of equilibria $\boldsymbol{\omega}_-$ (left) and $\boldsymbol{\omega}_+$ (right) at Euler level: plots of the real part of the eigenvalues versus the Mach number M_+; vertical asymptotes denote the instabilities occurring at critical values $M_+ = M_{2-}^{(E)}$ and $M_+ = M_{1+}^{(E)}$, set $\alpha = 0.1$ and $\chi = 0.25$

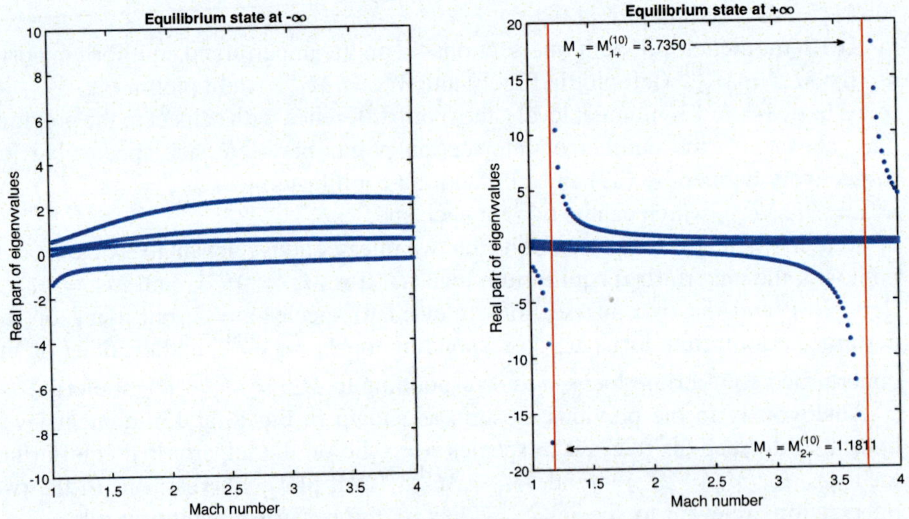

Fig. 10 Stability of equilibria $\boldsymbol{\omega}_-$ (left) and $\boldsymbol{\omega}_+$ (right) at 10–moment level: plots of the real part of the eigenvalues versus the Mach number M_+; vertical asymptotes denote the instabilities occurring at critical values $M_+ = M_{2+}^{(10)}$ and $M_+ = M_{1+}^{(10)}$, set $\alpha = 0.1$ and $\chi = 0.25$

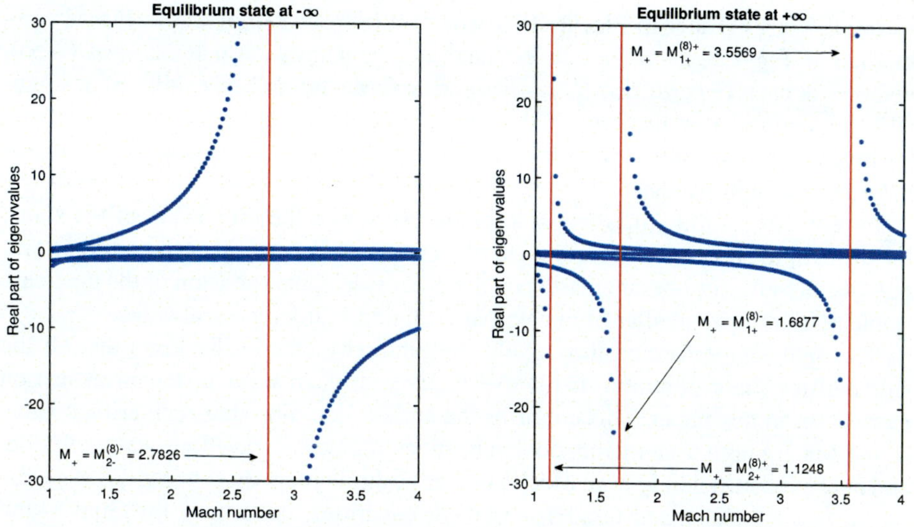

Fig. 11 Stability of equilibria $\boldsymbol{\omega}_-$ (left) and $\boldsymbol{\omega}_+$ (right) at 8–moment level: plots of the real part of the eigenvalues versus the Mach number M_+; vertical asymptotes denote the instabilities occurring at critical values $M_+ = M_{2-}^{(8)-}$, $M_+ = M_{2+}^{(8)+}$, $M_+ = M_{1+}^{(8)-}$ and $M_+ = M_{1+}^{(8)+}$, set $\alpha = 0.1$ and $\chi = 0.25$

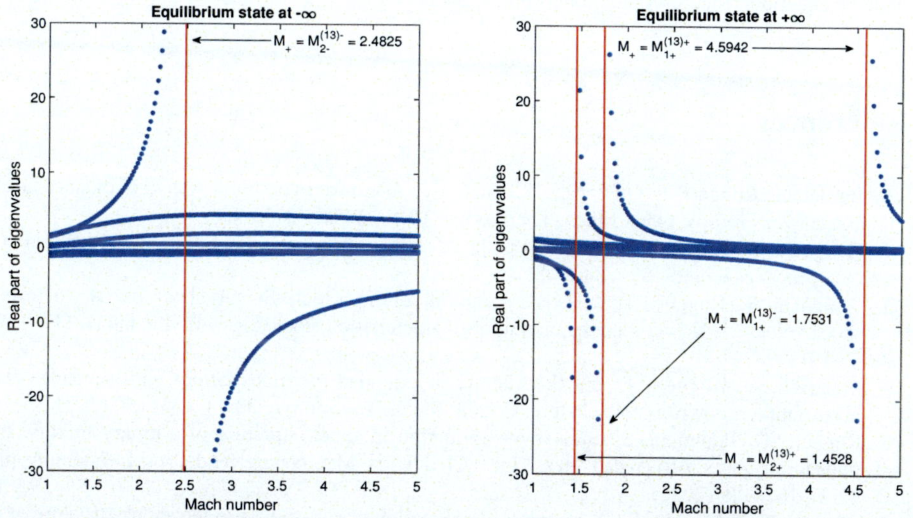

Fig. 12 Stability of equilibria $\boldsymbol{\omega}_-$ (left) and $\boldsymbol{\omega}_+$ (right) at 13–moment level: plots of the real part of the eigenvalues versus the Mach number M_+; vertical asymptotes denote the instabilities occurring at critical values $M_+ = M_{2-}^{(13)-}$, $M_+ = M_{2+}^{(13)+}$, $M_+ = M_{1+}^{(13)-}$ and $M_+ = M_{1+}^{(13)+}$, set $\alpha = 0.1$ and $\chi = 0.25$

In all figures (Euler in Fig. 9; 10–moment in Fig. 10; 8–moment in Fig. 11; 13–moment in Fig. 12), it can be noticed that one eigenvalue, evaluated at $+\infty$ ($-\infty$), diverges at each unperturbed (perturbed) critical Mach number, exhibiting a singularity; in particular, at the perturbed state (left panels) such eigenvalue passes from positive to negative values, whereas, the opposite change in sign occurs at the unperturbed state (right panels).

In terms of stability, it is worth to be observed that the dimension of the stable manifold of the downstream equilibrium get larger of one unit at each perturbed singularity as Mach number increases; on the other side, the dimension of the upstream stable manifold get smaller of one unit at each unperturbed critical value.

The analysis performed throughout the hierarchy clearly discloses that all the singularities share common distinctive features at each level of the investigation carried out in this paper. This confirms the authors' opinion that each critical value of the Mach number can generate a sub–shock along the travelling wave solution, and hence for sufficiently high Mach number multiple jump discontinuities can arise (up to two for Euler and Grad 10–moment equations, up to four for Grad 8–and 13–moment systems).

Acknowledgements This work is performed in the frame of activities sponsored by INdAM-GNFM and by the Universities of Messina and Parma (Italy). G. Martalò is grateful to GNFM for the financial support of the research project *Equazioni cinetiche, modelli mascroscopici e applicazioni* (Kinetic equations, macroscopic models and applications).

References

1. Boillat, G., Ruggeri, T.: On the shock structure problem for hyperbolic system of balance laws and convex entropy. Cont. Mech. Thermodyn. **10**(5), 285–292 (1998)
2. Taniguchi, S., Ruggeri, T.: On the sub-shock formation in extended thermodynamics. Int. J. Non Linear Mech. **99**, 69–78 (2018)
3. Taniguchi, S., Ruggeri, T.: A 2 × 2 simple model in which the sub-shock exists when the shock velocity is slower than the maximum characteristic velocity. Ric. di Mat. **68**(1), 119–129 (2019)
4. Mentrelli, A., Ruggeri, T.: Shock structure in extended thermodynamics with second-order maximum entropy principle closure. Contin. Mech. Thermodyn. **33**(1), 125–150 (2021)
5. Ruggeri, T., Taniguchi, S.: Sub–shock formation in shock structure of a binary mixture of polyatomic gases. Atti Accad. Naz. Lincei Cl. Sci. Fis. Mat. Natur. Rend. Lincei Matem. Appl. **32**(2), 167–179 (2021)
6. Ruggeri, T., Taniguchi, S.: A complete classification of sub-shocks in the shock structure of a binary mixture of Eulerian gases with different degrees of freedom. Phys. Fluids. **34**(6), 066116 (2022)
7. Madjarević, D., Pavić-Colić, M., Simić, S.: Shock structure and relaxation in the multi-component mixture of Euler fluids. Symmetry. **13**(6), 955 (2021)
8. Simić, S., Madjarević, D.: Shock structure and entropy growth in a gaseous binary mixture with viscous and thermal dissipation. Wave Motion. **100**, 102661 (2021)
9. Bisi, M., Groppi, M., Spiga, G.: Grad's distribution functions in the kinetic equations for a chemical reaction. Contin. Mech. Thermodyn. **14**(2), 207–222 (2002)
10. Pirner, M.: Kinetic Modelling of gas Mixtures. Wurzburg Univerisity Press, Wurzburg (2018)

11. Kosuge, S., Aoki, K., Takata, S.: Shock-wave structure for a binary gas mixture: finite-difference analysis of the Boltzmann equation for hard-sphere molecules. Eur. J. Mech. B-Fluids. **20**(1), 87–126 (2001)
12. Madjarević, D., Simić, S.: Shock structure in Helium-Argon mixture—a comparison of hyperbolic multi-temperature model with experiment. EPL. **102**(4), 44002 (2013)
13. Madjarević, D., Ruggeri, T., Simić, S.: Shock structure and temperature overshoot in macroscopic multi-temperature model of mixtures. Phys. Fluids. **26**(10), 106102 (2014)
14. Bisi, M., Martalò, G., Spiga, G.: Shock wave structure of multi-temperature Euler equations from kinetic theory for a binary mixture. Acta Appl. Math. **132**(1), 95–105 (2014)
15. Conforto, F., Mentrelli, A., Ruggeri, T.: Shock structure and multiple sub-shocks in binary mixtures of Eulerian fluids. Ric. di Mat. **66**(1), 221–231 (2017)
16. Bisi, M., Conforto, F., Martalò, G.: Sub-shock formation in Grad 10-moment equations for a binary gas mixture. Contin. Mech. Thermodyn. **28**(5), 1295–1324 (2016)
17. Artale, V., Conforto, F., Martalò, G., Ricciardello, A.: Shock structure and multiple sub-shocks in Grad 10-moment binary mixtures of monoatomic gases. Ric. di Mat. **68**(2), 485–502 (2019)
18. Artale, V., Conforto, F., Martalò, G., Ricciardello, A.: Shock structure solutions of Grad 13-moment equations for binary gas mixtures. Wave Motion. **115**, 103055 (2022)
19. Lax, P.D.: Hyperbolic systems of conservation laws II. Comm. Pure Appl. Math. **10**(4), 537–566 (1957)
20. Bisi, M., Martalò, G., Spiga, G.: Multi-temperature fluid-dynamic model equations from kinetic theory in a reactive gas: the steady shock problem. Comput. Math. with Appl. **66**(8), 1403–1417 (2013)
21. Bisi, M., Groppi, M., Macaluso, A., Martalò, G.: Shock wave structure of multi-temperature Grad 10-moment equations for a binary gas mixture. EPL. **133**(5), 54001 (2021)
22. Boillat, G., Ruggeri, T.: Hyperbolic principal subsystems: entropy convexity and subcharacteristic conditions. Arch. Ration. Mech. Anal. **137**(4), 305–320 (1997)
23. Currò, C., Fusco, D.: Discontinuous travelling wave solutions for a class of dissipative hyperbolic models. Atti Accad. Naz. Lincei Cl. Sci. Fis. Mat. Natur. Rend. Lincei Matem. Appl. **16**(1), 61–71 (2005)

A Conservative a-Posteriori Time-Limiting Procedure in Quinpi Schemes

Giuseppe Visconti, Silvia Tozza, Matteo Semplice, and Gabriella Puppo

Abstract The superior stability properties of implicit time schemes allow to avoid small time steps required to satisfy restrictive stability conditions for stiff hyperbolic systems. In Puppo et al. (Commun Appl Math Comput 2022) an implicit third order finite volume scheme based on a third order DIRK combined with a third order CWENO reconstruction for the space-limiting was proposed. The originality of the proposed method, named Quinpi, lies in the computation of a first order implicit predictor which is used to fix the nonlinear weights of the space reconstruction, thus simplifying considerably the non-linearity of the scheme. However, the time-limiting in the above mentioned paper, which is necessary to control spurious oscillations in the implicit time integration, requires a conservative correction. In this work, we address this problem and we propose a conservative a-posteriori time-limiting procedure inspired by the MOOD method. The numerical experiments show the reliability of the proposed scheme and include both linear and nonlinear scalar conservation laws.

Keywords Conservation laws · High-order schemes · Time-limiting · Implicit methods · CWENO reconstruction

---

G. Visconti (✉) · G. Puppo
Department of Mathematics, Sapienza University of Rome, P.le Aldo Moro 5, 00185 Rome, Italy
e-mail: giuseppe.visconti@uniroma1.it

G. Puppo
e-mail: gabriella.puppo@uniroma1.it

S. Tozza
Department of Mathematics, University of Bologna, Piazza di Porta S. Donato 5, 40126 Bologna, Italy
e-mail: silvia.tozza@unibo.it

M. Semplice
Dipartimento di Scienza e Alta Tecnologia, Università degli Studi dell'Insubria, Via Valleggio 11, 22100 Como, Italy
e-mail: matteo.semplice@uninsubria.it

G. Albi et al. (eds.), *Advances in Numerical Methods for Hyperbolic Balance Laws and Related Problems*, SEMA SIMAI Springer Series 32,
https://doi.org/10.1007/978-3-031-29875-2_9

1 Introduction

Many interesting propagation phenomena can be described by systems of one-dimensional partial differential equations (PDEs) of the form

$$\partial_t u(x,t) + \partial_x f(u(x,t)) = 0, \tag{1}$$

where $u : \mathbb{R} \times [0, +\infty) \to \mathbb{R}^m$, $m \geq 1$, is the unknown solution, and $f : \mathbb{R}^m \to \mathbb{R}^m$ is the vector of the flux functions. System (1) is said *hyperbolic* if the eigenvalues $\{\lambda_j(u)\}_{j=1}^m$ of the Jacobian of f are real, i.e. the propagation speeds of the waves are finite, and there exists a complete set of eigenvectors. In addition, system (1) is said *stiff* if $\frac{\max_{j=1,\dots,m} |\lambda_j(u)|}{\min_{j=1,\dots,m} |\lambda_j(u)|} \gg 1$, namely if the solution u is characterized by waves with very different speeds. This case is relevant in many physical applications, such as Low-Mach problems in gas dynamics, kinetic problems close to equilibrium.

The stiffness of the system introduces numerical difficulties when using explicit schemes which require a very small time step to meet the restrictive Courant-Friedrichs-Levy (CFL) stability condition. A huge literature has been developed for the numerical treatment of these problems. Here we refer to the non-exhaustive list [1, 6, 16–18, 33] for Low-Mach problems and to [19, 26, 29] for kinetic equations.

We will focus on the high-order implicit finite volume numerical approximation of stiff hyperbolic equations, specifically on the recent work [30] where a third order implicit scheme for scalar conservation laws was proposed. The numerical scheme in [30] was named Quinpi. The third order accuracy was achieved by using a third order Diagonally Implicit Runge-Kutta (DIRK) for the time integration and a third order Central Weighted Essentially Non Oscillatory (CWENO) reconstruction, cf. [14], for the space discretization. However, the use of CWENO to perform space-limiting, necessary to prevent spurious oscillations typical of high order schemes, introduces a source of non-linearity which becomes computationally challenging in the implicit time integration. The novel idea of [30] is to simplify considerably the non-linearity of the space-limiting procedure through the computation of a first order implicit predictor, which is used to freeze the nonlinear weights of the CWENO reconstruction. This approach makes the Quinpi scheme linear on linear problems. Instead, on nonlinear problems the only source of non-linearity is due to the nonlinear flux function, which is in turn due to the physical structure of the model, and therefore is unavoidable. The first order implicit scheme is based on a composite backward Euler, which is evaluated at the abscissae of the DIRK, naturally combined with a piecewise constant (i.e. non-limited) reconstruction in space. As noted in [4, 30], the space-limiting is not sufficient to control the appearance of spurious oscillations in implicit integration, especially for large Courant numbers. A time-limiting procedure is required, which is achieved by a nonlinear blending of the first order predictor and of the third order solution at each time level. In particular, in [30] the blending of the cell averages of the two solutions is performed in a CWENO framework using a combination of space and time regularity indicators.

Similar approaches to the Quinpi schemes have been developed in the literature. We mention, for instance, the fifth order implicit Weighted Essentially Non Oscillatory (WENO) schemes in [24], where a predictor-corrector technique is also used. The predictor used in [24] is based on an explicit first order scheme and therefore it does not allow to use large time steps. A fully nonlinear implicit scheme, based on a third order RADAU time integrator and a third order WENO reconstruction, which is limited in both space and time simultaneously, can be found in [4]. Fully implicit, semi-implicit, implicit-explicit and local time-stepping treatments of stiff hyperbolic systems have been investigated, e.g., in [5, 10–13, 22, 23].

In this work, we address the problem of the time-limiting in Quinpi schemes. In particular, the nonlinear blending proposed in [30] has the drawback of being non-conservative and, thus, of requiring a conservative correction at each time step. Here, we propose a different approach to control the spurious oscillations through a conservative a-posteriori time-limiting procedure which draws inspiration from the Multi-dimensional Optimal Order Detection (MOOD) method introduced in [8, 9] in order to reduce the order of the space reconstruction on problematic cells. A similar idea has been investigated in [21] where a MOOD-inspired technique controls unphysical oscillations of high order implicit numerical solutions of the water flow in district heating networks. MOOD has been also extended to other contexts, as for instance in [28, 31, 34]. Furthermore, in this work we also investigate the use of a first order predictor based on the continuous extension [35] of the backward Euler time integrator.

The chapter is organized as follows: In Sect. 2 we first review the third order Quinpi algorithm proposed in [30]. Next, we discuss the two modifications which motivate this work, namely the computation of the predictor using the first order continuous extension of backward Euler and the conservative a-posteriori time-limiting. In Sect. 3 we test the schemes on linear and nonlinear scalar conservation laws, comparing the results with the Quinpi scheme in [30]. Finally, in Sect. 4 we summarize the contributions of this work and discuss future perspectives.

2 Quinpi Scheme for Hyperbolic Conservation Laws

In order to introduce the third order Quinpi scheme, we briefly recall the finite volume setting for the numerical approximation of (1).

We consider a uniform discretization of the compact computational domain $\Omega \subset \mathbb{R}$ made by N cells $\Omega_j = [x_j - {}^{h}/{2}, x_j + {}^{h}/{2}]$ of amplitude $h > 0$. The computational domain Ω is thus covered by the cells Ω_j in such a way $\cup_j \Omega_j = \Omega$. We define the cell averages of the exact solution on the cells Ω_j as $\overline{u}_j(t) = \frac{1}{h}\int_{\Omega_j} u(x,t)\mathrm{d}x$. Using the method of lines (MOL), conservation law (1) can be written in the semi-discrete form

$$\frac{\mathrm{d}\overline{u}_j(t)}{\mathrm{d}t} = -\frac{1}{h}\left[f\left(u\left(x_j + \tfrac{h}{2}, t\right)\right) - f\left(u\left(x_j - \tfrac{h}{2}, t\right)\right)\right]. \tag{2}$$

Notice that (2) is still exact since it describes the exact evolution of the cell averages as the difference of the fluxes at the cell boundaries.

To transform system (2) in a numerical scheme for the approximation of the cell averages, one has to introduce several ingredients.

Space reconstruction with CWENO.

The solution of system (2), which evolves cell averages, requires the knowledge of point values of the solution at the cell interfaces. The problem of computing these values using the information of the cell averages is called *the reconstruction problem* and can be solved by a reconstruction algorithm $\mathcal{R} : \{\overline{u}_j(t)\} \mapsto \mathcal{R}\left(\{\overline{u}_j(t)\}\right)$. Classical examples are the WENO and CWENO reconstruction procedures, see [14, 25, 27, 32], and their developments, where $\mathcal{R}$ is suitably defined in order to prevent spurious oscillations in space.

In CWENO type reconstructions, the operator $\mathcal{R}$ provides a space limited approximation of the exact solution $u(\cdot, t)$, $t \geq 0$, as $u(x, t) \approx \sum_j R_j(x; t)\chi_{\Omega_j}(x)$, where χ_{Ω_j} is the characteristic function of the cell Ω_j, and $R_j(x; t)$ is the reconstruction polynomial of degree d on $x \in \Omega_j$, which depends on time through the time-dependent cell averages. With the reconstruction algorithm $\mathcal{R}$ we can estimate the values $u\left(x_j \pm {}^h/_2, t\right)$, $t \geq 0$, from the knowledge of the cell averages only. More precisely, the reconstruction algorithm provides the following values at each cell interface

$$u^-_{j+1/2}(t) = R_j\left(x_j + \tfrac{h}{2}; t\right), \quad u^+_{j+1/2}(t) = R_{j+1}\left(x_j + \tfrac{h}{2}; t\right), \tag{3}$$

which are named *boundary extrapolated data* (BED). We recall that with CWENO it is possible to reconstruct uniformly accurate and non-oscillatory point values of the solution at any point $x \in \Omega$.

Following [30], we focus on third order CWENO schemes in which the restriction of the operator $\mathcal{R}$ on a cell Ω_j can be defined as

$$\mathcal{R}_{|\Omega_j} : \{\overline{u}_k(t)\}_{k=j-1}^{j+1} \mapsto R_j(x; t) = \omega_{j,0}P_{j,0}(x) + \omega_{j,L}P^{(1)}_{j,L}(x) + \omega_{j,R}P^{(1)}_{j,R}(x).$$

Here, $P^{(1)}_{j,L}$ and $P^{(1)}_{j,R}$ are linear polynomials interpolating $\{\overline{u}_{j-1}(t), \overline{u}_j(t)\}$ and $\{\overline{u}_j(t), \overline{u}_{j+1}(t)\}$, respectively, in the sense of cell averages. Whereas, $P_{j,0}$ is the second degree second order accurate polynomial

$$P_{j,0}(x) = \frac{1}{d_0}\left(P^{(2)}_j(x) - d_L P^{(1)}_{j,L}(x) - d_R P^{(1)}_{j,R}(x)\right)$$

where $d_0, d_L, d_R \in (0, 1)$ with $d_0 + d_L + d_R = 1$ are the so-called linear or optimal coefficients, and $P^{(2)}_j(x)$ is the polynomial of degree 2 interpolating $\{\overline{u}_{j-1}(t), \overline{u}_j(t), \overline{u}_{j+1}(t)\}$ in the sense of cell averages. Observe that all these polynomials are time dependent, but we have suppressed the time dependencies for a better readability. For a detailed formulation of all these polynomials we refer the interested reader to [30, Sect. 3.1]. Finally, $\omega_{j,0}, \omega_{j,L}, \omega_{j,R} \in (0, 1)$ are the nonlinear weights which depend non-linearly on the cell averages. They are suitably defined, with the help of nonlinear regularity indicators, such as the Jiang-Shu smoothness indicators [25], in order to select the highest order of accuracy of the reconstruction on smooth flows and to downgrade to lower order non-oscillatory reconstructions on discontinuous solutions. Although several definitions of the nonlinear weights have been studied in the literature [2, 7, 15], we use the classical WENO type definition as in [30].

Notice that, in principle, the two BED in (3) obtained with the reconstruction algorithm are different, although computed at the same interface $x_{j+h/2}$, with $\left|u^+_{j+1/2}(t) - u^-_{j+1/2}(t)\right| = O(h^{d+1})$ on smooth flows. Therefore, in order to approximate the flux function at the interfaces, one introduces a consistent and monotone numerical flux $\mathcal{F}(a, b)$ such that $f(u(x_{j+1/2}, t)) \approx \mathcal{F}_{j+1/2}(t) = \mathcal{F}(u^-_{j+1/2}(t), u^+_{j+1/2}(t))$. The function $\mathcal{F}$ may be any approximate or exact Riemann solver. Finally, one obtains the system of ordinary differential equations (ODEs)

$$\frac{\mathrm{d}\overline{u}_j(t)}{\mathrm{d}t} = -\frac{1}{h}\left[\mathcal{F}_{j+\frac{1}{2}}(t) - \mathcal{F}_{j-\frac{1}{2}}(t)\right], \tag{4}$$

which provides the approximation of the PDE solution. From now on, we omit the dependence on t in order to lighten the notation.

Time integration with DIRK.

The semi-discrete approach with the MOL completely defines the system of ODEs (4). It is worth to notice that now $\overline{u}_j$ is an approximation of the exact cell average on Ω_j. In order to obtain a fully discrete scheme, one needs to approximate (4) with a time integration scheme. If we focus on Runge-Kutta (RK) schemes with s stages and general Butcher tableau

$$\begin{array}{c|cccc} c_1 & a_{11} & a_{12} & \dots & a_{1s} \\ c_2 & a_{21} & a_{22} & \dots & a_{2s} \\ \vdots & \vdots & \vdots & \ddots & \\ c_s & a_{s1} & a_{s2} & \dots & a_{ss} \\ \hline & b_1 & b_2 & \dots & b_s \end{array} \tag{5}$$

the time discretization of (4) leads to

$$\begin{aligned} \overline{u}_j^{n+1} &= \overline{u}_j^n - \frac{\Delta t}{h}\sum_{k=1}^{s} b_k\left[\mathcal{F}^{(k)}_{j+\frac{1}{2}} - \mathcal{F}^{(k)}_{j-\frac{1}{2}}\right], \quad n \geq 0, \ \forall j, \\ \mathcal{F}^{(k)}_{j+\frac{1}{2}} &= \mathcal{F}\left(u^{-,(k)}_{j+\frac{1}{2}}, u^{+,(k)}_{j+\frac{1}{2}}\right) \end{aligned} \tag{6}$$

where Δt is the time step, and now $\overline{u}_j^n$ denotes an approximation of the exact cell average at time $n\Delta t$, namely $\frac{1}{h}\int_{\Omega_j} u(x, n\Delta t)\mathrm{d}x$. Finally, $u^{\mp,(k)}_{j\pm 1/2}$ are BED from the stage values

$$\overline{u}_j^{(k)} = \overline{u}_j^n - \frac{\Delta t}{h} \sum_{\ell=1}^{s} a_{k\ell} \left[\mathcal{F}^{(\ell)}_{j+\frac{1}{2}} - \mathcal{F}^{(\ell)}_{j-\frac{1}{2}} \right], \quad \ell = 1, \ldots, s, \tag{7}$$

approximations of the solution at times $t^{(k)} = (n + c_k)\Delta t$. Typical assumption is that $c_k = \sum_{\ell=1}^{s} a_{k\ell}$, $k = 1, \ldots, s$, and one has $\sum_{k=1}^{s} b_k = 1$ for consistency. Clearly, one uses RK schemes with order matching the order of the space approximation.

It is well known that for explicit schemes, in the case of RK when $a_{k\ell} = 0$, $\ell \geq k$, the time step Δt must satisfy the CFL stability condition $\lambda := \Delta t / h \leq 1/ \max_{i=1,\ldots,m} |\lambda_i(u)|$. We recall that the $\lambda_i(u)$ are the eigenvalues of the Jacobian of the flux function f, namely the propagation speeds of the waves. Thus, if the system is stiff, i.e. characterized by fast waves, the CFL condition becomes very restrictive imposing the use of a very small time step for stability problems and not for accuracy. A way to overcome this restriction is to build implicit schemes with unconditional stability, e.g. based on DIRK schemes, for which $a_{k\ell} = 0$, $\ell > k$. The advantage of DIRK schemes is that the implicit computation of a given stage value (7) is independent from the following ones. Therefore, for each time step one has to solve the s nonlinear systems of equations

$$G_j \left(\overline{u}^{(k)} \right) := \overline{u}_j^{(k)} + \frac{a_{kk}\Delta t}{h} \left[\mathcal{F}^{(k)}_{j+\frac{1}{2}} - \mathcal{F}^{(k)}_{j-\frac{1}{2}} \right] - \overline{u}_j^n + \frac{\Delta t}{h} \sum_{\ell=1}^{k-1} a_{k\ell} \left[\mathcal{F}^{(\ell)}_{j+\frac{1}{2}} - \mathcal{F}^{(\ell)}_{j-\frac{1}{2}} \right] = 0.$$

We have highlighted the difference of the k-th numerical fluxes to emphasize that this makes the system $G(u) = 0$ nonlinear. In fact, G has two sources of non-linearity:

1. From the physics, if the phenomenon under study is described by a nonlinear flux function f in (1), and cannot be avoided;
2. Due to the high-order space discretization, since the numerical fluxes $\mathcal{F}^{(k)}_{j\pm 1/2}$ are computed on the BED of $\overline{u}_j^{(k)}$, obtained by the reconstruction procedure, which is nonlinear through the nonlinear weights even for linear problems. Therefore, even for linear PDEs, the implicit scheme requires a nonlinear solver to find the solution of $G(u) = 0$. This results in a prohibitive computational cost.

The Quinpi approach provides a way to circumvent the non-linearity determined by the high-order reconstruction procedure as described in the next subsection.

2.1 The Quinpi Approach

The Quinpi idea [30] is explained in the case of scalar conservation laws, i.e. taking $m = 1$ in (1). The name Quinpi stands for implicit CWENO and it is based on a predictor-corrector approach to tackle the second source of non-linearity discussed previously. To this end, an approximation of the solution at the s intermediate times $t^{(k)} \in [n\Delta t, (n+1)\Delta t]$, $k = 1, \ldots, s$, defined by the abscissae of the DIRK method is computed with a linear implicit low order scheme. The predictor solutions are then used to avoid the non-linearities of the high order method, the corrector, which are induced by the reconstruction procedure. In fact, it is possible to write the BED (3) at stage k as

$$u^{-,(k)}_{j+\frac{1}{2}} = \sum_{\ell=-1}^{1} W_{j,\ell}\left(x_{j+1/2}\right)\overline{u}^{(k)}_{j+\ell}, \quad u^{+,(k)}_{j+\frac{1}{2}} = \sum_{\ell=-1}^{1} W_{j+1,\ell}\left(x_{j+1/2}\right)\overline{u}^{(k)}_{j+1+\ell}, \tag{8}$$

where the weights $\{W_{j,\ell}(\cdot), W_{j+1,\ell}(\cdot)\}_{\ell=-1,0,1}$ contain the nonlinear part of the reconstruction. Then, if these weights are computed on the predictor scheme, they become constant with respect to the cell averages $\overline{u}^k$. In this way, the complete scheme is linear with respect to the space reconstruction in the sense that the solution of $G(u) = 0$ is obtained by solving sequentially s systems which are nonlinear only through the flux function f.

2.1.1 The Predictor: Composite Backward Euler Vs Continuous Extension

In order to obtain a predictor of the solution, we approximate system (2) with an implicit first order scheme. Specifically, in [30] the composite backward Euler method is employed providing s approximations, say $\overline{u}^{\mathsf{BE},(k)}$, $k = 1, \ldots, s$, within the time step $[n\Delta t, (n+1)\Delta t]$ of the high order scheme. Without loss of generality, assume that the abscissae of the DIRK method are ordered. Then, the k-th first order solution $\overline{u}^{\mathsf{BE},(k)}$ advances the solution from $t^{(k-1)} = (n + c_{k-1})\Delta t$ to $t^{(k)} = (n + c_k)\Delta t$, with the notation convention $c_0 = 0$. Overall, one has

$$\begin{aligned}
\overline{u}^{\mathsf{BE},n+1}_j &= \overline{u}^n_j - \frac{\Delta t}{h}\sum_{k=1}^{s}\theta_k\left[\mathcal{F}^{\mathsf{BE},(k)}_{j+\frac{1}{2}} - \mathcal{F}^{\mathsf{BE},(k)}_{j-\frac{1}{2}}\right], \quad n \geq 0,\ \forall j,\\
\mathcal{F}^{\mathsf{BE},(k)}_{j+\frac{1}{2}} &= \mathcal{F}\left(\overline{u}^{\mathsf{BE},(k)}_j, \overline{u}^{\mathsf{BE},(k)}_{j+1}\right)\\
\overline{u}^{\mathsf{BE},(k)}_j &= \overline{u}^{\mathsf{BE},(k-1)}_j - \frac{\theta_k\Delta t}{h}\left[\mathcal{F}^{\mathsf{BE},(k)}_{j+\frac{1}{2}} - \mathcal{F}^{\mathsf{BE},(k)}_{j-\frac{1}{2}}\right], \quad k = 1, \ldots, s,
\end{aligned} \tag{9}$$

where $\theta_k := c_k - c_{k-1}$ and $\overline{u}^{\mathsf{BE},(0)}_j := \overline{u}^n_j$.

Notice that the numerical flux function $\mathcal{F}$ in (9) is now computed on piecewise constant reconstructions from the cell averages. In fact, first order predictors do not require space-limiting, because they are unconditionally Total Variation Diminishing. Therefore, despite of the high order scheme, the first order predictor (9) is characterized by a single non-linearity, that is the one induced by the flux function f, which requires to solve the nonlinear system

$$G_j(\overline{u}^{\mathsf{BE},(k)}) := \overline{u}_j^{\mathsf{BE},(k)} + \frac{\theta_k \Delta t}{h}\left[\mathcal{F}_{j+\frac{1}{2}}^{\mathsf{BE},(k)} - \mathcal{F}_{j-\frac{1}{2}}^{\mathsf{BE},(k)}\right] - \overline{u}_j^{\mathsf{BE},(k-1)} = 0 \tag{10}$$

at each stage.

However, the solution of (10) requires the use of a numerical method, such as the Newton's method, s times within a single time step and this may be computationally expensive. In this work we explore a way to reduce this complexity which relies on the continuous extensions of RK schemes [35]. More precisely, we compute the predictor solution only at time $(n+1)\Delta t$ solving one nonlinear system in each time step. Then, we recover the first order approximations at the times $t^{(k)} = [n\Delta t, (n+1)\Delta t]$ as linear interpolation between $\overline{u}_j^n$ and $\overline{u}_j^{\mathsf{BE},n+1}$. Overall, we have

$$\begin{aligned}
\overline{u}_j^{\mathsf{BE},n+1} &= \overline{u}_j^n - \frac{\Delta t}{h}\left[\mathcal{F}_{j+\frac{1}{2}}^{\mathsf{BE},n+1} - \mathcal{F}_{j-\frac{1}{2}}^{\mathsf{BE},n+1}\right], \quad n \geq 0, \ \forall j, \\
\mathcal{F}_{j+\frac{1}{2}}^{\mathsf{BE},n+1} &= \mathcal{F}\left(\overline{u}_j^{\mathsf{BE},n+1}, \overline{u}_{j+1}^{\mathsf{BE},n+1}\right) \\
\overline{u}_j^{\mathsf{BE},(k)} &= \overline{u}_j^n - \frac{c_k \Delta t}{h}\left[\mathcal{F}_{j+\frac{1}{2}}^{\mathsf{BE},n+1} - \mathcal{F}_{j-\frac{1}{2}}^{\mathsf{BE},n+1}\right], \quad k = 1, \dots, s.
\end{aligned} \tag{11}$$

This approach has the advantage to re-use the numerical fluxes at time level $(n+1)\Delta t$ to compute the $\overline{u}_j^{\mathsf{BE},(k)}$, and thus it results in a lower computational cost compared to (9).

Once the first order predictions are obtained, either with (9) or by (11), they are used to evaluate the nonlinear terms $\{W_{j,\ell}(\cdot), W_{j+1,\ell}(\cdot)\}_{\ell=-1,0,1}$ of the high order BED (8). Finally, from (6)–(7) one obtains the high order solution $\overline{u}_j^{n+1}$ at time level $(n+1)\Delta t$.

2.1.2 A Conservative a-Posteriori Time-Limiting

The Quinpi scheme ends with the limiting of the high order solution. In fact, when using a large Δt, the space limiting is not enough to prevent the spurious oscillations of high order schemes. In this case, as noticed in [4, 20, 30], limiting in time is also required. In particular, in [30] the time limiting is performed in a CWENO-like framework. There, the limiting is applied on the computed high order solution, blending it with the low order predictor which is a reliable, stable and non-oscillatory solution. However, the procedure in [30] does not have mass conservation property and must be followed up by a suitable conservative correction.

Here, we address this issue and propose a conservative a-posteriori time-limiting inspired by the MOOD technique proposed by Diot et al. [8, 9]. MOOD is originally designed as an a-posteriori space-limiting technique for multi-dimensional finite volume schemes. Instead, we use the typical MOOD detectors to limit the high order solution of the Quinpi approach at time level $(n+1)\Delta t$. Specifically, on the cells where the MOOD criteria detect an oscillatory behavior, we replace the fluxes at the cell interfaces with a convex combination of the high order and the low order numerical fluxes. We point out that a similar idea has been investigated in [21], where an a-posteriori limiting for fully implicit finite volume schemes on transport networks is proposed.

Whenever the a-posteriori limiting finds that Ω_ℓ is a problematic cell, for some $\ell \in \{1, \ldots, N\}$, the high order numerical fluxes $\mathcal{F}^{(k)}_{\ell\pm 1/2}$ are replaced by the limited fluxes

$$\mathcal{F}^{\mathsf{TL},(k)}_{\ell\pm 1/2} = (1-\alpha_{\mathsf{TL}})\theta_k \mathcal{F}^{\mathsf{BE},(k)}_{\ell\pm 1/2} + \alpha_{\mathsf{TL}} b_k \mathcal{F}^{(k)}_{\ell\pm 1/2}, \quad \alpha_{\mathsf{TL}} \in [0, 1], \tag{12}$$

where the acronym TL stands for Time-Limited. Of course, if neither Ω_ℓ nor $\Omega_{\ell+1}$ are problematic, $\mathcal{F}^{\mathsf{TL},(k)}_{\ell+1/2} = b_k \mathcal{F}^{(k)}_{\ell+1/2}$. Thus, the high order numerical fluxes at the interfaces of the problematic cell Ω_ℓ are blended with the low order numerical fluxes reducing locally the order of the solution recomputing

$$\overline{u}^{n+1}_j = \overline{u}^n_j - \frac{\Delta t}{h} \sum_{k=1}^{s} \left[\mathcal{F}^{\mathsf{TL},(k)}_{j+\frac{1}{2}} - \mathcal{F}^{\mathsf{TL},(k)}_{j-\frac{1}{2}} \right], \quad j = \ell-1, \ell, \ell+1. \tag{13}$$

The time-limited fluxes are computed as in (12) when the predictor is obtained with the composite backward Euler, which provides all the low order numerical fluxes $\mathcal{F}^{\mathsf{BE},(k)}_{\ell\pm 1/2}$, $k = 1, \ldots, s$. Instead, when the predictor is computed with the use of the continuous extension, the high order numerical fluxes $\mathcal{F}^{(k)}_{\ell\pm 1/2}$ are replaced by

$$\mathcal{F}^{\mathsf{TL},(k)}_{\ell\pm 1/2} = \frac{1}{s} \mathcal{F}^{\mathsf{BE},n+1}_{\ell\pm 1/2}. \tag{14}$$

In order to determine where the time limiting has to occur, we detect spurious oscillations using the following criteria, inspired by the ones proposed in [9].

Extrema Detector (ED). The ED criterion checks if a cell Ω_j has a local extremum of the solution at time $(n+1)\Delta t$, i.e. if $\overline{u}^{n+1}_j$ satisfies

$$h^2 < \min\left\{\overline{u}^{n+1}_{j-1}, \overline{u}^{n+1}_{j+1}\right\} - \overline{u}^{n+1}_j \quad \text{or} \quad h^2 < \overline{u}^{n+1}_j - \max\left\{\overline{u}^{n+1}_{j-1}, \overline{u}^{n+1}_{j+1}\right\}.$$

If Ω_j does not have an extremum, then the approximation $\overline{u}^{n+1}_j$ is assumed to be valid. Otherwise, one has to determine if $\overline{u}^{n+1}_j$ is a physical extremum or a spurious oscillation, determined either by the high order scheme or by floating point errors.

Let us define a fourth order approximation of the second derivative of the solution in x_j at time $t^{n+1} = (n+1)\Delta t$ from cell averages as

$$\frac{\mathrm{d}^2 u(x, t^{n+1})}{\mathrm{d}x^2}\Big|_{x=x_j} \approx C_j := \frac{-\overline{u}^{n+1}_{j-2} + 12\overline{u}^{n+1}_{j-1} - 22\overline{u}^{n+1}_j + 12\overline{u}^{n+1}_{j+1} - \overline{u}^{n+1}_{j+2}}{8h^2}.$$

Then, C_j provides a discrete approximation of the local curvature in Ω_j and based on the following indicators

$$\chi_{j,m} := \min\left\{C_{j-1}, C_j, C_{j+1}\right\}, \quad \chi_{j,M} := \max\left\{C_{j-1}, C_j, C_{j+1}\right\}$$

one distinguishes the type of the extremum with the subsequent criteria.

Plateau Detector (PD). If the total curvature $\max\left\{|\chi_{j,m}|, |\chi_{j,M}|\right\}$ is close to zero, then the found extremum in Ω_j is an artifact due to round off errors and the solution $\overline{u}_j^{n+1}$ is considered valid. The presence of the local plateau is numerically checked thanks to the following condition

$$\max\left\{|\chi_{j,m}|, |\chi_{j,M}|\right\} < h.$$

Instead, if Ω_j does not have a local plateau, then one investigates the nature of the extremum with the following criteria.

Local Oscillation Detector (LOD). A spurious oscillation occurs in Ω_j if the sign of the curvature changes, i.e., if one observes numerically

$$\chi_{j,m}\chi_{j,M} < 10^{-8}.$$

If a local oscillation is found, then the time-limiting occurs as in (12) or (14). Otherwise, one checks if the extremum is smooth with the following detector.

Smoothness Detector (SD). There is a smooth extremum if

$$\frac{1}{4} < \frac{\min\left\{|\chi_{j,m}|, |\chi_{j,M}|\right\}}{\max\left\{|\chi_{j,m}|, |\chi_{j,M}|\right\}} < 1.$$

If the extremum is not smooth, then the time-limiting occurs as in (12) or (14).

In the numerical experiments of this chapter we consider only linear transport and Burgers equations, for which the admissible states are the entire $\mathbb{R}$ and the computer code does not perform operations that could lead to the occurrence of NaN values in the numerical solutions. Thus, contrary to classical MOOD detectors, we do not use the Physical Admissibility Detection (PAD) criterion [8], which checks if the solution stays in physically meaningful regimes.

What makes the present time-limiting approach conservative is the blending of the numerical fluxes instead of the blending of the solutions as done in [30]. The parameter α_{TL} in (12) is problem dependent, since it depends on the type of equation, presence of stiffness, initial profile chosen, etc.

3 Numerical Tests

In this section, we consider standard tests which are commonly used in the literature on high-order methods for scalar conservation laws: linear advection of non-smooth waves, shock formation and interaction in the nonlinear Burgers equation.

All the numerical simulations are performed with the third order Quinpi scheme in [30] named Q3P1 and with the Quinpi schemes of this work based on the conservative a-posteriori time-limiting inspired by MOOD. In particular, we name Q3P1$_{\mathsf{MOOD}}$ the third order scheme with the composite backward Euler as predictor and Q3P1$_{\mathsf{CEMOOD}}$ the third order scheme where the predictor is computed with the continuous extension.

All the schemes are built using the third order CWENO reconstruction with the optimal coefficients $d_0 = 3/4$, $d_L = d_R = 1/8$, whereas the nonlinear weights $w_{j,k}$, $k = 0, L, R, \forall j$, are computed using the Jiang-Shu regularity indicators, see [25], with parameter $\epsilon = h^2$. For the purposes of this work it is sufficient to consider the Lax-Friedrichs numerical flux

$$\mathcal{F}(u^-, u^+) = \frac{1}{2}\left(f(u^+) + f(u^-) - \alpha(u^+ - u^-)\right), \tag{15}$$

with $\alpha = \max_u |f'(u)|$. The time integrator is the three stage third order L-stable DIRK scheme of [3] with Butcher tableau

$$\begin{array}{c|ccc}
\lambda & \lambda & 0 & 0 \\
\frac{(1+\lambda)}{2} & \frac{(1-\lambda)}{2} & \lambda & 0 \\
1 & -\frac{3}{2}\lambda^2 + 4\lambda - \frac{1}{4} & \frac{3}{2}\lambda^2 - 5\lambda + \frac{5}{4} & \lambda \\
\hline
 & -\frac{3}{2}\lambda^2 + 4\lambda - \frac{1}{4} & \frac{3}{2}\lambda^2 - 5\lambda + \frac{5}{4} & \lambda
\end{array} \tag{16}$$

where $\lambda = 0.4358665215$. The Courant numbers (Cou) and the values α_{TL} of the conservative a-posteriori time-limiting are specified in each numerical test.

3.1 Test 1: Experimental Order of Convergence

First of all, we study the accuracy of the Quinpi scheme proposed in this work by numerically computing the rate of convergence. Similarly to what was done in [4, 30], we consider the nonlinear Burgers equation, that is

$$\partial_t u(x,t) + \partial_x \left(\frac{u^2(x,t)}{2}\right) = 0, \tag{17}$$

Table 1 Test 1. Comparisons of the orders of convergence of the Quinpi schemes with $\Delta t = h$

	Q3P1		Q3P1$_{\text{MOOD}}$		Q3P1$_{\text{CEMOOD}}$	
N	L^1 error	Rate	L^1 error	Rate	L^1 error	Rate
160	$1.83 \cdot 10^{-3}$	1.71	$7.63 \cdot 10^{-2}$	2.72	$7.63 \cdot 10^{-5}$	2.72
320	$4.27 \cdot 10^{-4}$	2.10	$1.04 \cdot 10^{-5}$	2.87	$1.04 \cdot 10^{-5}$	2.87
640	$7.28 \cdot 10^{-5}$	2.55	$1.34 \cdot 10^{-6}$	2.96	$1.34 \cdot 10^{-6}$	2.96
1,280	$1.00 \cdot 10^{-5}5$	2.86	$1.68 \cdot 10^{-7}$	2.99	$1.68 \cdot 10^{-7}$	2.99
2,560	$1.28 \cdot 10^{-6}6$	2.97	$2.10 \cdot 10^{-8}$	3.00	$2.10 \cdot 10^{-8}$	3.00
	Q3P1		Q3P1$_{\text{MOOD}}$		Q3P1$_{\text{CEMOOD}}$	
N	L^∞ error	Rate	L^∞ error	Rate	L^∞ error	Rate
160	$1.70 \cdot 10^{-2}$	1.09	$8.61 \cdot 10^{-4}$	2.01	$8.61 \cdot 10^{-4}$	2.01
320	$5.51 \cdot 10^{-3}$	1.63	$1.41 \cdot 10^{-4}$	2.61	$1.41 \cdot 10^{-4}$	2.61
640	$1.13 \cdot 10^{-3}$	2.29	$1.88 \cdot 10^{-5}$	2.90	$1.88 \cdot 10^{-5}$	2.90
1,280	$1.64 \cdot 10^{-4}$	2.78	$2.38 \cdot 10^{-6}$	2.98	$2.38 \cdot 10^{-6}$	2.98
2,560	$2.12 \cdot 10^{-5}$	2.95	$2.97 \cdot 10^{-7}$	3.00	$2.97 \cdot 10^{-7}$	3.00

on $(x, t) \in (0, 2) \times (0, 1]$, with the following simple smooth initial condition

$$u_0(x) = 0.5 - 0.25 \sin(\pi x). \tag{18}$$

The L^1 and L^∞ norms of the numerical errors, and the experimental order of convergence are computed at the final time $t = 1$ in order to avoid the shock formation. All the results are reported in Tables 1, 2 and 3 for different Courant numbers, where we observe third order convergence in both L^1 and L^∞ norms. With large Courant numbers, the proposed two modifications, Q3P1$_{\text{CEMOOD}}$ and Q3P1$_{\text{MOOD}}$, reach smaller errors and faster convergence with respect to the results in [4] for the same grid, and the results are similar to those of Q3P1 in [30]. Instead, considering $\Delta t = h$, the conservative a-posteriori time-limiting proposed here performs better than Q3P1 in [30]. This means that the new time-limiting allows to limit less the solution for moderate Courant numbers.
The results related to the proposed Q3P1$_{\text{MOOD}}$ scheme here reported are obtained with $\alpha_{\text{TL}} = 0.75$. We have not observed changes in the results choosing other values of α_{TL}.

3.2 *Test 2: Linear Transport Problem*

We consider the linear scalar conservation law

$$\partial_t u(x, t) + \partial_x u(x, t) = 0, \tag{19}$$

Table 2 Test 1. Comparisons of the orders of convergence of the Quinpi schemes with $\Delta t = 10\,h$

	Q3P1		Q3P1$_{\text{MOOD}}$		Q3P1$_{\text{CEMOOD}}$	
N	L^1 error	Rate	L^1 error	Rate	L^1 error	Rate
160	$1.93 \cdot 10^{-3}$	2.05	$1.73 \cdot 10^{-3}$	2.03	$1.72 \cdot 10^{-3}$	1.94
320	$3.75 \cdot 10^{-4}$	2.36	$3.43 \cdot 10^{-4}$	2.34	$3.43 \cdot 10^{-4}$	2.32
640	$6.00 \cdot 10^{-5}$	2.64	$5.57 \cdot 10^{-5}$	2.62	$5.58 \cdot 10^{-5}$	2.62
1,280	$8.25 \cdot 10^{-6}$	2.86	$7.76 \cdot 10^{-6}$	2.84	$7.77 \cdot 10^{-6}$	2.84
2,560	$1.05 \cdot 10^{-6}$	2.97	$1.00 \cdot 10^{-6}$	2.96	$1.00 \cdot 10^{-6}$	2.96
5,120	$1.32 \cdot 10^{-7}$	3.00	$1.26 \cdot 10^{-7}$	2.99	$1.26 \cdot 10^{-7}$	2.99
	Q3P1		Q3P1$_{\text{MOOD}}$		Q3P1$_{\text{CEMOOD}}$	
N	L^∞ error	Rate	L^∞ error	Rate	L^∞ error	Rate
160	$1.59 \cdot 10^{-2}$	1.33	$1.42 \cdot 10^{-2}$	1.26	$1.41 \cdot 10^{-2}$	1.14
320	$4.74 \cdot 10^{-3}$	1.74	$4.33 \cdot 10^{-3}$	1.72	$4.36 \cdot 10^{-3}$	1.69
640	$1.00 \cdot 10^{-3}$	2.24	$9.31 \cdot 10^{-4}$	2.22	$9.33 \cdot 10^{-4}$	2.22
1,280	$1.61 \cdot 10^{-4}$	2.64	$1.52 \cdot 10^{-4}$	2.61	$1.52 \cdot 10^{-4}$	2.61
2,560	$2.18 \cdot 10^{-5}$	2.89	$2.08 \cdot 10^{-5}$	2.87	$2.08 \cdot 10^{-5}$	2.87
5,120	$2.76 \cdot 10^{-6}$	2.98	$2.65 \cdot 10^{-6}$	2.97	$2.65 \cdot 10^{-6}$	2.97

Table 3 Test 1. Comparisons of the orders of convergence of the Quinpi schemes with $\Delta t = 50\,h$

	Q3P1		Q3P1$_{\text{MOOD}}$		Q3P1$_{\text{CEMOOD}}$	
N	L^1 error	Rate	L^1 error	Rate	L^1 error	Rate
1,280	$5.67 \cdot 10^{-4}4$	2.04	$5.66 \cdot 10^{-4}4$	2.03	$5.65 \cdot 10^{-4}4$	2.03
2,560	$9.74 \cdot 10^{-5}5$	2.54	$9.72 \cdot 10^{-5}5$	2.54	$9.73 \cdot 10^{-5}5$	2.54
5,120	$1.41 \cdot 10^{-5}5$	2.79	$1.41 \cdot 10^{-5}5$	2.79	$1.41 \cdot 10^{-5}5$	2.79
10,240	$1.90 \cdot 10^{-6}6$	2.89	$1.90 \cdot 10^{-6}6$	2.89	$1.90 \cdot 10^{-6}6$	2.89
20,480	$2.44 \cdot 10^{-7}7$	2.96	$2.44 \cdot 10^{-7}7$	2.96	$2.44 \cdot 10^{-7}7$	2.96
	Q3P1		Q3P1$_{\text{MOOD}}$		Q3P1$_{\text{CEMOOD}}$	
N	L^∞ error	Rate	L^∞ error	Rate	L^∞ error	Rate
1,280	$6.59 \cdot 10^{-3}3$	1.48	$6.58 \cdot 10^{-3}3$	1.48	$6.54 \cdot 10^{-3}3$	1.47
2,560	$1.54 \cdot 10^{-3}3$	2.10	$1.54 \cdot 10^{-3}3$	2.10	$1.54 \cdot 10^{-3}3$	2.09
5,120	$2.69 \cdot 10^{-4}4$	2.52	$2.68 \cdot 10^{-4}4$	2.52	$2.69 \cdot 10^{-4}4$	2.52
10,240	$3.92 \cdot 10^{-5}5$	2.78	$3.91 \cdot 10^{-5}5$	2.78	$3.91 \cdot 10^{-5}5$	2.78
20,480	$5.13 \cdot 10^{-6}6$	2.93	$5.13 \cdot 10^{-6}6$	2.93	$5.13 \cdot 10^{-6}6$	2.93

on $(x, t) \in (-1, 1) \times (0, 2]$, with periodic boundary conditions in space. As initial condition we will consider the following three profiles:

$$u_0(x) = \sin(\pi x) + \begin{cases} 3, & -0.4 \le x \le 0.4, \\ 0, & \text{otherwise}, \end{cases} \tag{20a}$$

$$u_0(x) = \begin{cases} 1, & -0.25 \le x \le 0.25, \\ 0, & \text{otherwise}, \end{cases} \tag{20b}$$

$$u_0(x) = \begin{cases} \frac{1}{6}\left(G(x, \beta, z - \delta) + G(x, \beta, z + \delta) + 4G(x, \beta, z)\right), & -0.8 \le x \le -0.6, \\ 1, & -0.4 \le x \le -0.2, \\ 1 - |10(x - 0.1)|, & 0 \le x \le 0.2, \\ \frac{1}{6}\left(F(x, \alpha, a - \delta) + F(x, \alpha, a + \delta) + 4F(x, \alpha, a)\right), & 0.4 \le x \le 0.6, \\ 0, & \text{otherwise}, \end{cases}$$

where

$$\begin{aligned} G(x, \beta, z) &= \exp(-\beta(x - z)^2), \\ F(x, \alpha, a) &= \sqrt{\max\{1 - \alpha^2(x - a)^2, 0\}}. \end{aligned} \tag{20c}$$

The initial conditions (20a) and (20b) are a discontinuous sinusoidal and a double-step profile, respectively, whereas the initial condition (20c) has been designed by Jiang and Shu in [25]. Fixing the constants in (20c) as $a = 0.5$, $z = -0.7$, $\delta = 0.005$, $\alpha = 10$, and $\beta = \log 2/36\delta^2$, this initial condition consists of smooth and non-smooth shapes. Precisely, from the left to the right side of the domain, there is a Gaussian, then a double-step, a sharp triangle and finally a half ellipse.

We use these numerical tests to study the properties of a scheme to transport different shapes with minimal dissipation, dispersion effects and oscillation effects, with different Courant numbers.

Let us focus on the results reported in Fig. 1, obtained using the first initial condition (20a) with two different Courant numbers, i.e. $\mathsf{Cou} = 3$ and $\mathsf{Cou} = 5$. On the left, with $\Delta t = 3\,h$, we note that the scheme $\mathsf{Q3P1}_{\mathsf{MOOD}}$ with $\alpha_{\mathsf{TL}} = 0$ performs better than the other two implicit schemes. In fact, $\mathsf{Q3P1}_{\mathsf{MOOD}}$ is closer to the exact solution exhibiting lower dissipation. The scheme $\mathsf{Q3P1}_{\mathsf{CEMOOD}}$ is more diffusive but outperforms $\mathsf{Q3P1}$. All the schemes do not oscillate around the discontinuities. Focusing on the right panel of Fig. 1, obtained with a bigger Courant number, $\mathsf{Cou} = 5$, we note that now $\mathsf{Q3P1}$ diffuses less than $\mathsf{Q3P1}_{\mathsf{CEMOOD}}$, but again $\mathsf{Q3P1}_{\mathsf{MOOD}}$, here used with two different values of α_{TL}, provides the best approximation. In particular, choosing $\alpha_{\mathsf{TL}} = 0$ leads to a more dissipative approximation. Instead, a bigger value, $\alpha_{\mathsf{TL}} = 0.75$, which weights less the predictor solution in the time-limiting procedure, allows us to reduce the dissipation error.

In Fig. 2 we observe the results obtained by using the initial condition (20b). We zoom-in the top part of the double-step, in which it is again clear that the $\mathsf{Q3P1}_{\mathsf{MOOD}}$ scheme with $\alpha_{\mathsf{TL}} = 0$ provides the best approximation. In this case, with

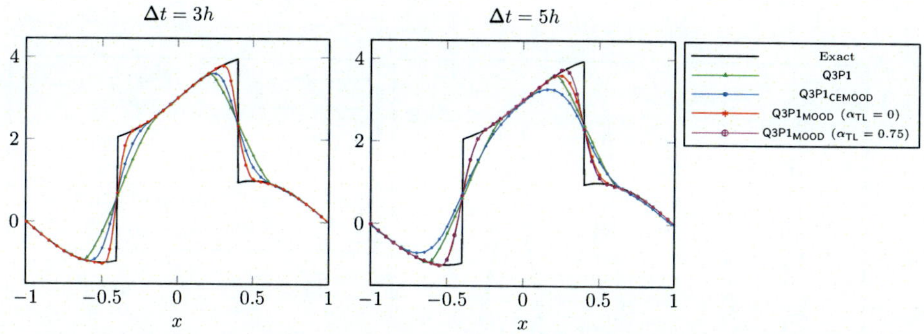

Fig. 1 Test 2. Linear transport equation (19) with initial condition (20a) on 400 cells. The markers are used to distinguish the schemes, and are drawn one out of 10 cells

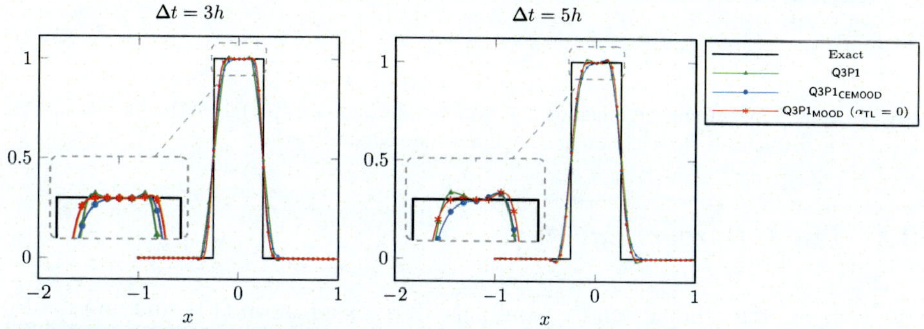

Fig. 2 Test 2. Linear transport equation (19) with initial condition (20b) on 400 cells. The markers are used to distinguish the schemes, and are drawn one out of 10 cells

a smaller Courant number, i.e. $\mathsf{Cou} = 3$, we do not observe oscillations. Instead, with $\mathsf{Cou} = 5$, the $\mathsf{Q3P1_{MOOD}}$ scheme shows a small undershoot on the bottom part and a small overshoot on the upper part, before the jump discontinuities. Whereas, the Q3P1 scheme shows in addition an undershoot and an overshoot also after the jump discontinuities (see the upper and the bottom parts after the jumps).

We conclude the numerical tests on the linear transport equation with the results reported in Fig. 3, obtained by using the initial condition (20c). Here we use $\mathsf{Cou} = 2.5$ as done in [21]. The scope of this test case is to study the behavior of the scheme $\mathsf{Q3P1_{MOOD}}$ for different values of the parameter that defines the mixing between predictor and corrector. Overall, the best choice is $\alpha_{\mathsf{TL}} = 0.75$. Both $\alpha_{\mathsf{TL}} = 0.75$ and $\alpha_{\mathsf{TL}} = 1$ provide a very high resolution of the half ellipse. On the contrary, the other values are very diffusive. However, the value $\alpha_{\mathsf{TL}} = 1$, which represents the unlimited solution, exhibits several oscillations in the bottom parts of the four shapes and in the upper part related to the square wave.

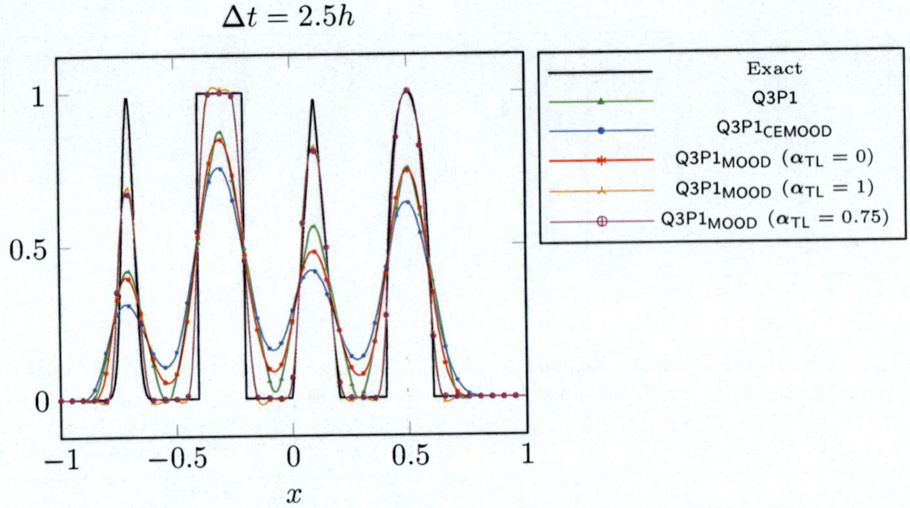

Fig. 3 Test 2. Linear transport equation (19) with initial condition (20c) on 400 cells. The markers are used to distinguish the schemes, and are drawn one out of 10 cells

3.3 Test 3: Burgers Equation

We compare the schemes on the nonlinear Burgers equation (17) studying the following situations.

Test 3a: Shock formation.

We consider again the smooth initial condition (18) on the domain $(x, t) \in (0, 2) \times (0, 2]$ with periodic boundary conditions in space. The final time is chosen in order to analyze the behavior of the schemes when the shock appears.

In Fig. 4 we show the results obtained with the three third order Quinpi schemes. For a better visualization of the behavior when a shock appears, we have reported the zoom in the bottom left part of the panes obtained with different Courant numbers, i.e. $\mathsf{Cou} = 3$ and $\mathsf{Cou} = 5$. In both cases, we note that the proposed $\mathsf{Q3P1}_{\mathsf{MOOD}}$ comes closest to the exact solution on the corners, as clearly visible in the zoom. The other two schemes are more diffusive. Whereas, for $\mathsf{Cou} = 5$ the classical Q3P1 shows a comparable resolution to $\mathsf{Q3P1}_{\mathsf{MOOD}}$, and both perform better than $\mathsf{Q3P1}_{\mathsf{CEMOOD}}$.

Test 3b: Shock formation and interaction.

Here, we consider the smooth initial condition

$$u_0(x) = 0.2 - \sin(\pi x) + \sin(2\pi x) \tag{21}$$

on the space domain $x \in [-1, 1]$ with periodic boundary conditions and fixed Courant number equal to 3, namely $\Delta t = 3\,\mathrm{h}$, studying the behavior of the con-

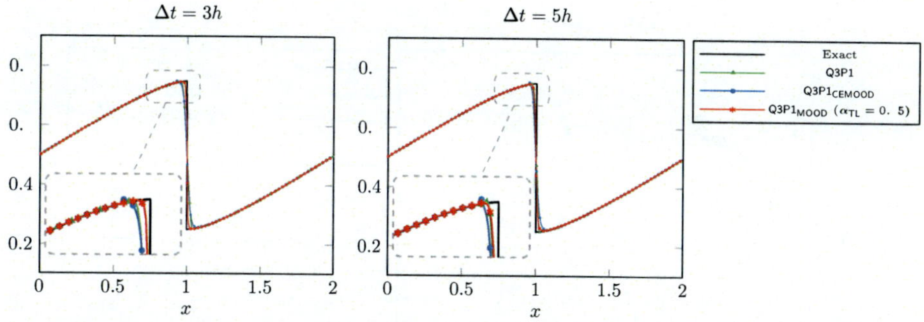

Fig. 4 Test 3a. Burgers equation (17) with initial condition (18) on $N = 400$ cells at time $t = 2$. Cou $= 3$ (on the left), Cou $= 5$ (on the right). The markers are used to distinguish the schemes, and are drawn each 5 cells

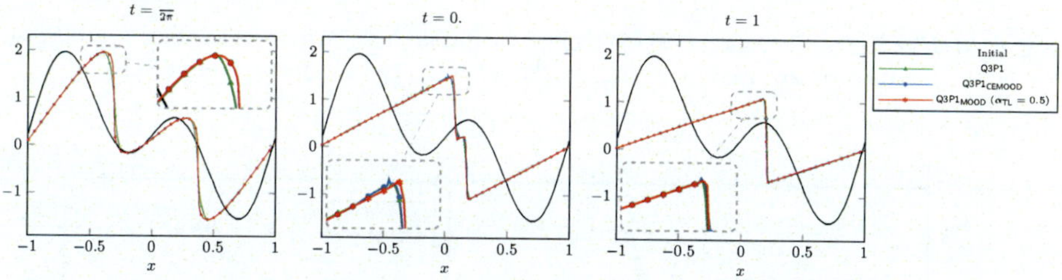

Fig. 5 Test 3b. Burgers equation (17) with initial condition (21) on 400 cells with $\Delta t = 3$h, at three different times. The markers are used to distinguish the schemes, and are drawn each 10 cells

sidered schemes at three different final times. The exact solution is characterized by the formation of two shocks which collide developing a single discontinuity. All the schemes do not produce spurious oscillations at time $t = 1/2\pi$, just before the two shocks appear. Starting from $t = 0.6$, Q3P1$_{CEMOOD}$ exhibits a small oscillation before the first shock, clearly visible in the zoom part on the left. Moreover, we observe that at time $t = 0.6$ the Q3P1$_{MOOD}$ scheme is less diffusive than the other two schemes. Similar considerations can be made at time $t = 1$, when the shocks collide (Fig. 5).

Test 3c: Rarefaction and shock waves.

Finally, we consider the discontinuous initial condition (20b) on the domain $(x, t) \in (-1, 1) \times (0, 0.5]$, with periodic boundary conditions in space. We reported below the results related to the case with two different Courant numbers.

In Fig. 6, with $\Delta t = 3$h, we observe again that the proposed Q3P1$_{MOOD}$ scheme provides the best approximation, followed by the classical Q3P1 proposed in [30], and then by Q3P1$_{CEMOOD}$. For the Q3P1$_{MOOD}$ scheme, we use the value $\alpha_{TL} = 0.75$.

Finally, in Fig. 7, obtained with $\Delta t = 5$h, we note that the three schemes are closer to each other in the zoomed part, even if Q3P1 and Q3P1$_{CEMOOD}$ perform worse

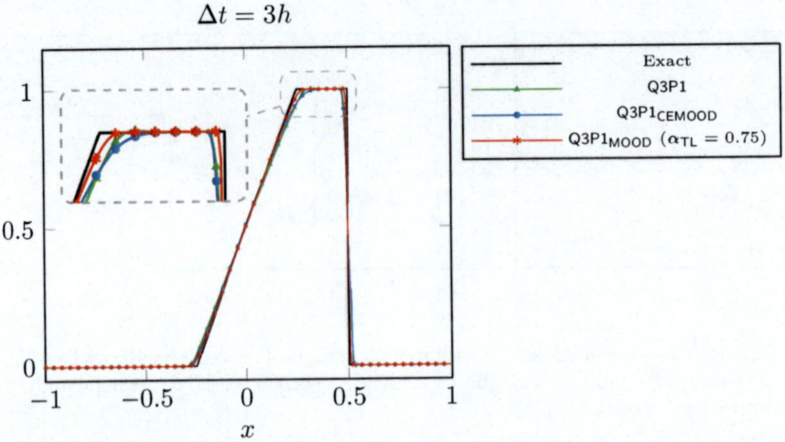

Fig. 6 Test 3c. Burgers equation (17) with initial condition (20b) on 400 cells at time $t = 0.5$, with $\Delta t = 3\,\mathrm{h}$. The markers are used to distinguish the schemes, and are drawn every 8 cells

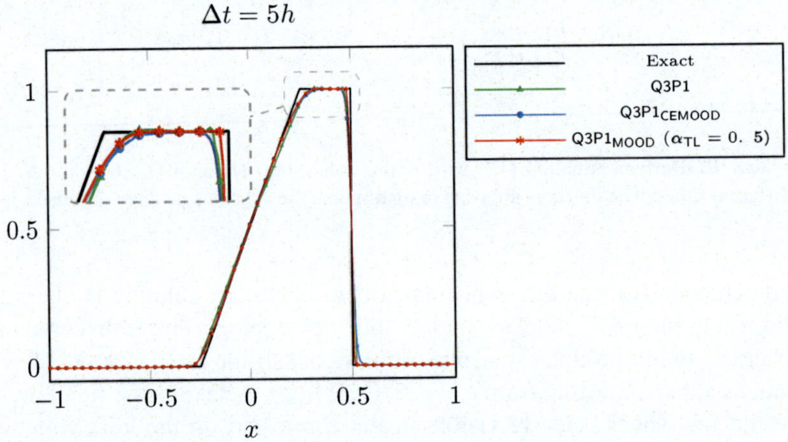

Fig. 7 Test 3c. Burgers equation (17) with initial condition (20b) on 400 cells at time $t = 0.5$, with $\Delta t = 5\,\mathrm{h}$. The markers are used to distinguish the schemes, and are drawn every 8 cells

in the bottom part, for example around $x = 0.5$, as expected since with a greater Courant number these schemes diffuse more. We do not observe the same for the Q3P1$_{\mathsf{MOOD}}$ scheme since we use here the value $\alpha_{\mathsf{TL}} = 0.45$.

3.4 Computational Performance of the Quinpi Schemes

In the following, we compare the computational CPU times required by each time step of the explicit third order SSP Runge-Kutta scheme and of the Quinpi schemes. In Table 4 we report the results obtained both on the linear equation (19) on $[-1, 1]$ and on the Burgers' equation (17) on $[0, 2]$ with initial condition (18). To make the test reliable and fair, the parameters are chosen to have a comparable and large enough total time of execution of all the methods. The results are obtained on a quad-core Intel Core i7-6600U with clock speed 2.60 GHz.

Looking at Table 4, we note that the Q3P1$_{\text{CEMOOD}}$ scheme is the fastest third order scheme in terms of CPU time in all the three problems considered, for all the choices of cells N. Q3P1$_{\text{CEMOOD}}$ is followed by Q3P1$_{\text{MOOD}}$, the slowest one results to be the Q3P1 proposed in [30]. In terms of time ratios, we can note bigger values for Q3P1 scheme, followed by Q3P1$_{\text{MOOD}}$ and then by Q3P1$_{\text{CEMOOD}}$. Of course, the SSP-RK3 explicit scheme is the fastest one, as expected, requiring about a third of the time of the Quinpi schemes in most cases, arriving until more than a quarter of the time of the Quinpi schemes for the last refinement in the nonlinear problem after shock formation.

Table 4 CPU times (in seconds) for each step of the SSP-RK3 explicit scheme, the Q3P1, the Q3P1$_{\text{MOOD}}$ and the Q3P1$_{\text{CEMOOD}}$ implicit schemes, on linear and nonlinear problems. The time ratios are given in parenthesis, using as reference the time of the explicit scheme

Cells N	SSP-RK3	Q3P1	Q3P1$_{\text{MOOD}}$	Q3P1$_{\text{CEMOOD}}$
200	0.0023s	0.0099s (4.30)	0.0100s (4.35)	0.0092s (4.00)
400	0.0033s	0.0132s (4.00)	0.0132s (4.00)	0.0115s (3.48)
800	0.0068s	0.0223s (3.28)	0.0220s (3.23)	0.0195s (2.87)
1600	0.0095s	0.0433s (4.56)	0.0418s (4.40)	0.0389s (4.09)

(a) Linear problem

Cells N	SSP-RK3	Q3P1	Q3P1$_{\text{MOOD}}$	Q3P1$_{\text{CEMOOD}}$
200	0.0023s	0.0077s (3.35)	0.0075s (3.26)	0.0068s (2.96)
400	0.0032s	0.0105s (3.28)	0.0105s (3.28)	0.0092s (2.87)
800	0.0066s	0.0182s (2.76)	0.0179s (2.71)	0.0159s (2.41)
1600	0.0091s	0.0344s (3.78)	0.0327s (3.59)	0.0282s (3.10)

(b) Nonlinear problem before shock formation

Cells N	SSP-RK3	Q3P1	Q3P1$_{\text{MOOD}}$	Q3P1$_{\text{CEMOOD}}$
200	0.0023s	0.0099s (4.30)	0.0100s (4.35)	0.0092s (4.00)
400	0.0033s	0.0132s (4.00)	0.0132s (4.00)	0.0115s (3.48)
800	0.0068s	0.0223s (3.28)	0.0220s (3.23)	0.0195s (2.87)
1600	0.0095s	0.0433s (4.56)	0.0418s (4.40)	0.0389s (4.09)

(c) Nonlinear problem after shock formation

4 Conclusion and Perspectives

Several systems of hyperbolic conservation laws are characterized by waves with very different speeds, and in many applications the phenomenon of interest travels with slow speeds, whereas the fastest waves do not need to be accurately represented. Low Mach problems provide typical examples. In fact, in these cases the actual speed of the gas is much slower than the acoustic waves. However, fast waves impose the Courant-Friedrichs-Levy stability condition, and if one is interested in the movement of the gas, accuracy in the propagation of sound is irrelevant.

These problems require implicit numerical treatment. An implicit third order scheme based on CWENO, named Quinpi, has been recently introduced in [30]. There, the implicit formulation has been tackled using low order linear implicit schemes, the predictor, to avoid the nonlinearities of the high order space reconstruction. In this chapter we have revisited Quinpi schemes by proposing a conservative a-posteriori time-limiting procedure inspired by the MOOD method. The time-limiting blends the low order predictor and the third order solution to dampen possible spurious oscillations due to large time steps. Furthermore, we have built two types of predictors, one based on the composite backward Euler, as in [30], and another one based on the continuous extension of the backward Euler.

We have numerically compared the scheme in [30] with the schemes proposed in this chapter. We have shown that the Quinpi scheme with the conservative a-posteriori time-limiting and with composite backward Euler as predictor performs better than the other two third order Quinpi schemes. In terms of CPU times, the new Quinpi schemes of this work are faster than the classical Quinpi approach in [30]. In particular, the scheme with the predictor based on the continuous extension results to be the fastest one.

Future research will be focused on the extension of the Quinpi approach to systems of conservation laws and on the use of other time integration schemes, such as stage accurate Runge-Kutta and Backward Differentiation Formula solvers.

Acknowledgements The authors are members of the INdAM Research Group GNCS. This work was partially supported by Ateneo Sapienza project 2019 "Metodi numerici per problemi evolutivi, networks ed applicazioni", and carried out within the MUR (Ministry of University and Research) PRIN-2017 project "Innovative Numerical Methods for Evolutionary Partial Differential Equations and Applications" (number 2017KKJP4X), Ateneo Sapienza projects 2020 "Algoritmi e modelli per sistemi di natura iperbolica, networks e applicazioni", and 2021 "Evolutionary problems: analysis techniques and construction of numerical solutions", INdAM-GNCS project "Metodi numerici per l'imaging: dal 2D al 3D", Code CUP_E55F22000270001.

References

1. Abbate, E., Iollo, A., Puppo, G.: An all-speed relaxation scheme for gases and compressible materials. J. Comput. Phys. **351**, 1–24 (2017)
2. Acker, F., Borges, R.B.d.R., Costa, B.: An improved WENO-Z scheme. J. Comput. Phys. **313**, 726–753 (2016)
3. Alexander, R.: Diagonally implicit Runge-Kutta methods for stiff O.D.E.'s. SIAM J. Numer. Anal. **14**(6), 1006–1021 (1977)
4. Arbogast, T., Huang, C., Zhao, X., King, D.N.: A third order, implicit, finite volume, adaptive Runge-Kutta WENO scheme for advection-diffusion equations. Comput. Methods Appl. Mech. Engrg. **368** (2020)
5. Avgerinos, S., Bernard, F., Iollo, A., Russo, G.: Linearly implicit all Mach number shock capturing schemes for the Euler equations. J. Comput. Phys. **393**, 278–312 (2019)
6. Boscarino, S., Russo, G., Scandurra, L.: All Mach number second order semi-implicit scheme for the Euler equations of gas dynamics. J. Sci. Comput. **77**(2), 850–884 (2018)
7. Castro, M., Costa, B., Don, W.S.: High order weighted essentially non-oscillatory WENO-Z schemes for hyperbolic conservation laws. J. Comput. Phys. **230**(5), 1766–1792 (2011)
8. Clain, S., Diot, S., Loubère, R.: A high-order finite volume method for hyperbolic systems: multi-dimensional Optimal Order Detection (MOOD). J. Comput. Phys. **230**(10), 4028–4050 (2011)
9. Clain, S., Diot, S., Loubère, R.: Improved detection criteria for the Multi-dimensional Optimal Order Detection MOOD on unstructured meshes with very high-order polynomials. Comput. Fluids. **64**, 43–63 (2012)
10. Coquel, F., Nguyen, Q.L., Postel, M., Tran, Q.H.: Local time stepping with adaptive time step control for a two-phase fluid system. In: ESAIM: Proceedings, vol. 29, pp. 73–88 (2009)
11. Coquel, F., Nguyen, Q.L., Postel, M., Tran, Q.H.: Entropy-satisfying relaxation method with large time-steps for Euler IBVPs. Math. Comput. **79**, 1493–1533 (2010)
12. Coquel, F., Nguyen, Q.L., Postel, M., Tran, Q.H.: Local time stepping applied to implicit-explicit methods for hyperbolic systems. Multiscale Model. Simul. **8**(2), 540–570 (2010)
13. Coquel, F., Postel, M., Poussineau, N., Tran, Q.H.: Multiresolution technique and explicit-implicit scheme for multicomponent flows. J. Numer. Math. **14**(3), 187–216 (2006)
14. Cravero, I., Puppo, G., Semplice, M., Visconti, G.: CWENO: uniformly accurate reconstructions for balance laws. Math. Comput. **87**(312), 1689–1719 (2018)
15. Cravero, I., Semplice, M., Visconti, G.: Optimal definition of the nonlinear weights in multidimensional Central WENOZ reconstructions. SIAM J. Numer. Anal. **57**(5), 2328–2358 (2019)
16. Degond, P., Tang, M.: All speed scheme for the low Mach number limit of the isentropic Euler equations. Commun. Comput. Phys. **10**(1), 1–31 (2011)
17. Dellacherie, S.: Analysis of Godunov type schemes applied to the compressible Euler system at low Mach number. J. Comput. Phys. **229**(4), 978–1016 (2010)
18. Dimarco, G., Loubere, R., Vignal, M.H.: Study of a new asymptotic preserving scheme for the Euler system in the low Mach number limit. SIAM J. Sci. Comput. **39**(5), A2099–A2128 (2017)
19. Dimarco, G., Pareschi, L.: Numerical methods for kinetic equations. Acta Numerica. **23**, 369–520 (2014)
20. Duraisamy, K., Baeder, J.D.: Implicit scheme for hyperbolic conservation laws using non oscillatory reconstruction in space and time. SIAM J. Sci. Comput. **29**, 2607–2620 (2007)
21. Eimer, M., Borsche, R., Siedow, N.: Implicit finite volume method with a posteriori limiting for transport networks. Adv. Comput. Math. **48**(3), 21 (2022)
22. Frolkovic, P., Krisková, S., Rohová, M., Zeravý: Semi-implicit methods for advection equations with explicit forms of numerical solution (2022). arXiv:2106.15474
23. Frolkovic, P., Zeravý: Semi-implicit high resolution numerical scheme for conservation laws (2022). arXiv:2206.09425

24. Gottlieb, S., Shu, C., Tadmor, E.: Strong stability preserving high-order time discretization methods. SIAM Rev. **43**, 73–85 (2001)
25. Jiang, G.S., Shu, C.W.: Efficient implementation of weighted ENO schemes. J. Comput. Phys. **126**, 202–228 (1996)
26. Lemou, M., Mieussens, L.: A new asymptotic preserving scheme based on micro-macro formulation for linear kinetic equations in the diffusion limit. SIAM J. Sci. Comput. **31**, 334–368 (2008)
27. Levy, D., Puppo, G., Russo, G.: Compact central WENO schemes for multidimensional conservation laws. SIAM J. Sci. Comput. **22**(2), 656–672 (2000)
28. Loubère, R., Dumbser, M., Diot, S.: A new family of high order unstructured mood and ADER finite volume schemes for multidimensional systems of hyperbolic conservation laws. Commun. Comput. Phys. **16**, 718–763 (2014)
29. Pieraccini, S., Puppo, G.: Microscopically implicit-macroscopically explicit schemes for the BGK equation. J. Comput. Phys. **231**, 299–327 (2012)
30. Puppo, G., Semplice, M., Visconti, G.: Quinpi: integrating conservation laws with CWENO implicit methods. Commun. Appl. Math. Comput. **5**, 343–369 (2023)
31. Semplice, M., Loubère, R.: Adaptive-Mesh-Refinement for hyperbolic systems of conservation laws based on a posteriori stabilized high order polynomial reconstructions. J. Comput. Phys. **354**, 86–110 (2018)
32. Shu, C.W.: Essentially Non-Oscillatory and Weighted Essentially Non-Oscillatory Schemes for Hyperbolic Conservation Laws. NASA/CR-97-206253 ICASE Report No. 97–65 (1997)
33. Tavelli, M., Dumbser, M.: A pressure-based semi-implicit space-time discontinuous Galerkin method on staggered unstructured meshes for the solution of the compressible Navier-Stokes equations at all Mach numbers. J. Comput. Phys. **341**, 341–376 (2017)
34. Zanotti, O., Dumbser, M., Loubère, R., Diot, S.: A posteriori subcell limiting for discontinuous Galerkin finite element method for hyperbolic system of conservation laws. J. Comput. Phys. **278**, 47–75 (2014)
35. Zennaro, M.: Natural continuous extensions of Runge-Kutta methods. Math. Comput. **46**, 119–133 (1986)

Applications of Fokker Planck Equations in Machine Learning Algorithms

Yuhua Zhu

Abstract As the continuous limit of the gradient-based optimization algorithms, Fokker Planck (FP) equation can provide a qualitative description of the algorithm's behavior and give principled theoretical insight into many mysteries in machine learning (ML). This chapter provides a theoretical interpretation of certain ML algorithms using FP equations. Specifically, we summarize the applications of FP equations in asynchronous stochastic gradient descent algorithms, algorithmic fairness for imbalanced data, and Reinforcement Learning.

Keywords Fokker Planck equations · Machine learning algorithms · Stochastic gradient descent · Fairness · Reinforcement learning

1 Introduction

Deep learning and reinforcement learning (RL) have achieved spectacular empirical success, while a theoretical understanding of these subjects remains incomplete despite ongoing research efforts. The study of partial differential equations (PDEs) is an essential mathematical tool for understanding machine learning (ML) algorithms. In recent years, there has been rich literature on studying machine learning algorithms using PDEs. PDEs are viewed as continuous limits for the optimization algorithms as the learning rate goes to zero. It can provide insights and theoretical guides to the properties of the ML algorithms that is hard to explain in discretized version.

The stochastic differential equation (SDE) or PDE approach for approximating stochastic optimization methods can be traced in the line of work [6, 10, 11, 14–16, 25, 29], just to mention a few. When one tries to minimize the following objective function,

Y. Zhu (✉)
Department of Mathematics and Halicioğlu Data Science Institute, University of California-San Diego, San Diego, CA 92093, USA
e-mail: yuz244@ucsd.edu

G. Albi et al. (eds.), *Advances in Numerical Methods for Hyperbolic Balance Laws and Related Problems*, SEMA SIMAI Springer Series 32,
https://doi.org/10.1007/978-3-031-29875-2_10

$$\min_{\theta} \quad F(\theta) = \frac{1}{n}\sum_{i=1}^{n} f_i(\theta), \tag{1}$$

where θ represents the model parameters, $f_i(\theta)$ denotes the loss function at the i-th training sample and n is the size of the training sample set. The stochastic gradient descent (SGD) algorithm is commonly used in the literature, whose updates can be described by

$$\theta_{k+1} = \theta_k - \frac{\eta}{|\mathcal{B}|}\nabla \sum_{i\in\mathcal{B}} f_i(\theta_k)$$

where η is the learning rate, and $\mathcal{B}$ is a random batch from the total data set. When the learning rate is sufficiently small, one can approximate the SGD by an SDE. Such a SDE approximation involves a data-dependent covariance coefficient for the diffusion term and is justified in the weak sense with an error of order $O(\sqrt{\eta})$ [14]. More specifically, the dynamics can be approximated by

$$d\Theta = -\nabla F(\Theta)dt + \sqrt{\eta}\Sigma(\Theta)^{1/2}dB, \tag{2}$$

where $\Theta(t = k\eta) \approx \theta_k$, and $\Sigma(\Theta)$ is the covariance of the stochastic gradient at location Θ. Furthermore, the probability density function $p(\theta, t)$ (p.d.f.) of Θ satisfies the Fokker Plank (FP) equation [22],

$$\partial_t p(\theta, t) = \nabla\cdot[\nabla F(\theta)p(\theta, t)] + \frac{\eta}{2}\sum_{i=1}^{N}\sum_{j=1}^{N}\frac{\partial^2}{\partial\theta_i\,\partial\theta_j}\left[\Sigma_{ij}(\theta)p(\theta, t)\right].$$

In other words, the above FP equation is the limiting p.d.f. of the SGD algorithm as the learning rate goes to zero.

Based on the above connection, we summarize the applications of FP equations in the following three directions: algorithmic fairness for imbalanced data, asynchronous stochastic gradient descent (ASGD) algorithms, and RL. To be more specific, we provide theoretical analyses based on the FP equation to understand the following phenomena in ML:

- Why resampling outperforms reweighting in correcting biased data when stochastic gradient-type algorithms are used in training.
- How the delayed gradients in ASGD affect the training process.
- Why the borrowing-from-the-future (BFF) algorithm can alleviate the double sampling problem in model-free RL.

The results are summarized from the papers [2, 36, 37], respectively, and they are discussed in Sects. 2, 3 and 4, respectively.

2 Algorithmic Fairness for Imbalanced Data

A data set sampled from a certain population is called biased if the subgroups of the population are sampled at proportions that are significantly different from their underlying population proportions. Applying machine learning algorithms naively to biased training data can raise serious concerns and lead to controversial results [12, 18, 32]. In many domains such as demographic surveys, fraud detection, identification of rare diseases, and natural disasters prediction, a model trained from biased data tends to favor oversampled subgroups by achieving high accuracy there while sacrificing the performance on undersampled subgroups.

In order to mitigate the biases and discriminations against the undersampled subgroups, a common technique is to preprocess the data set by compensating the mismatch between population proportion and the sampling proportion. Among various approaches, two commonly-used choices are reweighting and resampling. In reweighting, one multiplies each sample with a ratio equal to its population proportion over its sampling proportion. In resampling, on the other hand, one corrects the proportion mismatch by either generating new samples for the undersampled subgroups or selecting a subset of samples for the oversampled subgroups. Both methods result in statistically equivalent models in terms of the loss function. However, it has been observed in practice that resampling often outperforms reweighting significantly, such as boosting algorithms in classification [9, 28], off-policy prediction in RL [27] and so on. The obvious question is why.

Main contributions. The main contribution here is to provide an answer to this question: resampling outperforms reweighting because of the stochastic gradient-type algorithms used for training. In [2], the theoretical analysis is based on two points of view, one from the dynamical stability perspective and the other from stochastic asymptotic, i.e., the FP equation. In this chapter, we summarize the perspective from FP equation. We show via analyzing the steady state of the asymptotic FP equation why resampling generates expected results while reweighting performs undesirably.

2.1 *Setting*

Let us consider a population that is comprised of two different groups, where a proportion a_1 of the population belongs to the first group, and the rest with the proportion $a_2 = 1 - a_1$ belongs to the second (i.e., $a_1, a_2 > 0$ and $a_1 + a_2 = 1$). In what follows, we shall call a_1 and a_2 the population proportions. Consider an optimization problem for this population over a parameter θ. For simplicity, we assume that each individual from the first group experiences a loss function $V_1(\theta)$, while each individual from the second group has a loss function of type $V_2(\theta)$. Here the loss function $V_1(\theta)$ is assumed to be identical across all members of the first group and the same for $V_2(\theta)$ across the second group, however it is possible to extend the formulation to allow for loss function variation within each group. Based on this setup, a minimization problem over the whole population is to find

$$\theta^* = \underset{\theta}{\operatorname{argmin}}\, V(\theta), \quad \text{where } V(\theta) \equiv a_1 V_1(\theta) + a_2 V_2(\theta). \tag{3}$$

For a given set Ω of N individuals sampled uniformly from the population, the empirical minimization problem is

$$\theta^* = \underset{\theta}{\operatorname{argmin}} \frac{1}{N} \sum_{r \in \Omega} V_{i_r}(\theta), \tag{4}$$

where $i_r \in \{1, 2\}$ denotes which group an individual r belongs to. When N grows, the empirical loss in (4) is consistent with the population loss in (3) as there are approximately a_1 fraction of samples from the first group and a_2 fraction of samples from the second.

However, the sampling can be far from uniformly random in reality. Let n_1 and n_2 with $n_1 + n_2 = N$ denote the number of samples from the first and the second group, respectively. It is convenient to define f_i, $i = 1, 2$ as the sampling proportions for each group, i.e., $f_1 = n_1/N$ and $f_2 = n_2/N$ with $f_1 + f_2 = 1$. The data set is biased when the sampling proportions f_1 and f_2 are different from the population proportions a_1 and a_2. In such a case, the empirical loss is $f_1 V_1(\theta) + f_2 V_2(\theta)$, which is clearly wrong when compared with (3).

Let us consider two basic strategies to adjust the model: reweighting and resampling. In reweighting, one assigns to each sample $r \in \Omega$ a weight a_{i_r}/f_{i_r} and the reweighting loss function is

$$V_w(\theta) \equiv \frac{1}{N} \sum_{r \in \Omega} \frac{a_{i_r}}{f_{i_r}} V_{i_r}(\theta) = a_1 V_1(\theta) + a_2 V_2(\theta). \tag{5}$$

In resampling, one either adds samples to the minority group (i.e., oversampling) or removing samples from the majority group (i.e., undersampling). Although the actual implementation of oversampling and undersampling could be quite sophisticated in order to avoid overfitting or loss of information, mathematically we interpret the resampling as constructing a new set of samples of size M, among which $a_1 M$ samples are of the first group and $a_2 M$ samples of the second. The resampling loss function is

$$V_s(\theta) \equiv \frac{1}{M} \sum_{s} V_{i_s}(\theta) = a_1 V_1(\theta) + a_2 V_2(\theta). \tag{6}$$

Notice that both $V_w(\theta)$ and $V_s(\theta)$ are consistent with the population loss function $V(\theta)$. This means that, under mild conditions on $V_1(\theta)$ and $V_2(\theta)$, a deterministic gradient descent algorithm from a generic initial condition converges to similar solutions for $V_w(\theta)$ and $V_s(\theta)$. For a SGD algorithm, the expectations of the stochastic gradients of $V_w(\theta)$ and $V_s(\theta)$ also agree at any θ value. However, as we shall explain below, the training behavior can be drastically different for a stochastic gradient algorithm. The key reason is that the variances experienced for $V_w(\theta)$ and $V_s(\theta)$ can be drastically different: computing the variances of gradients for resampling and reweighting reveals that

$$\begin{aligned}\mathbb{V}\left[\nabla\hat{V}_s(\theta)\right] &= a_1\nabla V_1(\theta)\nabla V_1(\theta)^T + a_2\nabla V_2(\theta)\nabla V_2(\theta)^T - (\mathbb{E}[\nabla\hat{V}_s(\theta)])^2,\\ \mathbb{V}\left[\nabla\hat{V}_w(\theta)\right] &= \frac{a_1^2}{f_1}\nabla V_1(\theta)\nabla V_1(\theta)^T + \frac{a_2^2}{f_2}\nabla V_2(\theta)\nabla V_2(\theta)^T - (\mathbb{E}[\nabla\hat{V}_w(\theta)])^2.\end{aligned} \tag{7}$$

These formulas indicate that, when f_1/f_2 is significantly misaligned with a_1/a_2, the variance of reweighting can be much larger. Without knowing the optimal learning rates a priori, it is difficult to select an efficient learning rate for reliable and stable performance for stiff problems, when only reweighting is used. In comparison, resampling is more favorable especially when the choice of learning rates is restrictive.

2.2 *Theorems*

Let us use a simple example to illustrate the main idea. Consider the following two loss functions,

$$V_1(\theta) = \begin{cases} |\theta+1|-1, & \theta \le 0 \\ \epsilon\theta, & \theta > 0 \end{cases}, \quad V_2(\theta) = \begin{cases} -\epsilon\theta, & \theta \le 0 \\ |\theta-1|-1, & \theta > 0 \end{cases},$$

with $0 < \epsilon \ll 1$. The population loss function is $V(\theta) = a_1V_1(\theta) + a_2V_2(\theta)$ with local minimizers $\theta = -1$ and $\theta = 1$. Note that the $O(\epsilon)$ terms are necessary. Without it, if the SGD starts in $(-\infty, 0)$, all iterates will stay in this region because there is no drift from $V_2(\theta)$. Similarly, if the SGD starts in $(0, \infty)$, no iterates will move to $(-\infty, 0)$. That means the result of SGD only depends on the initialization when $O(\epsilon)$ term is absent.

In Fig. 1, we present numerical simulations of the resampling and reweighting methods for the designed loss function $V(\theta)$. If $a_2 > a_1$, then the global minimizer of $V(\theta)$ is $\theta = 1$ (see the Fig. 1(1)). Consider a setup with population proportions $a_1/a_2 = 0.4/0.6$ along sampling proportions $f_1/f_2 = 0.9/0.1$, which are quite different. Figure 1(2) and (3) show the dynamics under the reweighting and resampling methods, respectively. The plots show that, while the trajectory for resampling is stable across time, the trajectory for reweighting quickly escapes to the (non-global) local minimizer $\theta = -1$ even when it starts near the global minimizer $\theta = 1$.

As we mentioned in Sect. 1, when the learning rate η is small, FP equation can be viewed as the limiting p.d.f. of the SGD algorithm. However, unlike the usual case where the drift term is usually assumed to be Lipschitz, the loss function considered in this section has jump discontinuous in the first derivative, the classical approximation

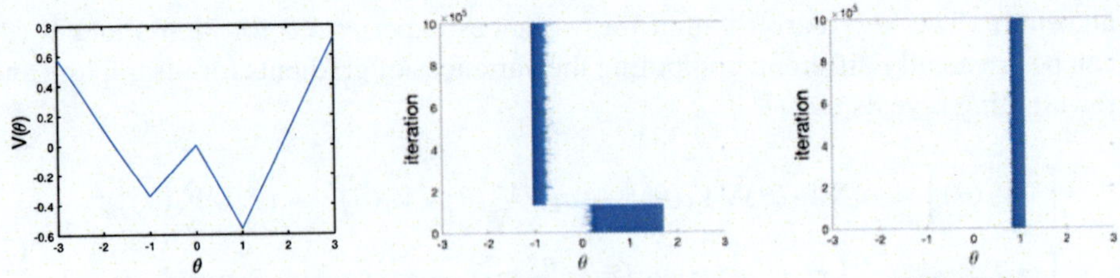

(1) Loss function $V(\theta)$ (2) Reweighting (3) Resampling

Fig. 1 Comparison of reweighting and resampling with learning rate $\eta = 0.12$. We set $a_1/a_2 = 0.4/0.6$, $f_1/f_2 = 0.9/0.1$ and $\epsilon = 0.1$. Both experiments start at $\theta_0 = 0.9$. The resampling strategy here is to randomly select the sub-population i with the probability a_i with replacement in each iteration. In **(2)** where reweighting is used, the trajectory skips to the local minimizer $\theta = -1$ later. In **(3)** where resampling is used, it stabilizes at the global minimizer $\theta = 1$ all the time

error results for the limiting SDE or PDE do not apply. In fact, the problem that the loss function $V(\theta) \notin C^1(\mathbb{R}^n)$ is a common issue in machine learning and deep neural networks, as many loss functions involves non-smooth activation functions such as ReLU and leaky ReLU [13]. To fill this gap, we give the following theorem as a justification of SDE approximation for the drift with jump discontinuities, based on the proof presented in [20].

Theorem 1 *For the continuous stochastic process Θ_t and discrete stochastic gradient decent updates θ_k satisfying the following equations:*

$$\begin{aligned} d\Theta &= b(\Theta)dt + \sigma(\Theta)dB, \\ \theta_{k+1} &= \theta_k + \eta b(\theta_k) + \sqrt{\eta}\sigma(\theta_k)Z_k, \quad Z_k \sim \mathcal{N}(0, 1), \end{aligned} \tag{8}$$

with the same initial data $\Theta_0 = \theta_0$, assume that the diffusion coefficient σ is Lipschitz continuous and non-degenerate, and the drift coefficient b is piecewise Lipschitz continuous, in the sense that b has finitely many discontinuity points $-\infty = \xi_0 < \xi_1 < \cdots < \xi_m < \xi_{m+1} = \infty$ and in each interval (ξ_{i-1}, ξ_i), b is Lipschitz continuous. Then, for all $k = 0, 1, 2, \ldots, K$, there exists $C > 0$ such that

$$\mathbb{E}[|\Theta_{k\eta} - \theta_k|] \le C\sqrt{\eta}, \tag{9}$$

where $\Theta_{k\eta}$ is the solution to SDE at time $k\eta$.

Next, by deriving the limiting FP equation and the steady state of the corresponding FP equation, one has the following two Lemmas. Denote $p_s(\theta)$ and $p_w(\theta)$ as the stationary distribution over θ under resampling and reweighting, respectively.

Lemma 1 *When $a_2 > a_1$, $V(1) < V(-1)$. The stationary distribution for resampling satisfies the relationship*

$$\frac{p_s(1)}{p_s(-1)} = \exp\left(-\frac{2}{a_1 a_2 \eta}(V(1) - V(-1))\right) + O(\epsilon) > 1.$$

Lemma 2 *With $a_2 > a_1$, $V(1) < V(-1) < 0$. Under the condition $\frac{f_2}{f_1} \leq \frac{a_2}{a_1}\sqrt{\frac{V(-1)}{V(1)}}$ for the sampling proportions, the stationary distribution for reweighting satisfies the relationship*

$$\frac{p_w(1)}{p_w(-1)} = \frac{a_1^2/f_1^2}{a_2^2/f_2^2} \exp\left(-\frac{2f_2/f_1}{a_2^2\eta}V(1) + \frac{2f_1/f_2}{a_1^2\eta}V(-1)\right) + O(\epsilon) < 1.$$

Lemma 1 shows that for resampling it is always more likely to find θ at the global minimizer 1 than at the local minimizer -1. Lemma 2 states that for reweighting it is more likely to find θ at the local minimizer -1 when $\frac{f_2}{f_1} \leq \frac{a_2}{a_1}\sqrt{\frac{V(-1)}{V(1)}}$. Together, they explain the phenomenon shown in Fig. 1.

To better understand the condition in Lemma 2, let us consider the case $a_1 = \frac{1}{2} - \epsilon$, $a_2 = \frac{1}{2} + \epsilon$ with a small constant $\epsilon > 0$. Under this setup, $V(-1)/V(1) \approx 1$. Whenever the ratio of the sampling proportions f_2/f_1 is significantly less than the ratio of the population proportions $a_2/a_1 \approx 1$, reweighting will lead to the undesired behavior. The smaller the ratio f_2/f_1 is, the less likely the global minimizer will be visited.

Piecewise convex results. The reason for constructing the above piecewise linear loss function is to obtain an approximately explicitly solvable SDE with a constant coefficient for the noise. One can further extend the results in 1D for piecewise strictly convex function with two local minima. Here we present the most general results in 1D, that is, piecewise strictly convex function with finite number of local minima. One may consider the population loss function $V(\theta) = \sum_{i=1}^{k} a_i V_i(\theta)$ with $V_i(\theta) = h_i(\theta)$ for $\theta_{i-1} < \theta \leq \theta_i$ and $V_i(\theta) = O(\epsilon)$ otherwise, where $h_i(\theta)$ are strictly convex functions and continuously differentiable, $O(\epsilon)$ term is sufficiently small and smooth. Here $\{\theta_i\}_{i=1}^{k-1}$ are $k-1$ disjoint points, and $\theta_0 = -\infty$, $\theta_k = \infty$. We assume that $V(\theta)$ has k local minimizers θ_i^* for $\theta_i^* \in (\theta_{i-1}, \theta_i)$. We present the following two lemmas with suitable assumptions (See Appendix B.3 of [2] for details of assumptions).

Lemma 3 *The stationary distribution for resampling at any two local minimizers θ_p^*, θ_q^* with $p > q$ satisfies the relationship*

$$\frac{p_s(\theta_p^*)}{p_s(\theta_q^*)} = \exp\left[\frac{2}{\eta}\int_{\theta_p^*}^{\theta_p} \frac{1}{h'_p(\theta)}d\theta \left(\frac{1}{1-a_p} - \frac{1}{1-a_q}\right)\right] + O(\epsilon) = \begin{cases} > 1, & \text{if } a_p > a_q; \\ < 1, & \text{if } a_p < a_q, \end{cases}$$

Lemma 4 *The stationary distribution for reweighting at any two local minimizers θ_p^*, θ_q^* with $p > q$ satisfies the relationship*

$$\frac{p_w(\theta_p^*)}{p_w(\theta_q^*)} = \exp\left[\frac{2}{\eta}\int_{\theta_p^*}^{\theta_p} \frac{1}{h'_p(\theta)}d\theta \left(\frac{f_p}{a_p(1-f_p)} - \frac{f_q}{a_q(1-f_q)}\right)\right] + O(\epsilon).$$

We first note that $\int_{\theta_p^*}^{\theta_p} \frac{1}{h'_p(\theta)}d\theta > 0$ due to the strictly convexity of h_p. Therefore, one can see from Lemma 3 that for resampling, the stationary solution always has the highest probability at the global minimizer. On the other hand, for the stationary solution of reweighting in Lemma 4, let us consider the case when $a_p > a_q$. In this case, $V(\theta_p^*) < V(\theta_q^*)$, therefore, one expects the above ratio larger than 1, which implies that $\frac{f_p}{a_p(1-f_p)} - \frac{f_q}{a_q(1-f_q)} > 0$. Note that if $f_p = a_p$, $f_q = a_q$, then this term is always larger than 0, but when f_p, f_q are significantly different from a_p, a_q in the sense that $f_p < f_q$ and $f_p < a_p$, $f_q > a_q$, then $\frac{f_p}{a_p(1-f_p)} - \frac{f_q}{a_q(1-f_q)} < 0$, which will lead to $\frac{p_s(\theta_p^*)}{p_s(\theta_q^*)} < 1$, i.e., higher probability of converging to θ_q^*, which is not desirable. To sum up, Lemma 4 shows that for reweighting, the stationary solution will not have the highest probability at the global minimizer if the empirical proportion is significantly different from the population proportion.

Multi-dimensional results. It is in fact not clear how to extend Lemmas 3 and 4 to multi-dimension. As far as we know, it is still an open problem how the stochastic process behaves when the covariance matrix of (2) depends on Θ in high dimensions. Instead, we focus on the case where the covariance matrix is piecewise constant. We divide the whole space into a finite number of disjoint convex regions $\mathbb{R}^d = \cup_{i=1}^k \Omega_i$. The loss function $V(\theta) = \sum_{i=1}^k a_i V_i(\theta)$ with $V_i(\theta) = \kappa_i \left\|\theta - \theta_i^*\right\|_1 - \beta_i$ for $\theta \in \Omega_i$ and $V_i(\theta) = O(\epsilon)$ otherwise, where $\|\theta\|_1 = \sum_{j=1}^d |\theta_j|$. The loss function has k local minimizers $\theta_i^* \in \Omega_i$. The following Lemma summarizes the results for the multi-dimensional case.

Lemma 5 *The stationary distribution for resampling and reweighting at any two local minimizers θ_p^*, θ_q^* satisfies the relationship*

$$\frac{p_s(\theta_p^*)}{p_s(\theta_q^*)} = \exp\left[\frac{2}{\eta}\left(\frac{\beta_p}{(1-a_p)\kappa_p^2} - \frac{\beta_q}{(1-a_q)\kappa_q^2}\right)\right] + O(\epsilon),$$

$$\frac{p_w(\theta_p^*)}{p_w(\theta_q^*)} = \exp\left[\frac{2}{\eta}\left(\frac{f_p\beta_p}{a_p(1-f_p)\kappa_p^2} - \frac{f_q\beta_q}{a_q(1-f_q)\kappa_q^2}\right)\right] + O(\epsilon),$$

respectively.

3 Asynchronous Stochastic Gradient Descent

Thanks to the availability of large datasets and modern computing resources, optimization-based machine learning has achieved state-of-the-art results in many applications of artificial intelligence. As the datasets continue to increase, distributed optimization algorithm have received more attention for solving large scale machine learning problems. ASGD is one of the most popular one among them. In ASGD, the local workers interact with the shared parameter independently without any synchronization, i.e., each local worker continues to compute the next gradient right after their own gradients have been added to the shared parameter. ASGD is efficient since the overall training speed is not affected by the slow local workers. However, it can potentially suffer from the problem of delayed gradients, i.e., the gradients that a local worker sends to the shared parameter are often computed with respect to the parameter of an older version of the model. Therefore, extra stochasticity is introduced in ASGD due to this delay. An interesting mathematical problem is how the delayed gradient affects the training process. There have been a few papers in the literature that analyze the convergence rate of ASGD. Most of them are from an optimization perspective [8, 17, 19, 24], while we take a perspective from a PDE.

Main contributions. The goal in this section is to study the convergence rate of p.d.f. of ASGD to its steady state distribution via the corresponding FP equation. The main focus is on the case where the loss function is a perturbed quadratic function. There are mainly two difficulties in this analysis. The first one is that asynchrony results in a degenerate diffusion operator in the corresponding PDE. No trivial analysis is able to give an exponential decay rate for the convergence to the steady state. Thanks to the association of a degenerate diffusion operator and a conservative operator, the decay rate can be recovered through "hypocoercivity" [33]. The key here is to construct a Lyapunov functional to prove the exponential decay of this functional. The second difficulty is to obtain a sharp convergence rate. Such a sharp rate is important to understand the influence of the asynchrony quantitatively. From the sharp rate, we prove that when the number of local workers is larger than the expected staleness, then ASGD is more efficient than stochastic gradient descent. Our theoretical result also suggests that longer delays result in slower convergence rate. Besides, the learning rate cannot be smaller than a threshold inversely proportional to the expected staleness.

3.1 Setting

We consider the minimization problem (1). In the ASGD algorithm, the parameter θ is updated with

$$\theta_{k+1} = \theta_k - \eta \nabla_\theta f_{\gamma_k}(\theta_{k-\tau_k}),$$

where γ_k is i.i.d. uniform random variable from $\{1, 2, \ldots, n\}$ and $\theta_{k-\tau_k}$ is the delayed read of the parameter θ used to update θ_{k+1} with a random staleness τ_k.

In [1], An et al. derived the modified stochastic differential equation for the algorithm, under the assumption that τ_k follows the geometric distribution, i.e., $\tau_k = l$ with probability $(1-\kappa)\kappa^l$ for $\kappa \in (0, 1)$. We call κ the staleness rate. Note that if κ is larger, then there is a longer delay. Besides, the expectation of the random staleness τ_k is $\frac{1}{1-k}$, so we call $\frac{1}{1-k}$ the expected staleness. By introducing

$$y_k = -\sqrt{\frac{\eta}{1-\kappa}} \mathbb{E}_{\tau_k} \nabla f(\theta_{\tau_k}),$$

when the learning rate η is small, (θ_k, y_k) can be approximated by time discretizations of a continuous time stochastic process $\left(\Theta_{k\delta_t}, Y_{k\delta_t}\right)$ for $\delta_t = \sqrt{\eta(1-\kappa)}$. (Θ_t, Y_t) satisfies the SDE

$$\begin{aligned} d\Theta_t &= Y_t dt + \tau dB_t, \\ dY_t &= -\nabla f(\Theta_t) dt - \gamma Y_t dt, \end{aligned} \tag{10}$$

where $\tau = \eta^{3/4}/(1-\kappa)^{1/4}\Sigma$, $\gamma = \sqrt{(1-\kappa)/\eta)}$, and Σ is the covariance matrix conditioned on τ_k, that is, Σ is the covariance between Θ_t and Y_t.

In what follows, we assume Σ is a constant for simplicity, and we consider the case where ∇f is a perturbed linear function,

$$\nabla f(\theta) = \omega_0^2 \theta + \varepsilon(\theta).$$

Denote $g(t, x, v)$ the joint p.d.f of (X_t, V_t) where

$$X_t = Y_t, \quad V_t = -\omega_0^2 \Theta_t - \gamma Y_t,$$

then it satisfies the following FP equation,

$$\partial_t g + v \cdot \nabla_x g - \omega_0^2 x \cdot \nabla_v g = \gamma \nabla_v \cdot \left(vg + \frac{1}{\beta} \nabla_v g \right) + \varepsilon \left(-\frac{1}{\omega_0^2}(v + \gamma x) \right) \cdot (\nabla_x g - \gamma \nabla_v g) \tag{11}$$

with $\beta = \frac{2\gamma}{\tau^2 \omega_0^4}$.

3.2 Theorems

When $\varepsilon = 0$, the steady state $M(x, v)$ of (11),

$$M(x, v) = M_x M_v := \left(\frac{1}{Z_1} e^{-\frac{\beta\omega_0^2}{2}|x|^2} \right) \left(\frac{1}{Z_2} e^{-\frac{\beta}{2}|v|^2} \right), \tag{12}$$

where Z_1, Z_2 are the normalization constants such that $\int M_v dv = \int M_x dx = 1$. However, for general $f(\theta)$, unfortunately there is no explicit form of the steady state. By denoting $F(x, v)$ as the steady state of (11), the weighted fluctuation function

$$h(t, x, v) = \frac{1}{M} [g(t, x, v) - F(x, v)]$$

satisfies the following equation,

$$\partial_t h + Th = Lh + Rh, \tag{13}$$

where

$$T = v \cdot \nabla_x - \omega_0^2 x \cdot \nabla_v \text{ is the transport operator;}$$
$$L = \frac{\gamma}{\beta}\frac{1}{M}\nabla_v \cdot (M\nabla_v) \text{ is the Linearized Fokker Planck operator;} \quad (14)$$
$$Rh = \varepsilon \cdot (\nabla_x h - \gamma \nabla_v h) - \beta\varepsilon \cdot \left(\omega_0^2 x - \gamma v\right) h \text{ is the perturbation terms.}$$

The above Eq. (13) is typically called the microscopic equation in the literature. It is also convenient for the forthcoming analysis to define the inner product $\langle \cdot, \cdot \rangle$ and the norm $\|\cdot\|_*$ as

$$\langle h, g\rangle_* = \int hgM\, dxdv, \quad \|h\|_*^2 = \langle h, h\rangle_* . \quad (15)$$

In addition, $\|\cdot\|^2$ is the standard L^2 norm with respect to the Lebesgue measure.

Under the above Gaussian measure M, the following Poincaré inequality holds

$$\|h\|_*^2 \le \frac{1}{d\beta \min\{\omega_0^2, 1\}} \left(\|\nabla_x h\|_*^2 + \|\nabla_v h\|_*^2\right), \quad \text{for } \forall h \ s.t. \int hMdxdv = 0 \quad (16)$$

The following key assumption ensures various bounds of the perturbation $\varepsilon(\theta)$.

Assumption 1 There exists a small constant $\epsilon_0 > 0$, such that,

$$\max_i \|\varepsilon_i\|_{L^\infty}, \|\varepsilon \cdot x\|_{L^\infty}, \|\varepsilon \cdot v\|_{L^\infty}, d \max_i \left\|\varepsilon_i'\right\|_{L^\infty}, \sum_i \left\|\varepsilon_i'\right\|_{L^\infty}, \left\|\varepsilon' \cdot x\right\|_{L^\infty}, \left\|\varepsilon' \cdot v\right\|_{L^\infty} \le \epsilon_0,$$

where$\varepsilon'(\theta)$ is the derivative of $\varepsilon(\theta)$.

The following theorem states an exponential decay bound for the fluctuation h.

Theorem 2 *Under Assumption 1 with ϵ_0 small enough, the fluctuation h decay exponentially as follows,*

$$\|h(t)\|_*^2 \lesssim e^{-2(\mu-\epsilon)t} H(0),$$

where $H(0) = \|\nabla_x h(0)\|_*^2 + C\,\|\nabla_v h(0)\|_*^2 + 2\hat{C}\,\langle \nabla_x h(0), \nabla_v h(0)\rangle_*$, $\epsilon = \epsilon_0 C_1$ for a constant C_1 depending on ω_0, γ and

$$\begin{cases} \text{when } \gamma < 2\omega_0: & \mu = \gamma, \quad C = \omega_0^2, \quad \hat{C} = \gamma/2; \\ \text{when } \gamma > 2\omega_0: & \mu = \gamma - \sqrt{\gamma^2 - 4\omega_0^2}, \quad C = \gamma^2/2 - \omega_0^2, \quad \hat{C} = \gamma/2; \\ \text{when } \gamma = 2\omega_0: & \forall \delta > 0, \text{ there exists } C(\delta), \hat{C}(\delta), \text{ such that the decay rate } \mu = \gamma - \delta. \end{cases} \quad (17)$$

More specifically $C_1 = \left(11 + 11C + 15\hat{C}\right)\epsilon_0 \cdot C_2^2 \cdot \frac{\max\{1,C\}}{C-\hat{C}^2}$, *where* $C_2 = \frac{\max\{1,\gamma,\gamma^2,\beta\gamma,\beta\omega_0^2\}}{\min\{1,\omega_0^2\}}$.

Remark 1 (*How the learning rate and staleness affect the convergence rate?*) When the perturbation ϵ_0 is small, the decay rate is dominated by $e^{-2\mu t}$. By the definition of δ_t, one has $t = k\delta_t = k\sqrt{\eta(1-\kappa)}$ with k the number of steps, η the learning rate and κ the staleness rate. Inserting the definition of $\gamma = \sqrt{(1-\kappa)/\eta)}$ into (17), the dominated decay rate $e^{-2\mu t}$ can also be written as,

$$\begin{cases} \text{when } \eta > \dfrac{1}{4\omega_0^2}(1-\kappa): & \mu t = (1-\kappa)k; \\ \text{when } \eta < \dfrac{1}{4\omega_0^2}(1-\kappa): & \mu t = (1-\kappa)k - \left(\sqrt{(1-\kappa)^2 - 4\omega_0^2(1-\kappa)\eta}\right)k. \end{cases} \tag{18}$$

From the above discussion, we make two observations:

- The learning rate should not be smaller than $\frac{1}{4\omega_0^2}(1-\kappa)$. For a fixed staleness rate κ, when the learning rate is larger than the threshold $\frac{1}{4\omega_0^2}(1-\kappa)$, the convergence rate is a constant only depending on $(1-\kappa)$. While the learning rate is smaller than this threshold, the convergence rate will become slower as the learning rate becomes smaller.
- Longer delays result in slower convergence rate. For a fixed learning rate, the optimal decay rate $e^{-2(1-\kappa)k}$ only relates to the staleness of the system. If the system has more delayed readings from the local workers, i.e., $(1-\kappa)$ is smaller, then the convergence rate is slower.

All the above discussion is based on the assumption that η is small enough so that the SME-ASGD is a good approximation for ASGD. In other words, we assume ω_0 is large here, hence the threshold $\frac{1}{4\omega_0^2}(1-\kappa)$ is still in the valid regime.

Remark 2 When is ASGD more efficient than SGD? Assume we have m local workers and the learning rate is larger than the threshold $\eta > \frac{1}{4\omega_0^2}(1-\kappa)$. When the perturbation ϵ_0 is small, for single batch SGD, the decay rate is e^{-2k} after k steps, while for ASGD, the decay rate is $e^{-2(1-\kappa)k}$ after calculating k gradients. Since now we have m local workers, for the same amount of time, the decay rate for ASGD becomes $e^{-2(1-\kappa)mk}$. Therefore, as long as $(1-\kappa)m > 1$, ASGD will be more efficient than SGD. Since the expectation of the random staleness τ_k is $\frac{1}{1-\kappa}$, in other words, when the number of local workers is larger than the expected staleness, then ASGD is more efficient than SGD.

We run a simple numerical experiment in Fig. 2 to verify the above conclusions.

(a) When $\kappa = 0.98$, the threshold for the learning rate is $(1-\kappa)/4 = 0.005$. One can see that the blue and red lines spend similar time to converge, which verifies that the convergence rate is the same when the learning rate is above the threshold. When the learning rate is below the threshold, as the learning rate becomes smaller, the convergence of ASGD becomes slower.
(b) When the learning rates are all above the threshold, as the staleness rate becomes larger, it takes a longer time for the ASGD to converge.
(c) When the staleness rate is 0.96, it takes 2 local workers for ASGD to converge faster than SGD in time. Remark 2 gives a conservative estimate for the number

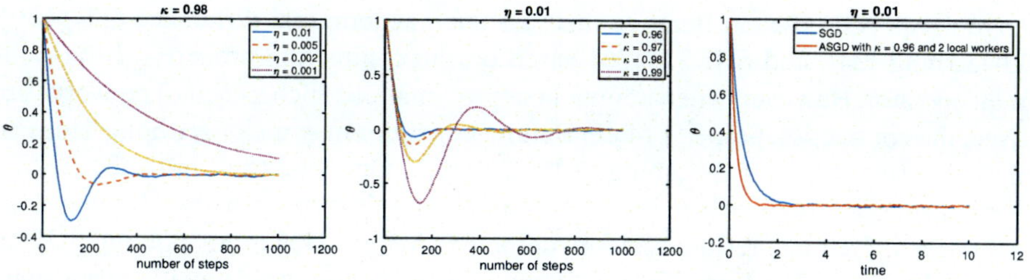

(a) (b) (c)

Fig. 2 Apply ASGD to minimize the quadratic function $f(\theta) = \theta^2$ with two components $f_1(\theta) = (\theta - 1)^2 - 1$ and $f_2(\theta) = (\theta + 1)^2 - 1$. All the plots are averaged results over 1000 simulations with initialization $\theta_0 = 1$. **a** Compare the convergence of ASGD with different learning rate when the staleness rate is $\kappa = 0.98$. **b** Compare the convergence of ASGD with different staleness rate when the learning rate is $\eta = 0.01$. **c** Compare the convergence of ASGD and SGD in time

of local workers. In practice, it actually requires fewer local workers for ASGD to be more efficient than SGD.

The proof of the theorem is given in Sect. 4 of [36]. The main ingredient of the proof is the following Lyapunov functional $H(t)$,

$$H(t) = \|\nabla_x h\|_*^2 + C\,\|\nabla_v h\|_*^2 + 2\hat{C}\,\langle \nabla_x h, \nabla_v h\rangle_* \tag{19}$$

where $C, \hat{C}$ are constants to be determined. Note that,

$$\frac{d}{dt}H(t) = \frac{d}{dt}\left(\|\nabla_x h\|_*^2 + C\,\|\nabla_v h\|_*^2\right) + 2\hat{C}\frac{d}{dt}\,\langle \nabla_x h, \nabla_v h\rangle_* . \tag{20}$$

The term $\frac{d}{dt}\,\|\nabla_v h\|_*^2$ will give the dissipation of $\|\nabla_v h\|_*^2$, and $\frac{d}{dt}\,\langle \nabla_x h, \nabla_v h\rangle_*$ will give the dissipation of $\|\nabla_x h\|_*^2$. The term $\|\nabla_x h\|_*^2$ and the constants $C, \hat{C}$ in the Lyapunov functional is to make sure the functional is always positive, so that one could eventually have

$$\frac{1}{2}\partial_t H(t) + \tilde{C}H(t) \le 0. \tag{21}$$

Finally, the exponential decay of $\|h(t)\|_*^2$ can be derived from this inequality and the relationship between $H(t)$ and $\|h(t)\|^2$.

4 Reinforcement Learning in Smooth Environment

The goal of RL is to find an optimal policy which maximizes the return of a Markov decision process (MDP) [31]. One of the most common ways of finding an optimal policy is to treat it as the fixed point of the Bellman operator. Researchers have developed efficient iterative methods such as temporal difference (TD) [30], Q-learning [34], and SARSA [26] based on the contraction property of the Bellman operator. However, when nonlinear approximations such as neural networks are used, the contraction property of the Bellman operator may no longer hold. This can in turn result in unstable training of the network. One way to stabilize RL with a nonlinear approximation is to formulate it as a minimization problem. This approach is known as Bellman residual minimization (BRM) [3]. However, applying SGD to BRM directly suffers from the so-called double sampling problem: at a given state, two independent samples for the next state are required in order to perform unbiased SGD. Such a requirement is often hard to fulfill in a model-free setting, especially for problems with a continuous state space. In order to alleviate the double-sampling issue, [37] proposed the BFF algorithm. The key idea is to borrow extra randomness from the future by leveraging the smoothness of the underlying RL problem. The main assumption is that the underlying dynamics of the MDP can be written as a discretized stochastic differential equation with a small step size. Note that knowledge of the dynamics is not required to implement the algorithm. Many RL applications in the physical environment are included under this assumption.

Contributions In this section, the FP equation is used to explain why BFF works and what elements affect the performance of BFF. The main theorem shows that for model-free control problems when the underlying dynamics change slowly with respect to actions, the training trajectory of the BFF algorithm is statistically close to the training trajectory of unbiased SGD. The difference between the two algorithms will first decay exponentially and eventually stabilize at an error of $O(\epsilon\delta_*)$, where δ_* is the smallest Bellman residual that unbiased SGD can achieve and ϵ is the size of the time step in the underlying SDE discretization. This theoretical result is based on the asymptotic behavior of the corresponding FP equation.

4.1 Setting

Working in the model-free RL setting, we consider a discrete-time MDP with a compact continuous state space $\Sigma \subset \mathbb{R}^{d_s}$. The action space $\mathbb{A} \subset \mathbb{R}^{d_a}$ can be a compact continuous set or a finite discrete set. The transition kernel of the MDP

$$P^a(s, s') = \mathbb{P}\left(s_{m+1} = s' | s_m = s, a_m = a\right). \tag{22}$$

denotes the likelihood of transferring from the current state $s_m = s$ under the current action $a_m = a$ to the next state $s_{m+1} = s'$. The immediate reward function $r(s', s, a)$ specifies the reward if one takes action a at state s and ends up at state s'. The immediate reward can also be random, in which case $r(s', s, a)$ represents the expected reward. A policy $\pi(a|s)$ gives the probability of taking action a at state s, i.e., $\mathbb{P}\{\text{take action } a \text{ at state } s\} = \pi(a|s)$. For a continuous state space, it is often convenient to rewrite the underlying transition in terms of the states:

$$s_{m+1} = s_m + \mu(s_m, a_m)\epsilon + \sqrt{\epsilon}\sigma Z_m. \tag{23}$$

where Z_m is a mean-zero noise. This form is particularly relevant when the MDP arises as a discretization of an underlying SDE, with ϵ as its discretized time step. We note that Z_m does not need to be i.i.d. Gaussian. Our theoretical and numerical results can be extended to any independent mean-zero noise with the same variance at each time step. In addition, the diffusion term σ can depend on state and action as well. The proof for this extension is given in Appendix B of [37]. Since the extension is trivial and does not affect the theoretical and numerical results, we stick to a constant diffusion term in this section. We note that this form is quite general: it encompasses both deterministic linear differential equations (e.g. [4, 7]) as well as cases with a nonlinear drift $\mu(s, a)$ and stochastic transitions, e.g. [23]. Throughout this chapter, we consider the case where for each state, the variation of the underlying drift $\mu(s, a)$ is a priori bounded in the action space, and for each action, the drift is smooth in the state space. Additionally, we assume the immediate reward $r(s', s, a)$ is continuous in $s', s \in \Sigma$ for each action.

Given a trajectory $\{s_m, a_m\}_{m\geq 0}$, the main object under study is the action-state pair value function $Q(s, a)$. Here are two types of problems: Q-evaluation and Q-control. Q-evaluation refers to the prediction of the value function $Q^\pi(s, a)$ when the policy π is given. For the Q-evaluation problem, the state space and action space can be continuous or discrete; while Q-control refers to finding the optimal policy π_* through the maximization of $Q^\pi(s, a)$ over all possible policies. For the Q-control problem, we mainly consider the case of finite discrete action space.

Q-evaluation Given a fixed policy π, the value function $Q^\pi(s, a)$ represents the expected return if one takes action a at state s and follows π thereafter, i.e.,

$$Q^\pi(s, a) = \mathbb{E}\left[\sum_{t\geq 0} \gamma^t r(s_{m+t+1}, s_{m+t}, a_{m+t}) \middle| s_m = s, a_m = a\right],$$

where $\gamma \in (0, 1)$ is a discount factor. The value function Q^π satisfies the Bellman equation [31] $Q^\pi(s, a) = T^\pi Q^\pi(s, a)$, with the Bellman operator T^π defined as

$$T^\pi Q^\pi(s, a) \equiv \mathbb{E}[r(s_{m+1}, s_m, a_m) + \gamma Q^\pi(s_{m+1}, a_{m+1})|(s_m, a_m) = (s, a)], \tag{24}$$

where the expectation is taken over (s_{m+1}, a_{m+1}) when the policy π is applied. In the nonlinear approximation setting, one seeks a solution to (24) from a family of functions $Q^\pi(s, a; \theta)$ parameterized by $\theta \in \Omega \subseteq \mathbb{R}^{d_\theta}$. One way to find $Q^\pi(s, a; \theta)$ is to solve the following BRM problem:

$$\min_{\theta\in\mathbb{R}^{d_\theta}} \mathop{\mathbb{E}}_{(s,a)\sim\rho(s,a)} \delta^2(s, a; \theta) \tag{25}$$

where $\rho(s, a)$ is a distribution over $\Sigma \times \mathbb{A}$ and

$$\delta(s, a; \theta) \equiv |T^\pi Q^\pi(s, a; \theta) - Q^\pi(s, a; \theta)| \tag{26}$$

is the absolute value of the Bellman residual. Note that the expectation in (25) can be taken with respect to different distributions ρ. For on-policy learning, it is often the stationary distribution of the Markov chain.

One approach for solving the Bellman minimization problem (25) is to directly apply SGD. The unbiased gradient estimate to the loss function is

$$F_m(\theta) = j(s_m, a_m, s_{m+1}; \theta)\nabla_\theta j(s_m, a_m, s'_{m+1}; \theta), \tag{27}$$

where (s_m, a_m) is a sample from the distribution $\rho(s, a)$ and

$$j(s_m, a_m, s_{m+1}; \theta) = r(s_{m+1}, s_m, a_m) + \gamma \int Q^\pi(s_{m+1}, a; \theta)\pi(a|s_{m+1})da \; - Q^\pi(s_m, a_m; \theta) \tag{28}$$

is an unbiased estimate for the Bellman residual $T^\pi Q^\pi(s_m, a_m) - Q^\pi(s_m, a_m)$. Here s_{m+1} is the next state in the trajectory, while s'_{m+1} is an independent sample for the next state according to the transition process. However, in model-free RL, as the underlying dynamics are unknown, another independent sample s'_{m+1} of the next state is unavailable. Therefore, this unbiased SGD, referred to as uncorrelated sampling (US), is impractical. Even if one can store the whole trajectory, it is impossible to revisit a certain state multiple times when the state space is either continuous or discrete but of high dimension. This is the so-called double sampling problem.

To address the double sampling problem, [35] introduced the BFF algorithm. The main idea of the BFF algorithm is to borrow the future difference $\Delta s_{m+1} = s_{m+2} - s_{m+1}$ and approximate the second sample s'_{m+1} with $s_m + \Delta s_{m+1}$. During SGD, the parameter θ is updated based on the following estimate of the unbiased gradient:

$$\hat{F}_m(\theta) = j(s_m, a_m, s_{m+1}; \theta)\nabla_\theta j(s_m, a_m, s_m + \Delta s_{m+1}; \theta), \tag{29}$$

where j is a sample of the Bellman residual defined in (28). When the difference between Δs_m and Δs_{m+1} is small, the new s'_{m+1} is statistically close to the distribution of the true next state.

Q-control The BFF algorithm mentioned above can be extended easily to Q-control, i.e., finding the value function Q^* of the optimal policy π_*. Q^* satisfies the Bellman equation $Q^*(s, a) = T^{\pi_*} Q^*(s, a)$, where T^{π_*} is the optimal Bellman operator,

$$T^{\pi_*} Q^*(s,a) = \mathbb{E}\left[r(s_{m+1}, s_m, a_m)] + \gamma \max_{a'} Q^*(s_{m+1}, a'; \theta) \middle| (s_m, a_m) = (s, a) \right], \tag{30}$$

where the expectation is taken over s_{m+1} when the optimal policy π_* is applied. The BRM problem is the same as (25) but with the absolute value of the Bellman residual $\delta(s, a; \theta)$ given by

$$\delta(s, a; \theta) = \left|T^{\pi_*} Q^*(s, a) - Q^*(s, a)\right|. \tag{31}$$

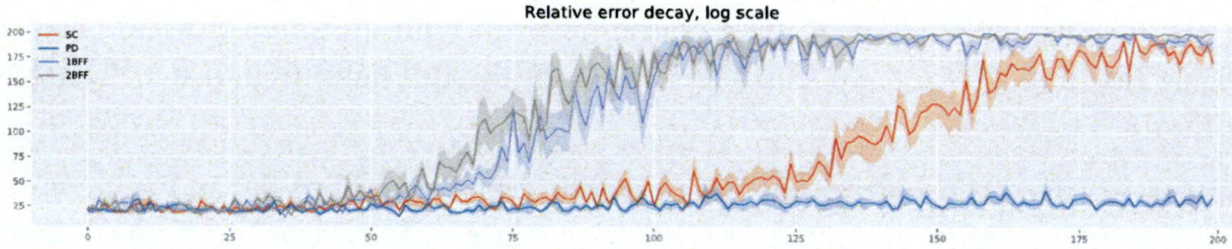

Fig. 3 Reward per training episode for the CartPole experiment. BFF is compared with sample cloning (SC) and primal-dual (PD) algorithms in the above plot. The shaded regions show the standard error of the mean over 5 trials. BFF is the first to reach the maximum reward and achieves it more consistently than sample-cloning. It achieves slightly better performance using 2 future steps (2BFF in the plot). Despite an extensive hyperparameter search, PD was not able to learn an effective policy

Rather than generating a trajectory offline with a fixed policy, we instead generate a training trajectory online using an ϵ-greedy policy. The BFF algorithm for Q-control is identical to (29), but with j replaced by $j(s_m, a_m, s_{m+1}; \theta) = r(s_{m+1}, s_m, a_m) + \gamma \max_a Q^*(s_{m+1}, a; \theta) - Q^*(s_m, a_m; \theta)$.

Figure 3 compares the performance of several methods on the CartPole problem (One can refer to [5] or the OpenAI Gym website [21] for the details of the CartPole problem setting.) We approximate the value function $Q(s, a; \theta)$ with a neural network. The parameters θ in the neural network is updated according to different methods.

- 1BFF refers to the standard BFF algorithm introduced in the above paragraph. One can refer to Algorithm 2 in [37] for the details of the algorithm.
- 2BFF refers to the modified BFF algorithm where one borrows two future steps to approximate the second sample for the next state. That is, the stochastic gradient is approximated by the following equation,

$$\frac{1}{2} j(s_m, a_m, s_{m+1}; \theta) \left[\nabla_\theta j(s_m, a_m, s_m + \Delta s_{m+1}; \theta) + \nabla_\theta j(s_m, a_m, s_m + \Delta s_{m+2}; \theta) \right].$$

- SC refers to the sample-cloning algorithm, where it assigns the same next sate's value to the second sample. That is, the stochastic gradient is approximated by the following equation,

$$j(s_m, a_m, s_{m+1}; \theta) \nabla_\theta j(s_m, a_m, s_{m+1}; \theta).$$

- PD refers to the primal-dual algorithm, where it avoids the double sampling problem by changing the minimization problem to a minimax problem. The objective function for PD is the following,

$$\min_\theta \max_\omega \quad \mathbb{E}\, \delta(s, a; \theta) y(s, a; \omega) - \frac{1}{2} y^2(s, a; \omega).$$

Here y is also approximated by a neural network with the same structure as the value function Q. (See Appendix E of [37] for the full definition of the PD algorithm.)

One can see that BFF reaches the max reward (200) faster than SC and achieves it with greater regularity throughout the training process. In contrast to both of these methods, the PD method fails to converge even after the extensive hyperparameter search described above. See Sect. 4.1 for more details of the experiment.

Now the question is why does it perform well and what are elements that affect the performance of BFF?

4.2 Theorems

This section states the main theoretical results which bound the difference between BFF and US on a continuous state space. Recall that the one-step transition is governed by the state dynamics (23), where $\mu(s, a)$ is the drift, Z_m is assumed to be normal $N(0, I_{d_s \times d_s})$, and σ is the diffusion coefficient. It is convenient to introduce $\Delta s_m := s_{m+1} - s_m = \mu(s_m, a)\epsilon + \sigma\sqrt{\epsilon}Z_m$. For a discrete action space $\mathbb{A}$, the drift term $\{\mu(s, a)\}_{a \in \mathbb{A}}$ is a family of continuous functions, while for a continuous action space, $\mu(s, a)$ is a continuous function in both state and action.

Error bound at each step. The following lemma bounds the difference between BFF and US at each step. That is, assuming the current parameters θ are the same, Lemma 6 bounds the expected difference between BFF and US for Q-evaluation and Q-control after one step.

Lemma 6 *Suppose that $Q^\pi(s, a; \theta)$ and $\max_{a \in \mathbb{A}} Q^*(s, a; \theta)$ are Lipschitz continuous in θ, and that $\partial_s \nabla_\theta Q^\pi(s, a; \theta)$ and $\partial_s \nabla_\theta \max_{a \in \mathbb{A}} Q^*(s, a; \theta)$ are continuous in the state and action space. Then the difference between the BFF gradient $\hat{F}$ and the unbiased gradient F defined in (27) and (29), respectively satisfies*

$$\mathbb{E}[\hat{F}_m(\theta)] - \mathbb{E}[F_m(\theta)] = \mathbb{E}\left[\delta(s_m, a_m; \theta)\left(C(s_m; \theta)\epsilon + o(\epsilon)\right)\right] = O(\mathbb{E}[\delta\epsilon]),$$

where δ is the absolute value of the Bellman residual defined in (26), (31) for Q-evaluation and Q-control respectively. For Q-evaluation, $C(s_m; \theta)$ is defined as

$$C(s_m; \theta) = \gamma\left(\partial_s \mathbb{E}_{a \sim \pi(a|s_m)}[\nabla_\theta Q^\pi(s_m, a; \theta)]\right) C_2(s_m), \tag{32}$$

and for Q-control, $C(s_m; \theta)$ is defined as

$$C(s_m; \theta) = \gamma\left(\partial_s \nabla_\theta \max_{a' \in \mathbb{A}} Q^*(s, a; \theta)\right) C_2(s_m) \tag{33}$$

with $C_2(s_m)$ an upper bound for the variation of the drift in the action space: $|\mu(s_m, a_{m+1}) - \mu(s_m, a_m)| \le C_2(s_m)$.

Note that the common factor $C_2(s)$ affects the magnitude of the difference between BFF and US. That is to say, when the drift changes more slowly with respect to the action, the difference is smaller and BFF performs better. The sizes of γ and ϵ play an important role in determining the statistical difference between BFF and US as well.

Differences of density evolutions. Next, we compare the p.d.f. for the parameters over the course of the complete BFF and US algorithms. The updates of the parameter θ_k by SGD follow

$$\text{US:} \quad \theta_{k+1} = \theta_k - \eta F_m(\theta_k)$$
$$\text{BFF:} \quad \theta_{k+1} = \theta_k - \eta \hat{F}_m(\theta_k)$$

where F_k, $\hat{F}_k$ are the estimates of the loss function's gradient defined in (27) and (29) for US and BFF algorithm respectively. When the learning rate η is small, the dynamics of SGD can be approximated by a continuous time SDE

$$\begin{aligned} \text{US:} \quad d\Theta_t &= -\mathbb{E}[F_m(\Theta_t)]dt + \sqrt{\eta}\mathbb{V}[F_m(\Theta_t)]dB_t \\ \text{BFF:} \quad d\Theta_t &= -\mathbb{E}[\hat{F}_m(\Theta_t)]dt + \sqrt{\eta}\mathbb{V}[\hat{F}_m(\Theta_t)]dB_t \end{aligned} \tag{34}$$

for $\Theta_{t=k\eta} \approx \theta_k$ with an error of $O(\sqrt{\eta})$, where $\mathbb{E}$ and $\mathbb{V}$ are expectation and variance taken over $\rho(s, a)$, the distribution defined in the loss function (25). Here $\mathbb{E}[F_m(\Theta_t)]$ denotes the true gradient of the population loss function used in US, and $\mathbb{E}[\hat{F}_m(\Theta_t)]$ denotes the biased gradient of the population loss used in BFF. For simplicity, we assume $\mathbb{V}[F_m] \equiv \xi$ is constant. Let $p(t, \theta)$ and $\hat{p}(t, \theta)$ be the p.d.f.s of the parameter θ at step $k = t/\eta$ for US and BFF, respectively. These p.d.f.s satisfy the following two equations [22]:

$$\text{US:} \quad \partial_t p = \nabla_\theta \cdot \left[\mathbb{E}[F_m]p + \frac{\eta}{2}\nabla_\theta \cdot (\mathbb{V}[F_m]p)\right]; \tag{35}$$
$$\text{BFF:} \quad \partial_t \hat{p} = \nabla_\theta \cdot \left[\mathbb{E}[\hat{F}_m]\hat{p} + \frac{\eta}{2}\nabla_\theta \cdot \left(\mathbb{V}[\hat{F}_m]\hat{p}\right)\right]. \tag{36}$$

In addition, we assume $\theta \in \Omega$ with Ω compact. As a result, we need a reflective boundary condition for the PDEs (35) and (36), i.e.

$$\left(\mathbb{E}[F_m]p + \frac{\eta}{2}\nabla\left(\mathbb{V}[F_m]p\right)\right) \cdot \mathbf{n}\Big|_{\partial\Omega} = 0, \quad \left(\mathbb{E}[\hat{F}_m]\hat{p} + \frac{\eta}{2}\nabla\left(\mathbb{V}[\hat{F}_m]\hat{p}\right)\right) \cdot \mathbf{n}\Big|_{\partial\Omega} = 0.$$

We define

$$\hat{d}(t, \theta) = p - \hat{p}$$

to be difference of the p.d.f.s for US and BFF. The goal in this section is to analyze how $\left\|\hat{d}\right\|_*^2$ evolves in time for some specific norm $\|\cdot\|_*$.

We introduce the following weighted norm to measure the difference between the p.d.f.s:

$$\left\|\hat{d}\right\|_*^2 := \int \hat{d}^2 \frac{1}{p^\infty} d\theta$$

where p^∞ is the steady distribution of US. p^∞ is obtained by setting the RHS of (35) to be zero. Since $\mathbb{E}[F_m] = \nabla_\theta \mathbb{E}_{(s,a)\sim\rho}[\delta^2(s, a; \theta)]$, where $\rho(s, a)$ is defined in the loss function (25) and $\delta(s, a; \theta)$ is the absolute Bellman residual defined in (26) for Q-evaluation and in (31) for Q-control, it is easy to check that the steady distribution of US is

$$p^\infty = \frac{1}{Z} e^{-\frac{2}{\eta\xi}\mathbb{E}[\delta^2]}, \tag{37}$$

where $Z = \int e^{-\frac{2}{\eta\xi}\mathbb{E}[\delta^2]} d\theta$ is a normalizing constant. We have reduced our problem to quantifying the evolution of $\left\|\hat{d}\right\|_*^2$ in time, which we accomplish with the following theorem.

Theorem 3 (short version) *For sufficiently small $\eta > 0$, the difference $\hat{d}$ of the p.d.f.s for US and BFF is bounded by*

$$\left\|\hat{d}(t)\right\|_* \leq C_1 e^{-C_2 t} + O\left(\epsilon\sqrt{\mathbb{E}[\delta_*^2]\eta^{C_3}}\right)\sqrt{1 - e^{-C_2 t}}, \tag{38}$$

where $\mathbb{E}[\delta_*^2] = \min_\theta \mathbb{E}[\delta^2]$ and C_1, C_2, C_3 are all positive constants.

The precise version of the above theorem is stated in Theorem C.5 in [37]. This theorem implies that as the algorithm moves on, the difference between BFF and US will decay exponentially. After running the algorithm for sufficiently many steps, the difference will eventually be $O\left(\epsilon\sqrt{\mathbb{E}[\delta_*^2]\eta^{C_3}}\right)$. As long as $\mathbb{E}[\delta_*^2]$ is small, BFF will achieve a minimizer close to US with an error much smaller than $O(\epsilon)$. Note that if $\mathbb{E}[\delta_*^2] = 0$, the difference still does not vanish. Instead, the leading order term of the last term in (38) becomes $O(\epsilon\eta^{C_3+1/2})$, which is shown in Corollary C.2 in [37]. The constant C_1 depends on the initial p.d.f. of the algorithm. The constant C_3 is related to the shape of $\mathbb{E}[\delta_*^2](\theta)$ in the parameter space. The flatter the shape at the minimizer is, the smaller C_3 is. The constant C_2 decreases as η decreases, so the first term increases as η decreases, while the last term $O(\epsilon^2\mathbb{E}[\delta_*^2]\eta^{C_3})$ does the opposite. This suggests that one should set the learning rate η large at first, making the exponential decay faster. As the training progresses, η should be reduced to make the final error smaller.

5 Conclusion

The study of partial differential equations is an essential mathematical tool to understand machine learning algorithms. The SGD algorithm can be approximated by a continuous stochastic process in which the resulting probability density function satisfies a Fokker Planck equation. Hence, Fokker Planck equation plays an important role in building the theoretical foundation of machine learning. There are still many open problems along this direction, including the asymptotic analysis when the diffusion is anisotropic or nonlinear, the limiting equation when the noise does not follow the normal distribution, etc, the behavior of the optimization process when the loss function is based on a deep neural network.

References

1. An, J., Lu, J., Ying, L.: Stochastic modified equations for the asynchronous stochastic gradient descent. Inf. Inference J. IMA **9**(4), 851–873 (2020)
2. An, J., Ying, L., Zhu, Y.: Why resampling outperforms reweighting for correcting sampling bias with stochastic gradients (2020). arXiv:2009.13447
3. Baird, L.: Residual algorithms: reinforcement learning with function approximation. In: Machine Learning Proceedings 1995, pp. 30–37. Elsevier (1995)
4. Bradtke, S.: Reinforcement learning applied to linear quadratic regulation. In: Advances in Neural Information Processing Systems, p. 5 (1992)
5. Brockman, G., Cheung, V., Pettersson, L., Schneider, J., Schulman, J., Tang, J., Zaremba, W.: Openai gym (2016). arXiv:1606.01540
6. Chaudhari, P., Soatto, S.: Stochastic gradient descent performs variational inference, converges to limit cycles for deep networks. In: 2018 Information Theory and Applications Workshop (ITA), pp. 1–10. IEEE (2018)
7. Doya, K.: Reinforcement learning in continuous time and space. Neural Comput. **12**(1), 219–245 (2000)
8. Duchi, J., Jordan, M.I., McMahan, B.: Estimation, optimization, and parallelism when data is sparse. In: Advances in Neural Information Processing Systems, 26 (2013)
9. Galar, M., Fernandez, A., Barrenechea, E., Bustince, H., Herrera, F.: A review on ensembles for the class imbalance problem: bagging-, boosting-, and hybrid-based approaches. IEEE Trans. Syst. Man Cybern. Part C (Appl. Rev.) **42**(4), 463–484 (2011)
10. Hu, W., Li, C. J., Li, L., Liu, J.-G.: On the diffusion approximation of nonconvex stochastic gradient descent. Ann. Math. Sci. Appl. **4**(1) (2019)
11. Jastrzebski, S., Kenton, Z., Arpit, D., Ballas, N., Fischer, A., Bengio, Y., Storkey, A.: Three factors influencing minima in sgd (2017). arXiv:1711.04623
12. Kay, M., Matuszek, C., Munson, S.A.: Unequal representation and gender stereotypes in image search results for occupations. In: Proceedings of the 33rd Annual ACM Conference on Human Factors in Computing Systems, pp. 3819–3828 (2015)
13. Krizhevsky, A., Sutskever, I., Hinton, G.E.: Imagenet classification with deep convolutional neural networks. Commun. ACM **60**(6), 84–90 (2017)
14. Li, Q., Tai, C., Weinan, E.: Stochastic modified equations and adaptive stochastic gradient algorithms. In: International Conference on Machine Learning, pp. 2101–2110 (2017)
15. Li, Q., Tai, C., Weinan, E.: Stochastic modified equations and dynamics of stochastic gradient algorithms i: mathematical foundations. J. Mach. Learn. Res. **20**, 40–1 (2019)
16. Mandt, S., Hoffman, M.D., Blei, D.M.: Stochastic gradient descent as approximate Bayesian inference (2017). arXiv:1704.04289
17. Mania, H., Pan, X., Papailiopoulos, D., Recht, B., Ramchandran, K., Jordan, M.I.: Perturbed iterate analysis for asynchronous stochastic optimization (2015). arXiv:1507.06970
18. Menon, S., Damian, A., Hu, S., Ravi, N., Rudin, C.: Pulse: Self-supervised photo upsampling via latent space exploration of generative models. In: Proceedings of the IEEE/CVF Conference on Computer Vision and Pattern Recognition, pp. 2437–2445 (2020)

19. Mitliagkas, I., Zhang, C., Hadjis, S., Ré, C.: Asynchrony begets momentum, with an application to deep learning. In: 2016 54th Annual Allerton Conference on Communication, Control, and Computing (Allerton), pp. 997–1004. IEEE (2016)
20. Müller-Gronbach, T., Yaroslavtseva, L., et al.: On the performance of the Euler–Maruyama scheme for sdes with discontinuous drift coefficient. In: Annales de l'Institut Henri Poincaré, Probabilités et Statistiques, pp. 1162–1178. Institut Henri Poincaré (2020)
21. OpenAI Gym. Toolkit for developing and comparing reinforcement learning algorithms. https://gym.openai.com/
22. Pavliotis, G.A.: Stochastic Processes and Applications: Diffusion Processes, the Fokker-Planck and Langevin Equations, vol. 60. Springer (2014)
23. Pedersen, M.L., Frank, M.J., Biele, G.: The drift diffusion model as the choice rule in reinforcement learning. Psychon. Bull. Rev. **24**(4), 1234–1251 (2017)
24. Recht, B., Re, C., Wright, S., Niu, F.: Hogwild!: A lock-free approach to parallelizing stochastic gradient descent. In: Advances in Neural Information Processing Systems, 24 (2011)
25. Rotskoff, G., Vanden-Eijnden, E.: Parameters as interacting particles: long time convergence and asymptotic error scaling of neural networks. In: Advances in Neural Information Processing Systems, pp. 7146–7155 (2018)
26. Rummery, G.A., Niranjan, M.: On-Line Q-Learning Using Connectionist Systems, vol. 37. Citeseer (1994)
27. Schlegel, M., Chung, W., Graves, D., Qian, J., White, M.: Importance resampling for off-policy prediction. In: Advances in Neural Information Processing Systems, pp. 1799–1809 (2019)
28. Seiffert, C., Khoshgoftaar, T.M., Van Hulse, J., Napolitano, A.: Resampling or reweighting: a comparison of boosting implementations. In: 2008 20th IEEE International Conference on Tools with Artificial Intelligence, vol. 1, pp. 445–451. IEEE (2008)
29. Shi, B., Du, S.S., Su, W., Jordan, M.I.: Acceleration via symplectic discretization of high-resolution differential equations. In: Advances in Neural Information Processing Systems, pp. 5744–5752 (2019)
30. Sutton, R.S.: Learning to predict by the methods of temporal differences. Mach. Learn. **3**(1), 9–44 (1988)
31. Sutton, R.S., Barto, A.G.: Reinforcement Learning: An Introduction. MIT press (2018)
32. Sweeney, L.: Discrimination in online ad delivery. Queue **11**(3), 10–29 (2013)
33. Villani, C.: Hypocoercivity (2006). arXiv preprint math/0609050
34. Watkins, C.J.C.H.: Learning from delayed rewards (1989)
35. Zhu, Y., Ying, L.: Borrowing from the future: an attempt to address double sampling. In: Mathematical and Scientific Machine Learning, pp. 246–268. PMLR (2020)
36. Zhu, Y., Ying, L.: A sharp convergence rate for the asynchronous stochastic gradient descent (2020). arXiv:2001.09126
37. Zhu, Y., Izzo, Z., Ying, L.: Borrowing from the future: Addressing double sampling in model-free control. In: Mathematical and Scientific Machine Learning, pp. 1099–1136. PMLR (2022)